Thermal Management Materials for Electronic Packaging

Thermal Management Materials for Electronic Packaging

Preparation, Characterization, and Devices

Edited by Xingyou Tian

WILEY-VCH

Editor

Prof. Xingyou Tian
Chinese Academy of Sciences
Institute of Solid State Physics
Hefei Institutes of Physical Science
350 Shushanhu Road
230031 Hefei
China

Cover Image: © Xuanyu Han/Getty Images

Library of Congress Card No.: applied for

British Library Cataloguing-in-Publication Data
A catalogue record for this book is available from the British Library.

Bibliographic information published by the Deutsche Nationalbibliothek
The Deutsche Nationalbibliothek lists this publication in the Deutsche Nationalbibliografie; detailed bibliographic data are available on the Internet at <http://dnb.d-nb.de>.

Print ISBN: 978-3-527-35242-5
ePDF ISBN: 978-3-527-84311-4
ePub ISBN: 978-3-527-84310-7
oBook ISBN: 978-3-527-84312-1

Typesetting Straive, Chennai, India
Printing and Binding CPI Group (UK) Ltd, Croydon, CR0 4YY

C9783527352425_061223

Contents

Overview of Works

Abstract: With high thermal conductive components → interface → structural design as the main line, the frontier exploration and research of high power density thermal management materials are carried out by combining experimental characterization, theoretical analysis, modeling simulation, and application verification. It will reveal the thermophysical response characteristics under the conditions of multiple, multi-dimensional, multi-level, and multi-field coupling and establish a full-chain common theory and technology platform through design and preparation, device integration, and service evaluation. Through the micro–nano interface layer, local dense stack distribution, spatial network distribution, orientation distribution, and other composite configurations, the substrate thermal management efficiency is significantly improved. This book prepares high thermal conductive substrate materials through theoretical guidance, structural control, and process screening.

Keywords: Devices; Electronic packaging; Preparation; Spatial nerwork structure; Thermal management Materials

The Key Technology

1. The design control and surface modification technology of substrates and high thermal conductive components can improve their physical and chemical compatibility.
2. Micro- and nanoscale interface layer design and fine control technology can realize the balanced matching of interface bonding strength and interface thermal conductivity.
3. The multi-scale design and control technology of composite configuration can improve the design margin of macro-mechanical and thermophysical properties.
4. The densification and near-net molding technology of substrate materials can improve the density and optimize the thermal flow conduction path.

This book has good guiding significance for the preparation, production, and device application of thermal management materials.

Description

Thermal management materials first introduce the basic principles and basic concepts of thermal conductivity of materials. Then the key issues of the design and preparation of thermal management materials with high thermal conductivity are discussed, including their characterization, performance, operation, and related research.

The latest methods, techniques, and applications are explained in detail regarding this rapidly developing area. The book will support the work of academic researchers and graduate students, as well as engineers and materials scientists working in industrial research and development.

In addition, it will be of great value to those directly involved in the industrial industry related to thermal management materials, as well as to researchers in the fields of materials science, polymer science, materials chemistry, nanomaterials, and others working on thermally conductive materials.

Key Features

This paper focuses on how to design and prepare thermal management materials with high thermal conductivity.

Includes new techniques for adjusting the thermal conductivity of thermal management materials.

Recent advances in the fields related to thermal management materials are reviewed, providing valuable insights into potential development pathways.

Next Work Assumption

1. Electronic packaging materials: Focus on the bottleneck problem of the electronic industry and the demands of national strategic for improving the anti-risk ability, promote the large-scale and stable mass production and application of achievements as soon as possible, and break the foreign technological monopoly. Strengthen cooperation and exchange with domestic downstream enterprises in the production and application of packaging materials, make full use of the existing pilot production line and first-class testing platform, timely control the changes in market demand, and develop world-class electronic packaging materials.
2. Packaging process and platform: The future work will be closely focused on the "14th Five-Year Plan" national strategic development needs. Focus on high-reliability ASIC design technology, third-generation semiconductor device driving and packaging technology related research work. Strive to make breakthroughs in the fields of new energy vehicle chip and application technology, aerospace intelligent sensing and interconnection technology, design, driving, packaging and testing technology of wide band gap semiconductor power devices, special motor drive and control technology, intelligent sensing and intelligent control technology, and high reliability packaging technology.

Acknowledgments

The author sincerely thanks Shuguang Bian and Bin Yang (High Technology Research and Development Center, Ministry of Science and Technology of China), Qing Yan and Guangyi Fu (Chinese Academy of Sciences), Tao Deng (Shanghai Jiao Tong University), Hongda Chen (Institute of Semiconductors, Chinese Academy of Sciences), and Daojun Song (Hefei High-Tech Zone) for fruitful discussions. The National Key R&D Program of China (No. 2017YFB0406200) and the Science and Technology Service Network Initiative of the Chinese Academy of Sciences (guide project for innovative and entrepreneurial) (No. KFJ-STS-SCYD-112) are acknowledged for financial support of our work in this area.

1

Physical Basis of Thermal Conduction

Xian Zhang[1], Ping Zhang[2], Chao Xiao[1], Yanyan Wang[1], Xin Ding[1], Xianglan Liu[1], and Xingyou Tian[1]

[1]*Chinese Academy of Sciences, Institute of Solid State Physics, Hefei Institutes of Physical Science, 350 Shushanhu Road, Hefei, 230031, Anhui, P. R. China*
[2]*School of Materials Science & Engineering, Anhui University, 111 Jiulong Road, Hefei, 230601, Anhui, P. R. China*

1.1 Basic Concepts and Laws of Thermal Conduction

1.1.1 Description of Temperature Field

The difference in temperature drives the heat transfer from the high-temperature zone to the low-temperature zone; it is therefore important to understand the temperature distribution of the object for studying the heat transfer. The temperature distribution can be expressed in mathematical equations and image forms using a scalar temperature field, which is a function of time (t) and space coordinates (x, y, z):

$$T = f(t, x, y, z) \tag{1.1}$$

According to the temperature variation with time, the temperature field can be divided into steady temperature field and unsteady temperature field. In the steady-state temperature field, the temperature of the object is only related to space but not to time change. It can be expressed by

$$T = f(x, y, z), \frac{\partial T}{\partial t} = 0 \tag{1.2}$$

According to the dimensional correlation with the spatial coordinates, the temperature field can be divided into one-dimensional, two-dimensional, and three-dimensional temperature fields, that is, the object temperature is only related to one, two, or three coordinates of space. For example, one-dimensional steady temperature field can be expressed by

$$T = f(x), \frac{\partial T}{\partial y} = \frac{\partial T}{\partial z} = \frac{\partial T}{\partial t} = 0 \tag{1.3}$$

When the temperature field is described by an image, the image formed by connecting the same points of temperature is called an isothermal surface. By analogy, a cluster of curves is formed in the two-dimensional temperature field, which is

Thermal Management Materials for Electronic Packaging: Preparation, Characterization, and Devices, First Edition. Edited by Xingyou Tian.

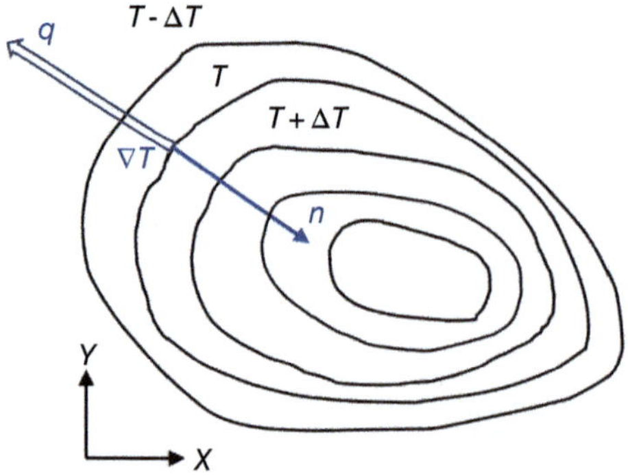

Figure 1.1 Temperature field and temperature gradient.

called an isotherm. As shown in Figure 1.1, there is no temperature difference on an isotherm, so heat can only transfer between different isotherms.

1.1.2 Temperature Gradient

As shown in Figure 1.1, the temperature difference between isotherms is the same. However, the ratio limit of the temperature difference to the distance of a point in different directions is different. The limit of the ratio of the temperature difference to the distance in the normal direction of the point is the maximum. The maximum limit value is defined as the temperature gradient, which is recorded as ∇T, it is a vector whose direction points to the direction of temperature rise along the isotherm normal and can be expressed by

$$\nabla T = \lim_{\Delta n \to \infty} \left(\frac{\Delta T}{\Delta n} \right) \tag{1.4}$$

1.1.3 Fourier's Law

In 1882, French scholar Fourier proposed the basic law of heat conduction process, namely Fourier's law. The expression is [1]

$$q = -k\nabla T \tag{1.5}$$

where q is the heat flow density in W/m^2 and k is the thermal conductivity in W/m K. Formula (1.5) shows that q is a vector, and its direction lies on the same normal of the temperature gradient isotherm, pointing to the direction of temperature reduction.

1.1.4 Heat Flux Density Field

In the space coordinate system, the heat flux expression is

$$q = q_x\overline{\boldsymbol{x}} + q_y\overline{\boldsymbol{y}} + q_z\overline{\boldsymbol{z}} \tag{1.6}$$

where $\overline{\boldsymbol{x}}, \overline{\boldsymbol{y}}, \overline{\boldsymbol{z}}$ are unit vectors in x-, y-, z-directions, respectively. The heat flux distribution inside the object constitutes a heat flux field, which is a vector field. As shown in Figure 1.2, the solid line represents the isotherm, and the dotted line represents the isochore. Isothermal streamline is a set of curves perpendicular to the isotherm everywhere. The heat flux density on the isothermic streamline is equal everywhere, the direction of the heat flux density at any point is always tangent to the isothermic

Figure 1.2 Isotherms and heat flux.

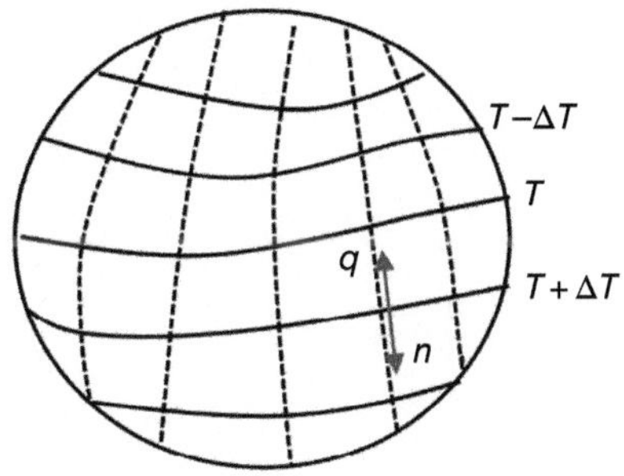

streamline, and the heat flux transmitted between two adjacent dotted lines is equal everywhere.

1.1.5 Thermal Conductivity

According to Formula (1.5),

$$k = -\frac{q}{\nabla T} = \frac{Q \cdot d}{S \cdot t \cdot \Delta T} = \frac{q \cdot d}{S \cdot \Delta T} \tag{1.7}$$

where S is the area, Q is the heat transferred through the area S in t time, and d is the distance of heat transfer. Thermal conductivity is a proportional coefficient in Fourier's law, which reflects the thermal conductivity of an object, that is, the heat flow that can be transferred through a unit area under the action of a unit temperature gradient.

The thermal conductivity of object usually varies with temperature. When the temperature range of the objective is small, the thermal conductivity is linearly related to the temperature:

$$k = k_0(1 + bT) \tag{1.8}$$

where k_0 is the thermal conductivity under a certain reference state and b is the constant determined experimentally. Due to the difference of heat conduction mechanism, the thermal conductivity of object in different forms is quite different. The heat conduction of objects is the collision and transfer of microscopic particles, including the thermal movement of molecules, the phonon movement formed by lattice vibration, and the movement of free electrons. Generally, the thermal conductivity of solid is the highest, while that of gas is the lowest.

1.2 Heat Conduction Differential Equation and Finite Solution

1.2.1 Heat Conduction Differential Equation

Figure 1.3 shows a cube element with side lengths of dx, dy, dz. Here, the density ρ, specific heat C_p, and thermal conductivity k are constants.

According to the law of conservation of energy, the sum of the net heat flowing into the cube unit in a certain time ΔQ_i and the heat generated by the cube unit itself ΔQ_p is equal to the increase in the enthalpy ΔE of the infinitesimal cube.

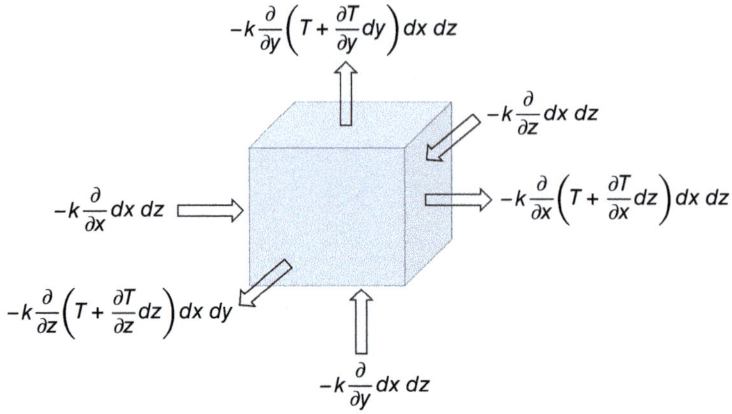

Figure 1.3 Cube element.

During $d\tau$ time, the total net inflow heat in the x-, y-, z-directions is

$$\Delta Q_i = k\left(\frac{\partial^2 T}{\partial x^2}+\frac{\partial^2 T}{\partial y^2}+\frac{\partial^2 T}{\partial z^2}\right)dx\,dy\,dz\,d\tau \tag{1.9}$$

The heat generated by the heat source in the cube unit in time $d\tau$ is

$$\Delta Q_p = q_1 dx\,dy\,dz\,d\tau \tag{1.10}$$

where q_1 is the calorific value of the heat source per unit time and volume, and the unit is W/m^3.

The increase in enthalpy of the heat source of the cube unit in time $d\tau$ is

$$\Delta E = \rho c\frac{\partial T}{\partial \tau}dx\,dy\,dz\,d\tau \tag{1.11}$$

According to the conservation of energy,

$$\rho c\frac{\partial T}{\partial \tau} = k\left(\frac{\partial^2 T}{\partial x^2}+\frac{\partial^2 T}{\partial y^2}+\frac{\partial^2 T}{\partial z^2}\right)+q_1 \tag{1.12}$$

It can also be written as

$$\frac{\partial T}{\partial \tau} = \alpha\left(\frac{\partial^2 T}{\partial x^2}+\frac{\partial^2 T}{\partial y^2}+\frac{\partial^2 T}{\partial z^2}\right)+\frac{q_1}{\rho c} = \nabla^2 T+\frac{q_1}{\rho c} \tag{1.13}$$

where ∇^2 is Laplace operator, and α is the thermal diffusion coefficient with unit of m^2/s. Formulas (1.12) and (1.13) are the differential equations of heat conduction, which describes the variation of temperature field in the heat conduction system with time and space.

Thermal diffusion coefficient ($\alpha = k/\rho c$) is a physical parameter related to the type of material, and the value depends on the thermal conductivity, density, and specific heat of the object. It reflects the ability of the object to transmit temperature changes.

For the steady-state temperature field with constant thermal conductivity, the thermal conductivity differential equation becomes

$$0 = \frac{\partial^2 T}{\partial x^2}+\frac{\partial^2 T}{\partial y^2}+\frac{\partial^2 T}{\partial z^2}+\frac{q_1}{k} = \nabla^2 T+\frac{q_1}{\rho c} \tag{1.14}$$

For steady-state temperature field with constant thermal conductivity and no internal heat source, the thermal conductivity differential equation is expressed as

$$0 = \frac{\partial^2 T}{\partial x^2} + \frac{\partial^2 T}{\partial y^2} + \frac{\partial^2 T}{\partial z^2} \tag{1.15}$$

The above equation is called Laplace equation, which is the most basic differential equation for studying the steady-state temperature field.

1.2.2 Definite Conditions

The differential equation of heat conduction is a universal equation for heat conduction problems. For specific heat conduction problems, the eigenvalue conditions of the corresponding problem must be given. These conditions include initial conditions (initial state of heat conduction system) and boundary conditions (interface characteristics, relationship between system and environment), where the initial condition expression is

$$T = f(x, y, z, 0) \tag{1.16}$$

If the initial temperature is uniform and constant ($T = T_i$), the initial condition is not required for the steady-state heat conduction problem. There are three common types of boundary problems.

The first boundary condition is that the boundary temperature distribution of the system is known:

$$T = f(x, y, z, 0) \tag{1.17}$$

The second boundary condition is that the heat flux distribution on the boundary is known:

$$q = q(x, y, z) \tag{1.18}$$

The third boundary condition is that the convective heat transfer coefficient and fluid temperature between the object and the surrounding fluid are known:

$$-k\frac{\partial T}{\partial n} = \alpha(T - T_\infty) \tag{1.19}$$

1.3 Heat Conduction Mechanism and Theoretical Calculation

The heat transfer of gases relies on the thermal motion and collision of molecules and atoms. The heat transfer in liquids relies on irregular elastic vibrations (or similar gases). In solids, heat energy is transferred through electrons and lattice vibrations. In metal solids, heat conduction is mainly realized by the interaction and collision of electrons. On the contrary, in dielectric solids including semiconductors and insulators, heat is mainly transferred through quantized lattice vibration. Among these heat conduction mechanisms, the speed of heat transfer through electron collision is the highest, while the one by molecular or atomic collision is the lowest.

1.3.1 Gases

The heat conduction mechanism of gases involves the heat transfer caused by the thermal movement of molecules and the collision between molecules. According to the theory of ideal gas molecular motion, the mathematical expression of molecular heat conduction mechanism can be deduced as [1]

$$k = \frac{1}{3}\overline{C_v}v\bar{l} \tag{1.20}$$

where k is the thermal conductivity, $\overline{C_v}$ is the heat capacity per unit volume of gas, v is the average velocity of gas molecules, and $\bar{l}$ is the average free path of gas molecules. Because the heat capacity of the gas and the average speed of molecular motion increase with the increase in temperature, the thermal conductivity of the gas also increases with the increase in temperature.

Chapman and Cowling [2] associate the thermal conductivity of simple gas with viscosity and specific heat at constant volume, and the expression is

$$k = f\eta C_v \tag{1.21}$$

where f is constant, 2.5 for smooth spherically symmetric molecules and 2.522 for rigid elastic spheres; η is the kinematic viscosity with a unit of kg/(m s); and c_v is the specific heat of constant volume with a unit of W/(kg K).

Eucken [3] correlated the thermal conductivity of monatomic gas with viscosity and specific heat at atmospheric pressure and 0 °C:

$$k = \frac{9r - 5}{4} \cdot \eta \cdot c_v \times 420(\mathrm{W/(m\,K)}) \tag{1.22}$$

Hirschfelder et al. [4] later modified the Euken equation:

$$k = \left(\frac{H}{v}\right)\eta/M(c_v + 9R/4) \tag{1.23}$$

where H and v are the complex interaction coefficients between molecular pairs, M is the molecular weight, and R is the ideal gas constant.

1.3.2 Solids

The lattice in solid is fixed and can only vibrate slightly near its equilibrium position. The heat conduction in solids is mainly realized by the lattice wave of lattice vibration and the movement of free electrons. According to quantum theory, the energy of lattice vibration is quantized, and the quantum of lattice vibration is usually called phonon [5]. Therefore, phonons can be regarded as free "gas" particles by analogy with gas heat conduction [6]. The contribution of phonons or electrons to heat conduction varies greatly depending on the type of solid.

1.3.2.1 Metals

There are a large number of unbound free electrons with light mass in metals. The electrons move like "electron gas." The interaction or collision between electrons is the main mechanism of metal heat conduction, that is, electronic heat conduction. Metals are also crystals; hence, lattice vibration (phonon) also contributes to the heat transfer in metals. Therefore, the thermal conductivity k of metal can be expressed by the following formula:

$$k = k_e + k_p \tag{1.24}$$

where k_e is the thermal conductivity representing the contribution of free electrons, and k_p is the thermal conductivity representing the contribution of phonon.

Based on the results of the kinetic theory of gas molecules, the mathematical expression of k_e is

$$k_e = \frac{1}{3}C_{ve} \cdot \overline{V_e} \cdot \bar{l}_e = \frac{1}{3}C_{ve} \cdot \overline{V_e}^2 \cdot \tau \tag{1.25}$$

where C_{ve} is the heat capacity of electrons per unit volume, $\overline{V_e}$ is the average velocity of electrons, and $\bar{l}_e$ is the average free path of the electron.

Free electrons in metals serve as both the carrier of heat and the carrier of electricity, so metal heat conduction and conductivity are closely related. The relationship between thermal conductivity and conductivity of metals follows Wiedman–Franz law:

$$\frac{k}{\sigma T} = \frac{\pi^2 {k_B}^2}{3e^2} \tag{1.26}$$

where σ is the electrical conductivity, e is the absolute value of electronic charge, and k_B is the Boltzmann constant. The law shows that the thermal conductivity of metal is proportional to the electrical conductivity.

The law of electron heat conduction in metal changing with temperature is shown in Figure 1.4. At very low temperatures, the electron thermal conductivity increases linearly with temperature; at medium temperatures, the electron thermal conductivity is almost constant and does not change with temperature; and at very high temperatures, the electron thermal conductivity decreases slightly with the increase in temperature.

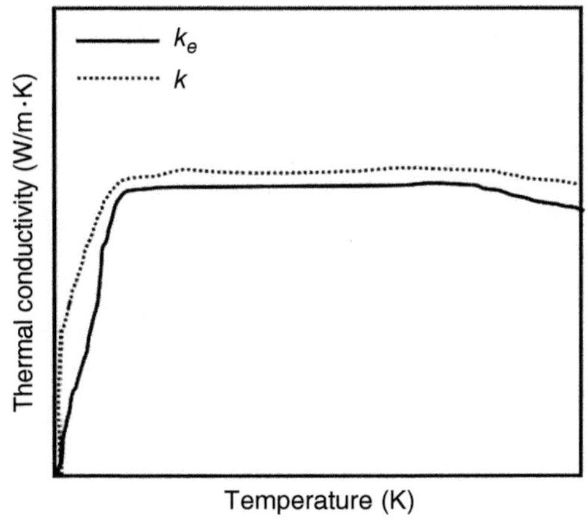

Figure 1.4 Relationship between metal thermal conductivity and temperature.

1.3.2.2 Inorganic Nonmetals

Crystal In dielectric crystals, heat energy is transferred by lattice vibration. Therefore, the propagation of lattice waves is regarded as the movement of phonons. The scattering encountered by lattice waves in the crystal is regarded as the collision between phonons, phonons and grain boundaries, and lattice defects. The thermal resistance in an ideal crystal is attributed to the collision between phonons.

According to Debye's assumption, by analogy with gases, it is considered that the mathematical expression of thermal conductivity in dielectric crystals should be the similar to that in gases. Therefore, the expression of phonon thermal conductivity can be expressed by

$$k_{pn} = \frac{1}{3} C_{vpn} \cdot \overline{V}_{pn} \cdot \overline{l}_{pn} \tag{1.27}$$

where C_{vpn}, $\overline{V}_{pn}$, and $\overline{l}_{pn}$ are the volume heat capacity, average velocity, and average free path of phonons, respectively.

According to the Klemens model, the lattice thermal conductivity of blocky solid is given by

$$k_{pn} = \frac{1}{3} \int \frac{\partial E_i(q,T)}{\partial T} v_i^2(q) \tau_i(q) dq \tag{1.28}$$

where i is the index of phonons, v is the group velocity of phonons, τ is the relaxation time, and q is the wave vector.

The curve of thermal conductivity of dielectric crystal changing with temperature is shown in Figure 1.5 [7]. At very low temperatures, the phonon-specific heat capacity is proportional to T^3. It means that heat conduction of crystal increases proportionally with the third power of temperature [8]. At higher temperatures, on the one hand, the phonon heat capacity does not change with temperature and is close to a constant 3R. On the other hand, the mean free path of phonons decreases gradually with the increase in temperature, and the mean free path of phonons is inversely proportional to the temperature ($l \propto 1/T$). Therefore, the thermal conductivity of the dielectric crystal at a higher temperature decreases [9].

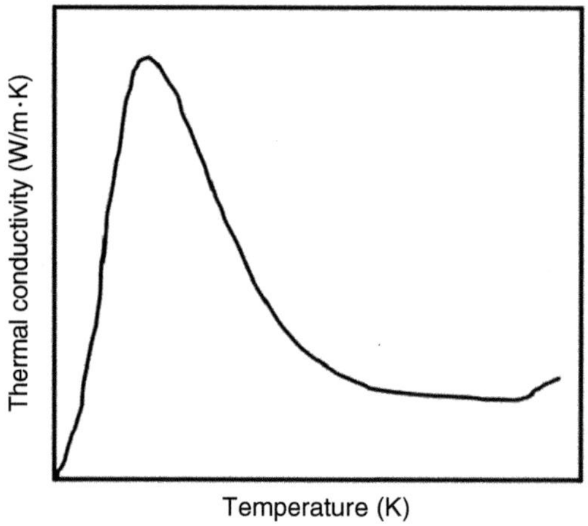

Figure 1.5 Relationship between thermal conductivity and temperature of dielectric crystals.

The formulas for calculating the theoretical thermal conductivity of dielectric crystals are summarized below:

Debye formula

$$k = \frac{1}{16\pi^3} \cdot \frac{3}{3\alpha^2 + 1} \cdot \frac{\rho v^5}{x_0 \nu^4} \cdot \frac{C}{k_B T} \tag{1.29}$$

where ρ is the density, v is the propagation velocity of phonons, C is the heat capacity, x_0 is the compression coefficient, ν is the frequency, k_B is the Boltzmann constant, T is the absolute temperature, and α is the correlation coefficient.

Compton formula

$$k = bn^2 dR\nu \frac{2 - C_0 T e^{-\frac{\beta\nu}{2T}}}{C_0{}^2 T} \tag{1.30}$$

where $b = 4m_1 m_2/(m_1 + m_2)$, m is the mass of different atoms, n^2 is the number of atoms per unit area perpendicular to the heat flow, d is the atomic spacing parallel to the heat flow, R is the gas constant of a single molecule, ν is the atomic vibration frequency, C_0 is the constant determined by thermal conductivity at a given temperature, T is the absolute temperature, and β is a constant.

Pell formula

$$\frac{1}{k} = AT + AT^B e^{-\frac{\Theta}{2T}} \tag{1.31}$$

where Θ is the Debye temperature, T is the absolute temperature, and A and B are constants.

Papapecchu formula

$$k = \frac{l_0 u_0 v}{3V} \left(\frac{C}{u} + \frac{u}{3RT^2} \right) \tag{1.32}$$

where V is the gram atomic volume; $l = l_0 u_0/u$ is the average free path; R is the gas constant; $u = 3LE$, E is the average energy of a wave and L is the Loschmidt constant; v is the average propagation speed; C is the specific heat of gram atom; and T is the absolute temperature.

Entuo formula

$$k = \frac{3nk\pi\nu Nd}{2} \left(\frac{h\nu}{kT} \right)^2 \left[\frac{e^{\frac{h\nu}{k_B T}}}{\left(e^{\frac{h\nu}{k_B T}} - 1 \right)^{\frac{3}{2}}} \right] \tag{1.33}$$

where k_B is the Boltzmann constant, n is a factor determined by the type of spatial lattice, N is the number of atoms per unit area, d is the atomic spacing, R is the gas constant, h is Planck constant, ν is the natural vibration frequency, and T is the absolute temperature.

Noncrystal Inorganic amorphous materials have a long-range disorder and short-range ordered structure. To analyze their heat conduction mechanism, they are generally treated as a "crystal" composed of very fine grains with only a few lattices spacing sizes. Allen and Feldman [10] studied the heat transfer mechanism in amorphous materials using lattice dynamics calculations and divided the vibration modes into three different categories: plane wave propagator, diffuser, and locator. The relative contribution of each mode determines the thermal conductivity of the material, including its value, dependence on sample size, and temperature [11]. In 1911, Einstein [12], first proposed the "amorphous limit" model to predict the minimum lattice thermal conductivity of amorphous solids. In 1979, Slack [13] further developed the "amorphous Limit" model, which equates the minimum mean free path of phonons with the wavelength of lattice waves and is called the "minimum thermal conductivity" model.

Cahill–Pohl model [14] simulates the thermal conductivity of amorphous solids by assuming that the average free path of each vibration mode is half of its wavelength. Agne et al. [15] simplified the Cahill–Pohl model results to

$$k = 1.21k_B\frac{1}{3}(2v_T + v_L)n^{2/3} \tag{1.34}$$

where k_B is the Boltzmann constant, n is the atomic number density, v_T is the lateral propagation velocity, and v_L is the longitudinal speed. In 1992, Cahill et al. [15] further modified the "minimum thermal conductivity" model.

For the relationship between thermal conductivity and temperature of inorganic amorphous, the contribution of photonic heat conduction should be considered at higher temperatures, and it will be mainly determined by the relationship between heat capacity and temperature in other temperature ranges. The thermal conductivity of inorganic amorphous materials changes with temperature as shown in Figure 1.6 [17]. At low temperatures, the thermal conductivity of inorganic amorphous increases monotonously with the increase in temperature. At this time, the change of phonon heat conduction with temperature is determined by

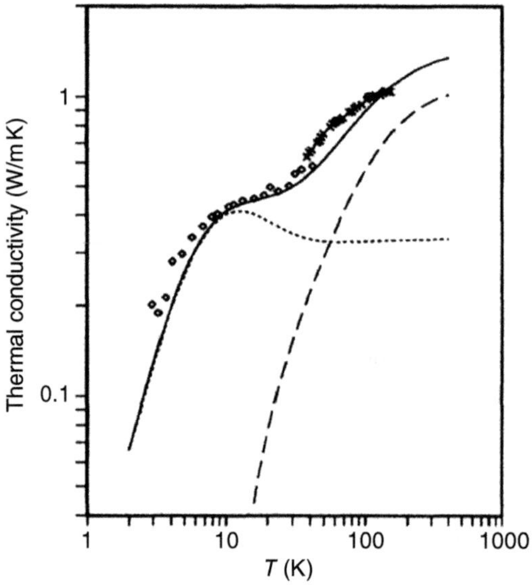

Figure 1.6 Relationship between thermal conductivity and temperature of inorganic amorphous solids. Source: Reproduced with permission from Ref. [16]; ©1993 American Physical Society.

the change rule of phonon heat capacity with temperature. As the temperature increases, the phonon heat capacity increases, so the thermal conductivity of the amorphous increases accordingly. From medium temperature to high temperature, although the temperature increases, the phonon heat capacity no longer increases but gradually becomes constant. Therefore, the phonon heat conduction no longer varies with temperature, and the thermal conductivity curve presents a straight line nearly parallel to the horizontal axis. Above the high temperature, the phonon heat conduction changes little with the further increase in temperature. As the average free path of photons increases significantly, the thermal conductivity of inorganic amorphous is proportional to the third power of temperature.

1.3.3 Liquids

The heat conduction mechanism of liquids is still controversial. One view is that the heat conduction mechanism of liquids is similar to that of gases. However, the unreasonable thing is that the distance between liquid molecules is relatively close, and the intermolecular force has a greater impact on the collision process than in gases. Another point is that the heat conduction mechanism of liquid is similar to that of non-metallic solids, mainly due to the effect of elastic waves.

The simplest equation applicable to all organic liquids is [18]

$$k = A\left[1 + B(1 - T/T_c)^{1/3} + C(1 - T/T_c)^{2/3} + D(1 - T/T_c)\right] \tag{1.35}$$

where T_c is the critical temperature, B, C, D are constants, A is the pseudo-critical thermal conductivity.

Bridgman [19] provides a simple formula that can accurately calculate the thermal conductivity of some liquids:

$$k = \frac{2k_B v_s}{d^2} = 2k_B v_s n^{2/3} \tag{1.36}$$

where v_s is the longitudinal speed of sound in a liquid, d is the mean distance between the centers of molecules and can be calculated from $d = n^{-1/3}$, and n is the number density of molecules in the liquid.

Chapman–Enskog [20] gives the thermal conductivity of liquids as follows:

$$k = \frac{k_0}{g(D)}\left[1 + \frac{2}{5}\pi\rho D^3 g(D)\right]^2 + \frac{3k_B}{2m}\text{II} \tag{1.37}$$

where D is the effective diameter of the molecule, m is the mass of a liquid molecule, ρ is the density of the liquid (kg/m^3), and T is the absolute temperature.

van Elk vet al. [21] developed an equation for the correlation between the components of binary mixed liquids and thermal conductivity, and verified various types of mixed liquids:

$$k_m = w_1k_1 + w_2k_2 - 0.72w_1w_2(k_2 - k_1) \tag{1.38}$$

The components are so selected that $k_2 \geq k_1$; the constant 0.72 may be replaced by an adjustable parameter when $k_2 < k_1$.

In 1974, Saksena and Harminder [22, 23] also successively proposed and improved the calculation formula for thermal conductivity of binary mixed liquids; Jamieson suggested that the thermal conductivity of binary mixture can be calculated by the following formula:

$$k = w_1 k_1 + w_2 k_2 - \alpha(k_2 - k_1)\left[1 - (w_2)^{1/2}\right] w_2 \tag{1.39}$$

The components are so selected that $k_2 \geq k_1$; α is an adjustable parameter that is set equal to unity if mixture data are unavailable for regression purposes. Equation (1.39) enables one to estimate m within about 7% for all types of binary mixtures with or without water.

In 1982, Rowley Method [24] proposed the thermal conductivity equation for multi-component mixed liquids using the two liquid theory modeling:

$$k_m = \sum_{i=1}^{n} w_i \sum_{j=1}^{n} w_{ji} k_{ji} \tag{1.40}$$

where k_m is the liquid mixture thermal conductivity, W/(m K); w_i is the weight fraction of component; w_{ji} is the local weight fraction of component j relative to a central molecule of component i; and k_{ji} is the characteristic parameter for the thermal conductivity that expresses the interactions between j and i, W/(m K).

In 1976, Li [25] provided a new idea for the calculation of thermal conductivity of multi-component mixed liquids:

$$k_m = \sum_{i=1}^{n} \sum_{j=1}^{n} \varphi_i \varphi_j k_{ji} \tag{1.41}$$

$$k_{ji} = 2\left(k_i^{-1} + k_j^{-1}\right)^{-1}$$

$$\varphi_i = \frac{X_i V_i}{\sum_{j=1}^{n} X_i V_j}$$

where X_i is the mole fraction of component i and φ_i is the superficial volume fraction of i. V_i is the molar volume of the pure liquid.

1.4 Factors Affecting Thermal Conductivity of Inorganic Nonmetals

1.4.1 Temperature

After summarizing the measurement results of the thermal conductivity of a large number of dielectric crystals and amorphous solids, Eueken [26] concluded that (i) the thermal conductivity of single crystals increases with the decrease of temperature, while the thermal conductivity of amorphous solids shows inverse trend; (ii) the thermal conductivity is inversely proportional to the absolute temperature of the crystal; and (iii) the thermal conductivity of amorphous solids is roughly proportional to the specific heat.

Except for lead oxide, the thermal conductivity of inorganic non-metallic crystals is almost proportional to the reciprocal of absolute temperature above room temperature [10]:

$$k = \frac{A}{T};\ k = \frac{A}{T} + B \tag{1.42}$$

where T is the absolute temperature, and A and B are constants determined experimentally.

For inorganic amorphous, the thermal conductivity increases approximately in direct proportion to the temperature [27]:

$$k = CT \tag{1.43}$$

where T is the absolute temperature, and C is the constant determined experimentally.

For the mixture of inorganic crystal and amorphous, the relationship between thermal conductivity and temperature can be expressed as

$$k = \frac{1}{AT + B + C/T} \tag{1.44}$$

1.4.2 Pressure

Under pressure, the correlation between thermal conductivity and pressure can be divided into increasing, decreasing, independent, and abnormal trends.

For example, the thermal conductivity of an object can increase sharply under pressure. This is because the strain generated by pressure enhances the atomic interaction and compresses the bond, changing the phonon dispersion, thus greatly increasing the propagation speed of phonons. Such enhancement is nonlinear in some cases, which can be attributed to the combined effect of decreasing phonon relaxation time and increasing phonon group velocity [28].

In addition, the thermal conductivity of the object can also be reduced under pressure, because phonon anharmonicity and phonon softening induced by pressure increase [29]. The first-principle calculation shows that the reduced thermal conductivity under pressure is mainly due to the stronger cubic anharmonic interaction, large mass ratio, and significant acousto-optic frequency gap [30].

In some cases, the thermal conductivity is independent of the pressure, which may be caused by the strong electron correlation effect driven by the electronic topological transition [31].

Several mechanisms have been proposed to understand abnormal thermal conductivity changes under pressure. The abnormal decrease of thermal conductivity of some materials under pressure can be ascribed to the large phonon frequency gap of materials under high pressure. The contribution of the inherent three-phonon scattering process is smaller than that of other cases, and the complex scattering process between more phonons dominates and increases the overall phonon scattering [32]. In the case of another non-monotonic behavior, the thermal conductivity first increases and then decreases with the increase in pressure. The possible mechanism of this behavior is the competitive scattering process of three phonons interacting with four phonons under high pressure [33] or the interaction between group velocity and phonon relaxation time under pressure [34]. When some rare earths show different pressure dependences of thermal conductivity, the pressure dependence is determined by the competition between the enhancement of phonon group velocity and the reduction of phonon relaxation time [35].

1.4.3 Crystal Structure

The more complex the crystal structure, the lower the thermal conductivity. For example, the structure of magnesia alumina spinel ($MgAl_2O_4$) is similar to that of alumina (Al_2O_3) and magnesium oxide (MgO), and their thermal solubility, thermal expansion coefficient, and elastic modulus are similar. However, due to the complex structure of the former, its thermal conductivity is much lower than the latter two. In addition, the mean free path of phonons of materials with complex structures is easy to approach or reach its minimum limit value, that is, the lattice size value, at high temperatures to obtain lower thermal conductivity.

The structural integrity and regularity of polycrystals are worse than that of single crystals. In addition, the influence of impurities and distortion on the grain boundary increases the phonon scattering, which is more significant at higher temperatures. In different crystal axis directions, the thermal conductivity of single crystals is also different, with anisotropy, and the phonon scattering is different. The difference of thermal conductivity of single crystals in different crystal axis directions decreases with the increase in temperature, because anisotropic crystals tend to improve or enhance their symmetry with the increase in temperature.

1.4.4 Thermal Resistance

As we all know, the smaller the thermal resistance, the greater the thermal conductivity. The main factor affecting thermal resistance is Debye temperature. The Debye temperature is closely related to atomic weight, theoretical density, and compressibility. The total thermal resistance in the crystal is

$$r = \frac{AT^m}{\left(e^{\frac{\Theta}{2T}} - 1\right)} + BNT^n + 1/(VLT^3) \tag{1.45}$$

where A is the atomic weight, Θ is the Debye temperature, ρ is the theoretical density, and X is the compression coefficient. Generally, the smaller the theoretical density is, the greater the Debye temperature is. The greater the Debye temperature, the smaller the thermal resistance. Therefore, lighter substances usually have higher thermal conductivity. The smaller the compression coefficient or the larger the Young's modulus, the higher the Debye temperature. The higher the Debye temperature, the smaller the thermal resistance. Therefore, materials with higher binding energy usually have higher thermal conductivity. The smaller the atomic weight, the higher the Debye temperature. Therefore, materials with large atomic weights have high thermal conductivity.

1.4.5 Others

Defects and impurities have a great influence on the thermal conductivity of materials, which is determined by the phonon heat conduction mechanism of dielectric crystals [10]. They are the center of phonon scattering, which will reduce the mean free path and thermal conductivity of materials [7]. Because the thermal conductivity of gas in the porosity is far lower than that of materials, porosity always reduces the thermal conductivity of the materials. Peierls-Boltzmann [36] revealed the process of strain affecting thermal conductivity and found the power law scaling relationship between thermal conductivity and temperature and strain. Strain first affects the speed of phonon propagation and then affects the relaxation time, while temperature only affects the relaxation time.

The experimental work of Krupskii and Manzhelli on the unconstrained argon sample revealed that the thermal conductivity has a deviation in addition to the inverse temperature dependence predicted theoretically [37]. They attributed the bias to the higher-order four-phonon interaction, while Clayton and Batchelder attributed it to the thermal expansion effect [38]. Dugdale and MacDonald assumed that micro-area lattice expansion would occur due to thermal expansion resulting from the temperature gradient [39]. For example, for a material with a positive thermal expansion coefficient, the relative expansion of the hotter region and the compression of the cooler region produces additional heat transfer sources, which further reduce the thermal conductivity.

References

1 Zhao, A.Z., Wingert, M.C., Chen, R. et al. (2021). Phonon gas model for thermal conductivity of dense, strongly interacting liquids. *Journal of Applied Physics* 129 (23): 235101.

2 Chapman, S. and Cowling, T.G. (1952). *The Mathematical Theory of Nonuniform Gases*. New York: Cambridge Univ. Press.

3 Hirschfelder, J.O., Bird, R.B., and Spotz, E.L. (1948). The transport properties for non-polar gases. *The Journal of Chemical Physics* 16 (10): 968–981.

4 Maloka, I.E., Hashim, E.T., and Ibrahim, S.Y. (2004). Effect of molecular weight on thermal conductivity of gases. *Petroleum Science and Technology* 22 (11–12): 1507–1511.

5 Peierls, R. (1929). The kinetic theory of thermal conduction in crystals. *Annalen der Physik* 3 (8): 1055–1101.

6 Lv, W. and Henry, A. (2016). Examining the validity of the phonon gas model in amorphous materials. *Scientific Reports* 6: 37675.

7 Chen, Y., Lukes, J.R., Li, D. et al. (2004). Thermal expansion and impurity effects on lattice thermal conductivity of solid argon. *The Journal of Chemical Physics* 120 (8): 3841–3846.

8 Klemens, P.G. and Simon, F.E. (1951). The thermal conductivity of dielectric solids at low temperatures (Theoretical). *Proceedings of the Royal Society of London. Series A: Mathematical and Physical Sciences* 208 (1092): 108–133.

9 Kittel, C. (1956). *Introduction to Solid State Physics*, 2e.

10 Kingery, W.D. and McQuarrie, M.C. (1954). Thermal conductivity: I, concepts of measurement and factors affecting thermal conductivity of ceramic materials. *Journal of the American Ceramic Society* 37 (2): 67–72.

11 Wingert, M.C., Zheng, J., Kwon, S. et al. (2016). Thermal transport in amorphous materials: a review. *Semiconductor Science and Technology* 31 (11): 113003.

12 Einstein, A. (1911). Über den Einfluß der Schwerkraft auf die Ausbreitung des Lichtes. *Annalen der Physik* 340 (10): 898–908.

13 Slack, G.A. (1979). The thermal conductivity of nonmetallic crystals. In: *Solid State Physics*, vol. 34 (ed. H. Ehrenreich, F. Seitz, and D. Turnbull), 1–71. Academic Press.

14 Cahill, D.G., Watson, S.K., and Pohl, R.O. (1992). Lower limit to the thermal conductivity of disordered crystals. *Physical Review B: Condensed Matter* 46 (10): 6131–6140.

15 Agne, M.T., Hanus, R., and Snyder, G.J. (2018). Minimum thermal conductivity in the context of diffuson-mediated thermal transport. *Energy & Environmental Science* 11 (3): 609–616.

16 Feldman, J.L., Kluge, M.D., Allen, P.B. et al. (1993). Thermal conductivity and localization in glasses: numerical study of a model of amorphous silicon. *Physical Review B* 48 (17): 12589–12602.

17 Kittel, C. (1949). Interpretation of the thermal conductivity of glasses. *Physical Review* 75 (6): 972–974.

18 Jamieson, D.T. (1979). Thermal conductivity of liquids. *Journal of Chemical and Engineering Data* 24 (3): 244–245.

19 Bridgman, P.W. (1923). The thermal conductivity of liquids. *Proceedings of the National Academy of Sciences* 9 (10): 341–345.

20 Huber, M.L., Perkins, R.A., Friend, D.G. et al. (2012). New international formulation for the thermal conductivity of H_2O. *Journal of Physical and Chemical Reference Data* 41 (3): 033102.

21 van Elk, E.P., Arendsen, A.R.J., and Versteeg, G.F. (2009). A new flowsheeting tool for flue gas treating. *Energy Procedia* 1 (1): 1481–1488. https://doi.org/10.1016/j.egypro.2009.01.194.

22 Harminder, M.P.S.a. (1974). Thermal Conductivity of binary liquid mixture. *Chemical Physics Letters* 25 (3): 445–446.

23 Harminder, M.P.S.a. (1974). Thermal conductivity of binary liquid mixtures. 27 (3): 448–449.

24 Rowley, R.L. (1982). A local composition model for multicomponent liquid mixture thermal conductivities. *Chemical Engineering Science* 37 (6): 897–904.

25 Maloka, I.E. (2007). Thermal conductivities of liquid mixtures. *Petroleum Science and Technology* 25 (8): 1065–1072.

26 Eucken, A. (1911). Über die Temperaturabhängigkeit der Wärmeleitfähigkeit fester Nichtmetalle. *Annalen der Physik* 339 (2): 185–221.

27 Berman, R. and Simon, F.E. (1951). The thermal conductivities of some dielectric solids at low temperatures (Experimental). *Proceedings of the Royal Society of London. Series A: Mathematical and Physical Sciences* 208 (1092): 90–108.

28 Sun, Z., Yuan, K., Zhang, X. et al. (2018). Pressure tuning of the thermal conductivity of gallium arsenide from first-principles calculations. *Physical Chemistry Chemical Physics* 20 (48): 30331–30339.

29 Zhou, Y., Dong, Z.-Y., Hsieh, W.-P. et al. (2022). Thermal conductivity of materials under pressure. *Nature Reviews Physics* 4 (5): 319–335.

30 Wang, L., Tian, F., Liang, X. et al. (2019). High-pressure phases of boron arsenide with potential high thermal conductivity. *Physical Review B* 99 (17).

31 Saha, P., Mazumder, A., and Mukherjee, G.D. (2020). Thermal conductivity of dense hcp iron: Direct measurements using laser heated diamond anvil cell. *Geoscience Frontiers* 11 (5): 1755–1761.

32 Lindsay, L., Broido, D.A., Carrete, J. et al. (2015). Anomalous pressure dependence of thermal conductivities of large mass ratio compounds. *Physical Review B* 91 (12): 121202(R).

33 Ravichandran, N.K. and Broido, D. (2019). Non-monotonic pressure dependence of the thermal conductivity of boron arsenide. *Nature Communications* 10 (1): 827.

34 Yuan, K., Zhang, X., Tang, D. et al. (2018). Anomalous pressure effect on the thermal conductivity of ZnO, GaN, and AlN from first-principles calculations. *Physical Review B* 98 (14): 144303.

35 Lan, G., Ouyang, B., and Song, J. (2015). The role of low-lying optical phonons in lattice thermal conductance of rare-earth pyrochlores: a first-principle study. *Acta Materialia* 91: 304–317.

36 Bhowmick, S. and Shenoy, V.B. (2006). Effect of strain on the thermal conductivity of solids. *The Journal of Chemical Physics* 125 (16): 164513.

37 Krupskii, I.N. and Manzhely, V.G. (1967). Thermal conductivity of solid argon. *Physica Status Solidi B* 24 (1): K53–K56.

38 Clayton, F. and Batchelder, D.N. (1973). Temperature and volume dependence of the thermal conductivity of solid argon. *Journal of Physics C: Solid State Physics* 6 (7): 1213.

39 Dugdale, J.S. and Macdonald, D.K.C. (1955). Lattice thermal conductivity. *Physical Review* 98 (6): 1751–1752.

2

Electronic Packaging Materials for Thermal Management

Xian Zhang[1], Ping Zhang[2], Chao Xiao[1], Yanyan Wang[1], Xin Ding[1], Xianglan Liu[1], and Xingyou Tian[1]

[1]*Chinese Academy of Sciences, Institute of Solid State Physics, Hefei Institutes of Physical Science, 350 Shushanhu Road, Hefei, 230031, Anhui, P. R. China*
[2]*School of Materials Science & Engineering, Anhui University, 111 Jiulong Road, Hefei, 230601, Anhui, P. R. China*

2.1 Definition and Classification of Electronic Packaging

Electronic packaging is a whole concept, including all technical contents from integrated circuit bare chip to integrated electronic assembly. Electronic packaging has developed from providing mechanical support, protection, electrical and thermal connection for chips in the early days to gradually integrating into chip manufacturing technology and system integration technology. Electronic packaging directly affects the electrical, thermal, optical, and mechanical properties of integrated circuits and devices. With the arrival of the "post Moore era," the competition in the electronic information industry is mainly reflected in the packaging of electronic products.

2.1.1 Definition of Electronic Packaging

Electronic packaging is a comprehensive design and manufacturing technology. Electronic packaging refers to the reasonable arrangement, assembly, bonding, and interconnection of various components and chips constituting electronic components and modules according to the specified circuit requirements [1]. In addition, the electronic components and modules are isolated from the external environment to achieve the purpose of protection. The scope of electronic packaging is very wide. In the narrow sense, it can generally refer to the welding, electrical interconnection, and encapsulation of devices, chips, or modules. In the broad sense, it can be extended to the assembly and debugging of electronic complete machines.

To be specific, electronic packaging in a narrow sense is the process of assembling integrated circuits into the final product of chips. To put it simply, the bare chips of integrated circuits produced by wafer foundry are placed on a substrate that plays a bearing role, the pins are led out, and then packaged as a whole. The broad sense

Thermal Management Materials for Electronic Packaging: Preparation, Characterization, and Devices, First Edition. Edited by Xingyou Tian.

of electronic packaging should be the sum of the narrow sense of packaging, real packaging engineering, and substrate technology. It refers to the transformation of the electronic and physical functions of semiconductors and electronic components into forms that can be applied to equipment or systems and make them serve human society.

2.1.2 Functions of Electronic Packaging

As electronic products are composed of semiconductor chips, miniaturized passive components, and interconnects, packaging and assembly technologies constitute a bridge between silicon chips and electronic systems. For electronic devices, electronic packaging has the following five basic functions:

(1) Providing mechanical support for the chip
    Silicon chip is brittle and does not tolerate stress. Packaging can prevent damage to the chip by external forces. It can also match the thermal expansion coefficient of the chip with that of the frame or substrate, which can alleviate the stress caused by changes in the external environment and the stress caused by the chip heating in the working state.
(2) Environmental protection
    The active devices of the chip are mainly concentrated in the area of several micrometers thick on the silicon surface, and these active devices are also connected through the interconnection of the chip surface. Electronic packaging can prevent these areas from physical and mechanical damage or invasion of external chemical substances, thus maintaining the performance of chips, electronic components, and functional components as much as possible.
(3) Electrical connection
    Electronic packaging is an important means to realize the electrical connection between the chip and its peripheral components. At present, the typical channel size of metal oxide semiconductor (MOS) tubes in the chip has reached below 32 nm; solder spot size of chip reaches 10 μm; external pin size of chip package reaches 100 μm; and the line width of the substrate reaches the mm level. The electrical signal transmission and chip power drive between them are realized through the interconnection technology and assembly in packaging. Here, electronic packaging can transform the circuit from small to large, from difficult to easy, and from complex to simple. Therefore, electronic packaging reduces the material and processing cost, improving the work efficiency and reliability.
(4) Standardization of integrated circuit appearance
    The general function of specification refers to the standard specification of package size, shape, number of pins, pitch, length, etc. The standardized distribution of I/O ports of the chip can obtain pin pitches that are easier to process in assembly and achieve the standardization of assembly interfaces, thus improving the processing capacity in subsequent assembly. Standardization is very convenient for packaging users, circuit board manufacturers, and chip manufacturers. Standardization is not only convenient for processing but also

convenient for matching with subsequent processes, meeting the universality requirements of the production line.

(5) Provide heat dissipation path
With the increase in the clock frequency and functions of the chip, the power consumption per unit area has also increased correspondingly, which also puts forward higher requirements for the heat dissipation of integrated circuits. The cooling and heat dissipation function should be strengthened. When the power consumption of the chip is more than 2–3 W, heat sink or heat sink shall be installed on the package; If the power consumption of the chip is above 5–10 W, forced cooling is required.

2.1.3 The Levels of Electronic Packaging

The electronic packaging can be divided into several levels, but the development trend of electronic packaging is to simplify and reduce the level of packaging [2]. The driving force of packaging technology innovation is to make packaging devices miniaturized, low cost, and high performance. Too complex electronic packaging level can no longer meet the future development needs of the electronic packaging field. Therefore, electronic packaging materials with integrated integration, simple structure, and excellent performance will be the favorite in this field in the future.

The packaging and interconnection of electronic devices can be divided into four levels:

(1) Zero-level package
Zero-level packaging refers to the interconnection within a semiconductor chip: the electrical connection between MOS tubes in a bare chip for signal transmission or power supply within the chip.

(2) Primary packaging
The primary packaging is the sealing protection of the bare chip. It is to seal the chip and package it into a microelectronic integrated circuit, including the interconnection between the semiconductor chip and the package shell, such as lead bonding, automatic tape welding, and solder bump bonding. Another type of primary-level package is about the assembly of multi-chip and wafer-level hybrid integrated circuits. Most of them are large-scale hybrid packages, multi-device integrated packages, etc. They include multi-chip modules and system in package, etc.

(3) Secondary packaging
Secondary packaging refers to the assembly of printed circuit boards. It refers to the welding of various microelectronic packaging products, other types of components, and a small number of bare chips on printed circuit boards. At this time, the whole circuit is no longer encapsulated in the secondary packaging. For high-frequency and high-density circuits, it also includes the impedance control of wiring, the fineness of wiring, electromagnetic compatibility, and the application of low dielectric constant dielectric materials.

(4) Tertiary packaging
Tertiary packaging is about the interconnection between plug-in interface, motherboard, and components. Tertiary packaging is actually a three-dimensional assembly technology, which is to insert each plug-in board or card after secondary packaging onto a larger motherboard.

2.2 Thermal Management in Electronic Equipment

With the increase in heat flux of electronic devices, thermal management needs to be given priority in electronic packaging design. Thermal management can maintain system performance and reliability, prevent catastrophic thermal failure, and extend the service life of electronic systems.

2.2.1 Thermal Sources

The improvement of power density and current intensity of microprocessors has made them become the main heat source, so it is necessary to conduct thermal management on-chip hotspots and Joule heat of packaging and interconnection.

(1) Power density
For microprocessors, power dissipation is proportional to the capacitance of the logic elements, the square of the operating voltage swing, and the operating frequency:

$$P \sim NCV^2 f \tag{2.1}$$

where P is the power dissipation (W), N is the number of devices per chip, C is the capacitance of the logic elements (F), V is the operating voltage (V), and f is the operating frequency (Hz).

(2) Density factor (DF)
DF can be used to quantify the impact of non-uniform chip heating on thermal management by the following equation:

$$\Psi_{jc} = R_{jc} \cdot DF \tag{2.2}$$

where Ψ_{jc} is the junction-to-case thermal resistance of the package, and the R_{jc} is the thermal interface material resistance. As the power density of hot spots increases, DF is increasing. Therefore, the thermal management solution must meet more stringent requirements for heat flux.

(3) Joule heat
Joule heat is generated when the current passes through the resistance, which follows Joule's law

$$P = I^2 R \tag{2.3}$$

where P is the power (W), I is the current (A), and R is the resistance (Ω). As the current increases, the Joule heat in the circuit will become more significant. In electronic packaging, smaller package size is required at the same time of high current, which will lead to high heat generation in packaging materials and interconnection systems.

2.2.2 Thermal Failure Rate

The failure rate of an electronic component can be estimated by the Arrhenius equation:

$$f = A\varepsilon^{-e_A/k_B T} \tag{2.4}$$

where A is the Arrhenius constant, f is the failure rate, e_A is the activation energy (eV), k_B is the Boltzmann's constant, and T is the junction temperature (K).

The activation energy varies according to the failure mechanism. When the activation energy is 0.65 eV, the failure efficiency will increase twice as the temperature increases from 50 °C to 60 °C. The activation energy value of 0.65 eV is usually used as a rule of thumb. For every 10 °C increase in temperature, the failure rate will be doubled. For example, the failure rate of bipolar digital devices increases with temperature as shown in Figure 2.1. The figure shows that the failure rate increases exponentially after 75 °C [3].

2.2.3 The Thermal Management at Different Package Levels

The primary packaging thermal management mainly involves the heat dissipation and thermal diffusion in semiconductor devices and chip packages. The thermal management technologies in packaging include traditional copper, coppertungsten, or copper-molybdenum heat sinks, as well as advanced thermal design, on-chip thermoelectric refrigeration, micro-heat pipes, micro-channel cooling, micro-freezers, and coolers. Secondary packaging thermal management is to transfer or diffuse heat to the chassis or electronic system through substrate materials or from chip packaging. Traditional technologies include copper and aluminum radiators. The tertiary packaging heat management refers to the heat dissipation through the system heat

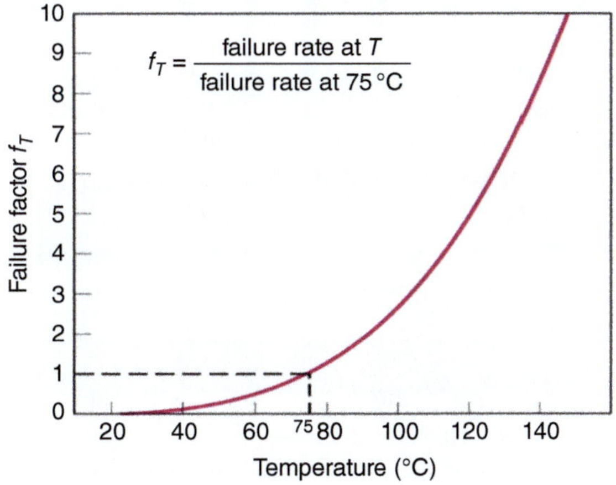

Figure 2.1 Graph of failure factor against device temperature. Source: Reproduced with permission from Ref. [3]; © 2020 Springer Nature.

exchanger and the heat diffusion from the chassis to the system heat exchanger. The traditional technologies include air-cooled heat sink and radiator, forced airflow, and liquid pump circulation. Although the thermal design of different package levels has some common features, each level has its own unique cooling requirements. For example, high heat flow and miniaturization should be considered in the cooling design of the primary package, the secondary package needs to transfer a large amount of heat with the minimum temperature gradient, and the heat exchanger design of the tertiary package needs to absorb most of the heat with the minimum volume and weight.

2.3 Requirements of Electronic Packaging Materials

As shown in Figure 2.2, electronic packaging materials play different functions in typical electronic packaging structure diagrams. It can be divided into thermal interface materials, substrate materials, and plastic packaging materials according to different application positions. These materials are described in detail below.

2.3.1 Thermal Interface Material

Due to the roughness in shape, large area of the interface of two contacting solid surfaces will be separated by an air-filled gap, leading to a very low real contact area [5]. This will result in large contact thermal resistance between heat spreaders because of the low thermal conductivity of air (0.023 W/m K, https://www.electronics-cooling.com/2001/02/the-thermal-conductivity-of-air/), thus rendering low heat dissipation efficiency [6]. As shown in Figure 2.3, this is shown by the temperature drop on the interface.

Thermal interface materials (TIM) are used to fill these voids because they can adapt to rough surfaces and have high thermal conductivity. This can exhaust the air with low thermal conductivity and establish an effective thermal conduction path

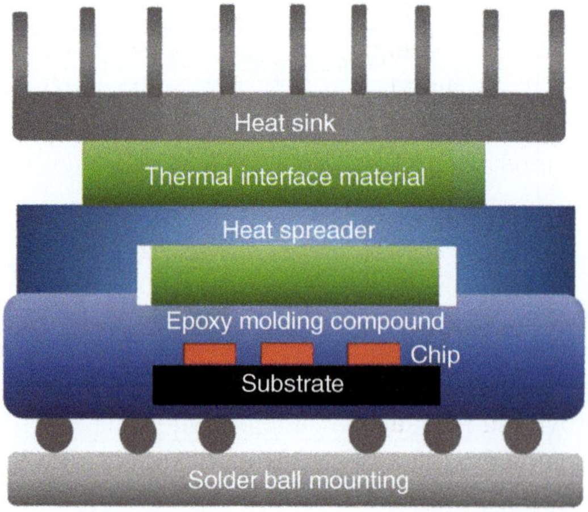

Figure 2.2 Schematic diagram of electronic packaging structure. Source: Reproduced with permission from Ref. [4]; © 2011 Springer: Berlin.

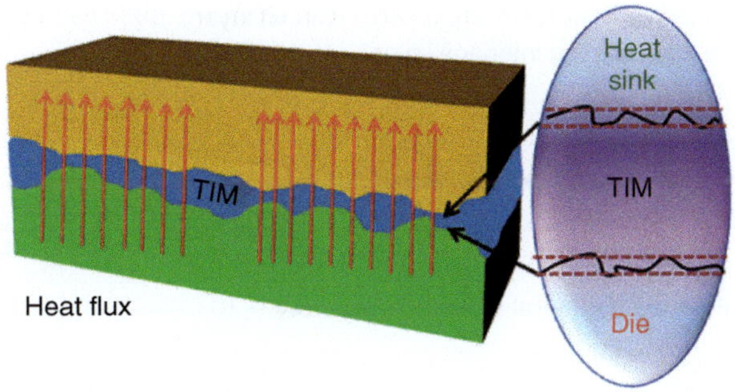

Figure 2.3 Thermal interface material. Source: Reproduced with permission from Ref. [7]; © 2012 American Chemical Society.

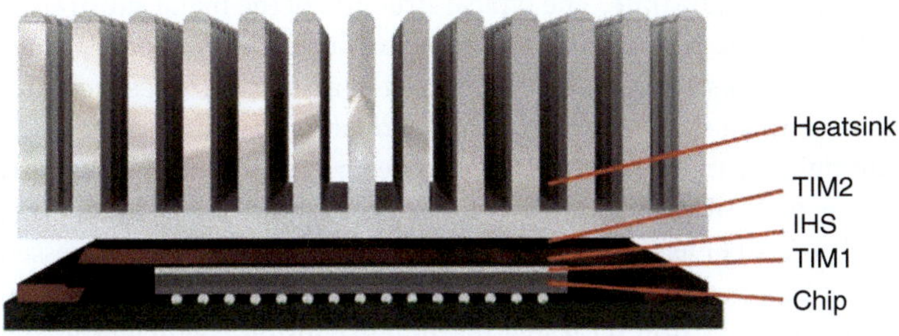

Figure 2.4 Classification of thermal interface material. Source: Reproduced with permission from Ref. [15]; © 2018 International Materials Reviews.

between the electronic components and the radiator. In turn, the contact thermal resistance of heat transfer is reduced to improve the heat dissipation performance of electronic components [8]. Traditional thermal interface materials include thermal conductive silicone grease, thermal conductive pads, and phase change materials [9]. The thermal conductivity is usually improved by filling silicone grease or polymer [7, 10] with high thermal conductivity fillers (silver, graphene [11], alkene [12], boron nitride [13], aluminum oxide [14], etc.).

In electronic packaging, thermal interface material can be divided into TIM1 and TIM2 according to its location, as shown in Figure 2.4. TIM1 is the thermal interface material between the chip and the package shell, while TIM2 is that between the package shell and the heat sink. TIM1 needs to meet the requirements of low thermal resistance and high thermal conductivity because it is in contact with the chip with large heat output, and the thermal expansion coefficient also needs to match with the silicon chip. More importantly, it must be electrically insulated to prevent short circuit of electronic components.

2.3.2 Heat Dissipation Substrate

As a key material of electrical and electronic equipment, the packaging substrate plays an important role, including providing current and signal connection between the chip and other components, providing mechanical support and protection for the chip, timely transferring the heat generated by the chip from the inside, and preventing the accumulation of heat [16].

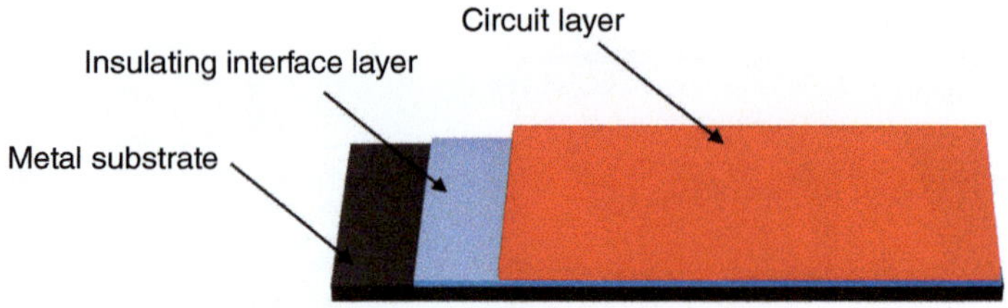

Figure 2.5 Metal substrate structure.

Traditionally, substrate materials are mainly divided into ceramic substrate, metal substrate, and organic resin substrate. The typical structure of the metal substrate is shown in Figure 2.5. Organic resin substrates are mainly used in civil applications, while metal and ceramic substrates are generally used in aerospace, aerospace, and military fields. At present, most integrated circuits use organic resin substrate, which can be divided into rigid organic resin substrate and flexible organic resin substrate. The rigid substrate is generally made of thermosetting resin as the base material, inorganic filler, and glass fiber as the reinforcement material, which are hot pressed and compounded with copper foil to form a rigid organic substrate. The flexible substrate is generally made of PI film with low thermal expansion coefficient as the dielectric layer and then compounded with copper foil.

In order to transmit high-frequency signals, the substrate material needs to adjust the module dielectric constant and dielectric loss factor to reduce the signal propagation delay signal loss [17, 18]. In order to ensure the mechanical properties, the coefficient of thermal expansion (CTE) of the substrate needs to be adjusted to match the CTE of the copper circuit (17 ppm/K) [19]. Most importantly, in order to overcome the heat dissipation problems caused by the miniaturization and integration of electronic components, it is necessary to develop high thermal conductivity substrate materials [20].

2.3.3 Epoxy Molding Compound

Epoxy molding compound (EMC), used for semiconductor chip packaging, not only protects the chip from external environment, especially from external mechanical and physical forces (such as impact and pressure) and external chemical forces (such as moisture, heat, and ultraviolet), but also provides a heat dissipation channel for the chip. Because EMC also ensures the electrical insulation of the chip, it provides a form of semiconductor packaging to make it easier to install on the printed circuit board. The structure of epoxy molding compound after molding is shown in Figure 2.6.

The main material of EMC is epoxy resin, and its components include epoxy resin, curing agent, curing accelerator, inorganic filler (silicon powder), coupling agent, low-stress agent, ion trapping agent, and release agent. Generally, the epoxy molding compound shall meet the following basic requirements: low thermal expansion coefficient, good thermal conductivity, good air tightness, good chemical stability,

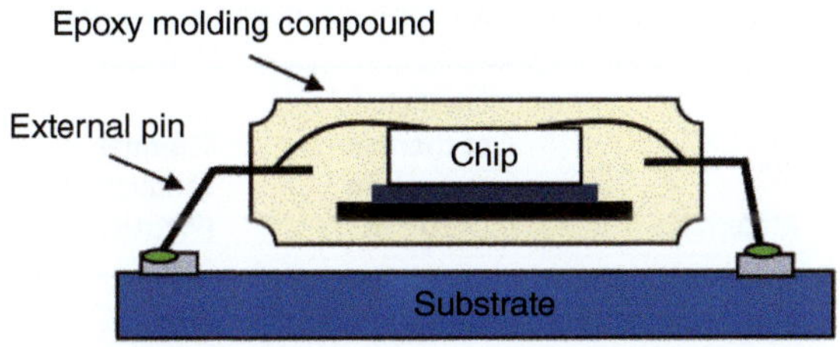

Figure 2.6 Structure of epoxy molding compound after molding.

acid and alkali resistance, high humidity resistance, high-temperature resistance, good mechanical strength, and good processing and molding performance [21].

The development of integrated circuit packaging technology has put forward strict requirements for the curability, fluidity, moldability, and reliability of EMC, including high heat resistance and low melt viscosity, high thermal conductivity and high insulation, low warpage and high melt fluidity, low dielectric constant and dielectric loss, and adaptation to the packaging and molding process of large-size devices.

2.4 Electronic Packaging Materials

A large number of materials are used in electronic packaging, such as semiconductors, ceramics, glass, composites, polymers, and metals. Zweben [22] proposed that electronic packaging materials are mainly divided into metal matrix composites, ceramic matrix composites, polymer matrix composites, and carbon–carbon composites. Metal matrix composites have high thermal conductivity, high strength, and high reliability, and are usually used for electronic packaging in military and aerospace applications. Ceramic matrix composites have low dielectric constant and thermal expansion coefficient, high mechanical strength and thermal conductivity, good thermal stability, and moisture resistance, showing excellent performance at high frequencies. Due to its easy processing, low cost, and excellent comprehensive performance, polymer matrix composites are gradually replacing metals and ceramics in civil electronic devices [23]. Carbon–carbon composites possess excellent mechanical properties that resulted from reinforced fibers and excellent high-temperature thermophysical properties that resulted from carbon materials.

2.4.1 Metal Matrix Packaging Materials

Metal matrix packaging materials are the earliest developed packaging materials. Traditional metal packaging materials mainly include Cu, Al, W, Mo, Kovar alloy (Fe Co Ni alloy), Invar alloy (Fe Ni alloy), steel, and stainless steel. Their main properties are shown in Table 2.1. Among them, Cu and A1 have high thermal conductivity and excellent heat dissipation performance. However, its coefficient of thermal expansion is relatively high, and it is difficult to match with the silicon chip, which will cause great stress due to the thermal cycle generated by the chip during operation. In addition, the density of Cu is high. The thermal expansion coefficients of Kovar alloy and Invar alloy can be well matched with silicon wafers. However, their

Table 2.1 Properties of common metal packaging materials.

Materials	Coefficient of thermal expansion ($\times 10^{-6}$ K^{-1})	Thermal conductivity (W/m K)	Density ($g\,cm^{-3}$)
Si	4.10	150	2.3
GaAs	5.80	39	5.3
Al	23.00	230	2.7
Cu	17.00	400	8.9
Mo	5.00	140	8.9
W	4.45	168	19.3
Kovar	5.90	17	8.3
Invar	1.60	10	–
Steel	12.6	65.2	7.9
Stainless steel	17.3	32.9	7.9

Source: Adapted from Ref. [24]/Springer Nature.

thermal conductivity is relatively low, which will affect the working performance in highly integrated package devices. In addition, compared with Al, the density of alloy is higher and the stiffness is poorer. The thermal conductivity of W and Mo is much better than that of Kovar alloy and Invar alloy, and the coefficient of thermal expansion is lower, but their poor wettability with silicon wafer restrict their application. To overcome this, Ag or Ni are usually coated on their surfaces. However, this will increase the complexity and cost of the process. In addition, W and Mo have higher densities, limiting their use in the aerospace field.

In order to reduce the thermal expansion coefficient of Cu, Cu can be compounded with materials with low thermal expansion coefficients, such as Mo and W, to obtain Cu/W and Cu/Mo metals. They have high conductivity and thermal conductivity as well as combine the low thermal expansion coefficient and high hardness of W and Mo. A plenty of work has been researched and developed various metal matrix composites on the basis of metal matrix packaging materials and Cu/W and Cu/Mo metals.

Metal matrix composite is a kind of composites with metal and intermetallic compound as the matrix and particles, whiskers, short fibers, or continuous fibers as the reinforcement phase. Al, Cu, Mg, and their alloys are usually selected as the matrix of metal matrix composites for electronic packaging, because of their good thermal conductivity, electrical conductivity, good processability, and weldability. The selection of reinforcements for metal matrix composites is critical. Generally, they should have low thermal expansion coefficient, high thermal conductivity, and good chemical stability. At the same time, the compatibility of reinforcements with the base metal should be considered. When ceramic materials are selected as reinforcements, the composites can not only obtain high thermal conductivity of the base metal materials but also possess the advantages of low thermal expansion coefficient of the reinforcement materials. Therefore, it presents the performance characteristics that single metal or ceramic materials cannot achieve. The performance of common reinforcing phases is shown in Table 2.2.

As shown in Table 2.3, the development of metal matrix composites has gone through several stages. The first generation is metal-based electronic packaging materials (W/Cu, Mo/Cu) with metal particles W and Mo as reinforcements, which are the most widely used metal electronic packaging materials at present. However, the thermal conductivity of this material can no longer meet the higher requirements of modern high-power devices, especially its high density, which is not

Table 2.2 Performance of common reinforcements.

Materials	Coefficient of thermal expansion ($\times 10^{-6}$ K^{-1})	Thermal conductivity (W/m K)	Density (g cm^{-3})
Boron fiber	~5.00	40	2.6
SiC fiber (NL200)	−3.10	12	
ThornelP-120	−1.45	640	2.18
SiCp	4.00	80–220	3.21
Al_2O_3	4.40	20–40	3.98
BeO	4.10	300	3.01
AlN	2.69	150–250	3.26
Si_3N_4	1.44	28	3.18
B_4C	3.38	29–67	2.52
TiB_2	4.60	25	4.50
Diamond	1.70	700–2000	3.52
SiC whisker	3.40	32	

Table 2.3 The development of metal matrix packaging materials.

	Materials	Coefficient of thermal expansion ($\times 10^{-6}$ K^{-1})	Thermal conductivity (W/m K)	Density (g cm^{-3})
Chip materials	Si	4.1	135	2.3
	GaAs	5.8	39	5.3
First generation	Cu/W	5.7–8.8	157–190	15–17
	Cu/Mo	7.0–7.1	183–197	9.9–10
Second generation	Al/SiC (≥50 vol%)	6–7	190–200	3.0
Third generation	Cu/diamond	5–6	≥500	5.5–6

conducive to the application in portable electronic equipment, aerospace, and other equipment with weight requirements. The second generation is a high-volume-fraction silicon carbide particle-reinforced aluminum matrix composite with three-dimensional bicontinuous structure, which has high thermal conductivity and low density under the condition of meeting low thermal expansion coefficient. Diamond has the highest thermal conductivity of all materials, and its thermal expansion coefficient is smaller than that of common semiconductor chip materials such as Si and GaAs. The composite of diamond and copper can not only obtain higher thermal conductivity but also obtain lower thermal expansion coefficient which is closer to the chip material. With the development of synthetic diamond technology, the price of diamond powder has dropped significantly. Therefore, diamond/Cu composite has become the key research object of the third generation of high-performance electronic packaging materials and heat dissipation materials. At present, the focus of research on metal matrix composite electronic packaging materials mainly focuses on silicon aluminum (Si–Al) alloy, silicon carbide aluminum (SiC–Al) alloy, copper aluminum, Cu–C fiber, diamond–copper (diamond–Cu) composite, diamond aluminum (amond–Al) composite, and Cu/ZrW208 (TiNi) negative expansion materials [25].

2.4.2 Ceramic Matrix Packaging Materials

Ceramic packaging is often multilayer ceramic substrate (MLC). This technology began with the tape-casting process patent invented by JL Park in 1961. Bemark Schwartz, the founder of ceramic packaging, researched and developed many packaging technology patents related to MLC. Ceramic packaging materials are mainly used as electronic packaging substrate materials because they have the following advantages: low dielectric constant, good high-frequency performance; high strength and good chemical stability; high thermal conductivity and low thermal expansion coefficient; and excellent air tightness and moisture resistance. At present, Al_2O_3 is a widely used ceramic packaging material, accounting for 90% of the total ceramic packaging materials. The remaining ceramic packaging materials mainly include AlN, BeO, BN, and Si_3N_4 [26]. The performance comparison of various ceramic packaging materials is summarized in Table 2.4.

Al_2O_3 ceramics are mainly made of Al_2O_3. All kinds of ceramics are made with α- Al_2O_3 as the main crystal phase and Al_2O_3 content above 75% (mass fraction). Al_2O_3 ceramics have a series of advantages, such as rich raw materials, low price, high mechanical strength and hardness, good insulation, thermal shock resistance and chemical corrosion resistance, high dimensional accuracy, and good adhesion with metal. In the actual preparation process, the content of Al_2O_3 should be increased as much as possible to improve the sintering density of the powder, so as to improve the comprehensive performance. However, this will also lead to higher sintering temperature and increase the cost. In order to reduce the sintering temperature and ensure the mechanical and electrical properties of Al_2O_3 ceramic substrate, it is often necessary to add a certain amount of sintering aids, such as B_2O_3, MgO, CaO, SiO_2, TiO_2, and other metal oxides. Although

Table 2.4 Performance of ceramic packaging materials.

Performance	Al_2O_3	AlN	BeO	SiC	BN	Si_3N_4
Coefficient of thermal expansion ($\times 10^{-6}\ K^{-1}$)	7.1	4.4	7.5	3.8	0.7–7.5	3.2
Thermal conductivity (W/m K)	20–30	180	285	270	20–60	10–40
Specific heat capacity (J g K)	—	0.75	1.046	—	—	—
Density ($g\ cm^{-3}$)	3.80	3.28	2.85	2.9	2.25	3.18
Resistivity (Ω cm)	>1015	>1013	>1015	>1015	>1013	>1014
Dielectric constant (1 MHz)	9.3	8.9	6.7	40	4.0	9.4
Dielectric loss (MHz)	0.0001	0.0005	0.001	0.005	0.0002–0.0006	—
Insulation strength ($KV{\cdot}mm^{-1}$)	10	15	10.6	0.07	300–400	100
Tensile strength ($N\ mm^2$)	323.8	186.0	232.2		40–80	980
Elastic modulus (GPa)	370	310	350	450	98	320
Hardness (GPa)	25	12	12	25	2	20
Toxicity	No	No	Yes	No	No	No

Source: Adapted from Ref. [34].

the application of Al_2O_3 ceramic packaging materials is mature, the thermal conductivity of Al_2O_3 ceramic substrate materials is relatively low, and the thermal expansion coefficient and dielectric constant are different from those of silicon wafers. Therefore, Al_2O_3 ceramic substrate is not suitable for use in high-frequency, high-power, and large-scale integrated circuits. In recent years, the main research direction of alumina ceramic substrate is to try new molding methods and optimize sintering aids.

A1N crystal belongs to hexagonal system and is a wurtzite-type covalent bond compound with AlN_4 tetrahedron as the structural unit. This structure endows them with excellent thermal, electrical, and mechanical properties. AIN ceramics have good thermal conductivity, low dielectric constant, and thermal expansion coefficient matching with semiconductor materials such as Si and GaAs. It is considered as the ideal material for a new generation of highly integrated semiconductor substrate and electronic device packaging and thus has drawn great attention from scholars. However, AlN ceramic substrate materials have high requirements for purity and sintering densification. The preparation process and cost of the materials are high, and the batch production is difficult. In recent years, the research focus of AIN ceramics is turning to further development and application, such as the preparation of easy sintering powder and the discussion of the types of sintering additives.

BeO crystal is the only hexagonal wurtzite structure in alkaline earth metal oxides, in which oxygen ions are arranged in a hexagonal dense manner to form hexagonal lattice. Because BeO has wurtzite-type and strong covalent bond structure, and the average molecular weight of Be is only 12, BeO has very high thermal conductivity and low density. In addition, BeO has low dielectric constant, low dielectric loss, and strong packaging process adaptability, which is suitable for high-power integrated circuit substrate materials. The melting point of BeO is as high as 2570 °C, which makes the sintering temperature of pure BeO ceramics reach above 1900 °C, resulting high production cost. In order to reduce the sintering temperature, Al_2O_3 and MgO are usually used as sintering aids. The biggest disadvantage of BeO is that it is highly toxic. Inhalation in large quantities will lead to acute pneumonia, and long-term inhalation will lead to chronic beryllium disease. This toxicity is more serious in the state of fine powder, and it is safer after sintering into ceramic substrate. For satellite communication and avionics equipment, BeO ceramic substrate is still used in pursuit of high thermal conductivity and high-frequency characteristics.

The thermal conductivity of SiC ceramics is very high and increases with the crystallinity. The coefficient of thermal expansion of SiC ceramics is low and close to that of Si. They show high hardness (only slightly inferior to diamond and cubic BN) and high mechanical strength. SiC ceramics have strong covalent bond characteristics and are difficult to sinter. Usually, a small amount of boron or aluminum oxides are added as sintering aids to improve the density. For example, adding a certain amount of BeO to the SiC material can not only improve the grain boundary resistance of SiC but also improve the insulation and dielectric properties. It can also improve the order of SiC particles in the sinter, enhance the symmetry of lattice vibration, and thus improve its thermal conductivity. The beryllium, boron, aluminum, and their compounds are the most effective additives, which can make the density of SiC ceramics reach more than 98%. In addition, SiC has low resistivity, high dielectric constant, large dielectric loss, and poor high-frequency characteristics, making it only suitable for low-density packaging.

BN has hexagonal crystal and cubic crystal. Both cubic and hexagonal BN are typical covalent bond crystals and should be prepared at high temperature and high pressure. Cubic BN has high hardness and high-temperature resistance and is suitable for super hard materials. Hexagonal BN is similar to graphite in structure, low in hardness, and anisotropic in thermal conductivity. BN material has the characteristics of high thermal stability, chemical stability, high thermal conductivity, low dielectric constant, and excellent insulation performance. It is widely used in radar windows, tube sockets, tube shells, heat sinks, and microwave output windows of high-power transistors. Cubic BN is expensive, so it is not suitable for general high thermal conductivity ceramic materials. The mismatch between thermal expansion coefficient and silicon also limits its application.

Si_3N_4 ceramics have been widely studied as structural materials before thanks to their excellent mechanical properties. Besides, Si_3N_4 ceramics have high electrical insulation, chemical stability, and excellent thermal stability. Therefore, they can be used to manufacture large-scale integrated circuit boards with high integration. In addition, Si_3N_4 ceramics have the greatest advantage of low thermal expansion coefficient, but their slightly poor dielectric property and high production cost limit the application as electronic packaging ceramic substrate.

2.4.3 Polymer Matrix Packaging Materials

Polymer matrix packaging materials have the advantages of low price, lightweight, good insulation, excellent processing performance, adhesion, mechanical property, and corrosion resistance. Hence, they are easy to realize the miniaturization and lightweight of electronic products. However, the polymer matrix packaging materials have some defects, such as poor air tightness and sensitivity to humidity, so the chips packaged with plastic need special attention in storage. Polymer matrix packaging materials cannot adapt to the harsh environment and can only be used in the civil field with low-reliability requirements.

The types of common polymer packaging materials are shown in Table 2.5. Polymers can be divided into thermoplastic polymers and thermosetting polymers according to different glass transition temperatures. There is no covalent cross-linking between the polymer chains of thermoplastic polymers, which can be repeatedly heated for molding. Thermosetting polymers are cross-linked polymers. They are cross-linked and cured under the action of heat or hot pressure to form covalent bonds. This cross-linking and curing is irreversible. High temperature will not melt the material but can soften or reduce the material's performance, even degrade it. The materials used for polymer matrix packaging materials are mainly thermosetting plastics, including polyester, epoxy resin, phenolic, and silicone polymer materials, among which epoxy resin is the most widely used. In addition, a large number of thermoplastics are also used in the preparation of polymer matrix packaging materials, such as methyl methacrylate, polycarbonate, nylon 66, and polyimide.

Polymer materials in the field of electronic packaging can be roughly divided into two categories: intrinsic and filled [27]. Intrinsic type is to improve the intrinsic thermal conductivity of polymer materials by finely regulating the arrangement and interaction of molecular chain segments through special synthesis and molding processes. Filled type refers to the polymer as the matrix material, in which thermal conductivity filler with high thermal conductivity is added to enhance the thermal conductivity. Filled polymer matrix composite is the most commonly used process for preparing polymer matrix electronic packaging materials due to its low cost, easy processing, and other advantages [28].

Table 2.5 Thermal conductivity of common polymers at room temperature.

Type	Material	Thermal conductivity (W/m K)
Thermoplasticity polymer	Epoxy resin	0.19
	PI	0.11
	PDMS	0.25
	EVA	0.35
Thermosetting polymer	PA6	0.22–0.33
	PA66	0.24–0.33
	PMMA	0.16–0.25
	PC	0.19–0.21
	PP	0.11–0.17
	HDPE	0.33–0.53
	LDPE	0.30–34
	PPS	0.30
	PBT	0.25–0.29
	PTFE	0.27
	ABS	0.15–0.20
	PVDF	0.19
	PET	0.15
	PS	0.10–0.15

Source: Adapted from Ref. [35].

The low thermal conductivity of the intrinsic polymer is mainly attributed to the serious phonon scattering in the amorphous region. The reasons for phonon scattering are as follows: highly curled entangled polymer chains are not conducive to the fast transmission of phonons; the loose polymer structure in the amorphous region will form voids and cause phonon scattering; and most of the interactions between chain segments are weak non-bonding forces (van der Waals force and dipole force) [29]. In order to improve the intrinsic thermal conductivity of polymers, the above shortcomings need to be overcome. At present, the most commonly used method to improve the crystallinity and directional arrangement of polymer chains is super stretched polymer fibers. Physical stretching, electrospinning, injection molding, nano-template, and strong magnetic field can be used to achieve orderly alignment of polymer chain segments in a certain direction, which can effectively improve the intrinsic thermal conductivity of polymer [30]. The weak coupling between multiple molecular segments in polymers is also considered to be one of the bottlenecks limiting the improvement of their thermal conductivity. Many studies have found that strengthening the interaction between chains can effectively improve the intrinsic thermal conductivity of polymers [31].

The molecular structure of polymer is directly related to the entanglement and rotation of chain segments, and also has a great impact on the intrinsic thermal conductivity. Building more conjugated carbon–carbon double bonds in the polymer to form π–π interaction force can further enhance the phonon transport between chains, which is expected to significantly improve the phonon transport efficiency [32, 33].

For filled polymer materials, thermal conductive fillers can generally be divided into metal fillers, ceramic fillers, and carbon materials. The thermal conductivity of common fillers is shown in Table 2.6. The excellent conductivity, expensive price, and biased quality of metal materials often determine the large amount of filler, poor mechanical properties, and high cost of composites. Ceramic materials have excellent electrical insulation and thermal conductivity, making them ideal fillers for preparing filled insulating and thermal conductive polymer materials. Carbon materials have super high thermal conductivity, lightweight, and good mechanical properties, which are better than metal materials. It is an ideal filler for filled thermal conductive polymer materials. However, carbon materials with excellent conductivity are disadvantageous to the insulation requirements of polymers. It is often necessary to modify the structure of composites or fillers to achieve the goal of electrical insulation [27].

Table 2.6 Thermal conductivity of filler.

Type	Material	Thermal conductivity (W/m K)
Metal	Al	204
	Ni	158
	Ag	450
	Cu	483
Ceramic	Al_2O_3	20–29
	MgO	36
	BeO	370
	SiC	80–120
	Si_3N_4	50
	BN	250–300
	AlN	150–220
Carbon material	Graphite	100–400
	CB	6–174
	CF	8–70
	Graphene	5000
	Diamond	2000
	CNT	3000–3500

Source: Adapted from Ref. [27].

2.4.4 Carbon–Carbon Composite

Carbon–carbon composites refer to a kind of composite materials with carbon fiber (or graphite) as reinforcement and carbon (or graphite) as matrix. Its discovery originated from a laboratory accident at the American Chance Sound Company in 1958. They show low density, good thermal conductivity, low coefficient of thermal expansion, high friction coefficient, strong thermal shock resistance, high specific heat capacity, and strong designability. They can be used not only in high-temperature environment but also in the situation of rapid temperature change. Therefore, they are widely used in aerospace field. The preparation of carbon–carbon composite packaging materials mainly includes carbonization of carbon precursor—thermal decomposition into carbon and other products in an air-isolated environment; and graphitization—the orderly transformation of thermodynamic unstable carbon atoms from disordered layer structure to graphite crystal structure by thermal activation.

The high thermal conductivity of carbon–carbon composites makes them a good candidate for the thermal management of electronic packaging. In addition, carbon–carbon composites have very low thermal expansion coefficients and can maintain dimensional stability in a wide range of temperatures. Carbon–carbon composites retain their original mechanical properties even at the temperature (non-oxidizing atmosphere) above 2000 °C. They demonstrate high resistance to thermal shock caused by rapid and extreme temperature changes.

References

1 Rasmussen, F.E. (2003). *Electrical Interconnections Through CMOS Wafers*. Hellerup: Technical University of Denmark.

2 Tong, X.C. (2011). *Thermal Management Fundamentals and Design Guides in Electronic Packaging, Advanced Materials for Thermal Management of Electronic Packaging* (ed. X.C. Tong), 1–58. New York, NY: Springer New York.

3 Khan, J., Momin, S.A., and Mariatti, M. (2020). A review on advanced carbon-based thermal interface materials for electronic devices. *Carbon* 168: 65–112.

4 Tong, X.C. (2011). *Advanced Materials for Thermal Management of Electronic Packaging*, vol. 30, 39–65. Berlin: Springer.

5 Grujicic, M., Zhao, C.L., and Dusel, E.C. (2005). The effect of thermal contact resistance on heat management in the electronic packaging. *Applied Surface Science* 246 (1–3): 290–302.

6 Uetani, K., Ata, S., Tomonoh, S. et al. (2014). Elastomeric thermal interface materials with high through-plane thermal conductivity from carbon fiber fillers vertically aligned by electrostatic flocking. *Advanced Materials* 26 (33): 5857–5862.

7 Shahil, K.M.F. and Balandin, A.A. (2012). Graphene-multilayer graphene nanocomposites as highly efficient thermal interface materials. *Nano Letters* 12 (2): 861–867.

8 Xu, J., Munari, A., Dalton, E. et al. (2009). Silver nanowire array-polymer composite as thermal interface material. *Journal of Applied Physics* 106 (12): 124310.

9 Zeng, J.L., Cao, Z., Yang, D.W. et al. (2010). Thermal conductivity enhancement of Ag nanowires on an organic phase change material. *Journal of Thermal Analysis and Calorimetry* 101 (1): 385–389.

10 Chen, L., Zhao, P., Xie, H. et al. (2016). Thermal properties of epoxy resin based thermal interfacial materials by filling Ag nanoparticle-decorated graphene nanosheets. *Composites Science and Technology* 125: 17–21.

11 Zhang, P., Zeng, J., Zhai, S. et al. (2017). Thermal properties of graphene filled polymer composite thermal interface materials. *Macromolecular Materials and Engineering* 302 (9): 1700068.

12 Naghibi, S., Kargar, F., Wright, D. et al. (2020). Noncuring graphene thermal interface materials for advanced electronics. *Advanced Electronic Materials* 6 (4): 1901303.

13 Hu, Q., Bai, X., Zhang, C. et al. (2022). Oriented BN/silicone rubber composite thermal interface materials with high out-of-plane thermal conductivity and flexibility. *Composites Part A: Applied Science and Manufacturing* 152: 106681.

14 Feng, C.-P., Chen, L.-B., Tian, G.-L. et al. (2020). Robust polymer-based paper-like thermal interface materials with a through-plane thermal conductivity over $9\ W\,m^{-1}\,K^{-1}$. *Chemical Engineering Journal* 392: 123784.

15 Hansson, J., Nilsson, T.M.J., Ye, L. et al. (2018). Novel nanostructured thermal interface materials: a review. *International Materials Reviews* 63 (1): 22–45.

16 Kim, Y.H., Lim, Y.W., Kim, Y.H. et al. (2016). Thermally stable siloxane hybrid matrix with low dielectric loss for copper-clad laminates for high-frequency applications. *ACS Applied Materials & Interfaces* 8 (13): 8335–8340.

17 Gao, Y.W., Gu, A.J., Jiao, Y.C. et al. (2012). High-performance hexagonal boron nitride/bismaleimide composites with high thermal conductivity, low coefficient of thermal expansion, and low dielectric loss. *Polymers for Advanced Technologies* 23 (5): 919–928.

18 Ge, M., Zhang, J., Zhao, C. et al. (2019). Effect of hexagonal boron nitride on the thermal and dielectric properties of polyphenylene ether resin for high-frequency copper clad laminates. *Materials & Design* 182: 108028.

19 Apeldorn, T., Keilholz, C., Wolff-Fabris, F. et al. (2013). Dielectric properties of highly filled thermoplastics for printed circuit boards. *Journal of Applied Polymer Science* 128 (6): 3758–3770.

20 Lee, G.W., Park, M., Kim, J. et al. (2006). Enhanced thermal conductivity of polymer composites filled with hybrid filler. *Composites Part A: Applied Science and Manufacturing* 37 (5): 727–734.

21 Phansalkar, S.P., Kim, C., and Han, B. (2022). Effect of critical properties of epoxy molding compound on warpage prediction: a critical review. *Microelectronics Reliability* 130: 114480.

22 Zweben, C. (2001). Advanced composites and other advanced materials for electronic packaging thermal management, Proceedings International Symposium on Advanced Packaging Materials Processes, Properties and Interfaces (IEEE Cat.

No.01TH8562). Braselton, GA, USA, 2001, pp. 360-365, https://doi.org/10.1109/ISAOM.2001.916602.
23 Wan, Y.-J., Li, G., Yao, Y.-M. et al. (2020). Recent advances in polymer-based electronic packaging materials. *Composites Communications* 19: 154–167.
24 Zweben, C. (1992). Metal-matrix composites for electronic packaging. *JOM* 44 (7): 15–23.
25 Tong, X.C. (2011). *High Thermal Conductivity Metal Matrix Composites, Advanced Materials for Thermal Management of Electronic Packaging* (ed. X.C. Tong), 233–276. New York, NY: Springer New York.
26 Tong, X.C. (2011). *Thermally Conductive Ceramic Matrix CompositesAdvanced Materials for Thermal Management of Electronic Packaging* (ed. X.C. Tong), 277–304. New York, NY: Springer New York.
27 Zhang, L., Deng, H., and Fu, Q. (2018). Recent progress on thermal conductive and electrical insulating polymer composites. *Composites Communications* 8: 74–82.
28 Tong, X.C. (2011). *Thermally Conductive Polymer Matrix Composites, Advanced Materials for Thermal Management of Electronic Packaging* (ed. X.C. Tong), 201–232. New York, NY: Springer New York.
29 Huang, C., Qian, X., and Yang, R. (2018). Thermal conductivity of polymers and polymer nanocomposites. *Materials Science and Engineering: R: Reports* 132: 1–22.
30 Ma, H., Ma, Y., and Tian, Z. (2019). A simple theoretical model for thermal conductivity of crystalline polymers. *ACS Applied Polymer Materials* 1 (10): 2566–2570.
31 Xu, X., Chen, J., Zhou, J. et al. (2018). Thermal conductivity of polymers and their nanocomposites. *Advanced Materials* 30 (17): 1705544.
32 Zhang, T., Wu, X.F., and Luo, T.F. (2014). Polymer nanofibers with outstanding thermal conductivity and thermal stability: fundamental linkage between molecular characteristics and macroscopic thermal properties. *The Journal of Physical Chemistry C* 118 (36): 21148–21159.
33 Frieden, E. (1975). Non-covalent interactions: key to biological flexibility and specificity. *Journal of Chemical Education* 52 (12): 754.
34 Chen, H., Ginzburg, V.V., Yang, J. et al. (2016). Thermal conductivity of polymer-based composites: fundamentals and applications. *Progress in Polymer Science* 59: 41–85.
35 Han, Z. and Fina, A. (2011). Thermal conductivity of carbon nanotubes and their polymer nanocomposites: a review. *Progress in Polymer Science* 36 (7): 914–944.

3

Characterization Methods for Thermal Management Materials

Kang Zheng and Xingyou Tian

Chinese Academy of Sciences, Institute of Solid State Physics, Hefei Institutes of Physical Science, 350 Shushanhu Road, Hefei, 230031, Anhui, P. R. China

3.1 Overview of the Development of Thermal Conductivity Test Methods

With the rapid-developing miniaturization and integration of microelectronics packaging materials, ultrahigh-voltage electrical devices, and light-emitting diodes (LEDs) [1–6], highly thermally conductive polymeric composites with excellent thermal stabilities are promising candidates to transfer the heat to the outside of the products, in order to effectively avoid substantial overheating and prolong their working life [7–10]. The conduction of heat is one of the fundamental energy transport mechanisms in nature. The vibrational modes of lattices, also known as phonons, have critical impacts on heat control and management in the real world. Thermal conductivity is the ability of an object to conduct heat. Also known as thermal conductivity, it is one of the thermophysical parameters of materials and the most important parameter for solids. It is defined as the heat transferred through $1\ \mathrm{m}^2$ in one second (1 s) under the condition of stable heat transfer with a temperature difference of 1 degree (K, or °C) between the surfaces of 1-meter-thick material. The unit is watt/meter · degree (W/(m K)), where K can be replaced by °C. There are several methods that have been developed for the determination of thermal conductivity. They have different application fields, measurement ranges, accuracy, and sample size requirements. The measurement results of the same sample by different methods may vary greatly. Therefore, to select an appropriate test method is the first consideration.

As nonmetals, heat transfer in polymer and ceramic filler-filled composites involves lattice vibrations, of which are phonons. The heat absorbed in the hot region increases the amplitudes of the lattice vibrations which is the same as generating more phonons. These new phonons transfer toward the cold regions and transport the lattice energy from the hot to cold region. The thermal conductive coefficient λ measures the rate at which heat can be transported through a medium per unit area per unit temperature gradient. It is proportional to the rate at which a

Thermal Management Materials for Electronic Packaging: Preparation, Characterization, and Devices, First Edition. Edited by Xingyou Tian.

medium can absorb energy, that is, λ is proportional to the heat capacity. A rigorous classical treatment gives λ as $\lambda = \rho\alpha Cp$, where α and Cp are the thermal diffusivity and specific heat capacity, and ρ is the density of the specimen.

The mean phonon velocity is constant and approximately independent of temperature so that λ is governed primarily by phonon velocity and mean free path. The mean free path depends on various processes that can scatter the phonons and hinder their propagation along the direction of heat flow. Phonons can impact other phonons, crystal defects, impurities, and crystal surfaces.

At present, the methods measuring thermal conductivity can be divided into steady-state method and transient method (also called unsteady-state method).

3.2 Test Method Classification and Standard Samples

There are many measurement methods for the thermal conductivity meter, and the applicable methods vary according to the different measurement objects and measurement ranges.

According to the heat transfer mechanism, methods for λ measurements are mainly divided into the two following methods: steady-state measurement method and non-steady-state measurement method.

Steady-state method can be divided into plate method, protective hot plate method, heat flow meter method, etc. Non-steady-state method, also known as transient method, includes hot-wire method, hot disk method, and laser method.

According to the shape of the sample, the measuring methods of the thermal conductivity meter can be divided into plate method, cylinder method, ball method, hot-wire method, etc. In addition, there are classifications of direct method and indirect method. For example, the laser method needs to obtain the thermal diffusivity first, and then calculate the thermal conductivity according to the density and specific volume, while the hot-wire method and the plate method directly obtain the thermal conductivity. Generally, the steady-state method is used as the basis for testing the accuracy of other methods and is often used as the arbitration scheme of other methods.

The thermal conductivity is only applicable when the heat transfer form is in heat-conducting state. When there exist other forms of heat transfer, such as radiation, convection, mass transfer, and other heat transfer forms, the complex heat transfer relationship is usually referred to as the apparent thermal conductivity, dominant thermal conductivity, or effective thermal conductivity.

In addition, the thermal conductivity is mainly referred to as those of homogeneous materials. In fact, for porous, multilayer, multi-structure, and anisotropic materials, the thermal conductivity measured by the thermal conductivity meter is actually a performance of comprehensive thermal conductivity, also known as the mean thermal conductivity.

The thermal conductivity of materials is not only related to the type of material but also closely related to its microstructure, filler content, etc. In scientific experiments and engineering design, the thermal conductivity of all materials needs to be accurately measured by experimental methods.

Thermal conductivity is an intrinsic parameter of the material, independent of its shape. At present, there are a lot of thermal conductivity testing methods, but no one method can meet all requirements of products in all conditions. The characteristics of products, the test standards, and test environment will largely affect the results of thermal conductivity. The thermal conductivity of materials obtained by different test methods cannot be compared with each other. In order to obtain accurate and meaningful results, it is necessary to select appropriate test methods for measurement.

We have listed some standards for measuring thermal conductivities.

3.2.1 Steady-State Measurement Method

ASTM C177-2010 Standard Test Method for Steady-State Heat Flux Measurements and Thermal Transmission Properties by Means of the Guarded-Hot-Plate Apparatus

ASTMC182-19 Standard Test Method for Thermal Conductivity of Insulating Firebrick

ASTM C201-1993(2009) Standard Test Method for Thermal Conductivity of Refractories

ASTM C202-93(2013) Standard Test Method for Thermal Conductivity of Refractory Brick

ASTM C335-05a Standard Test Method for Steady-State Heat Transfer Properties of Pipe Insulation

ASTM C518-04 Standard Test Method for Steady-State Thermal Transmission Properties by Means of the Heat Flow Meter Apparatus

ASTM C680-19 Standard Practice for Estimate of the Heat Gain or Loss and the Surface Temperatures of Insulated Flat, Cylindrical, and Spherical Systems by Use of Computer Programs

ASTM C687-18 Standard Practice for Determination of Thermal Resistance of Loose-Fill Building Insulation

ASTM C1043-19 Standard Practice for Guarded-Hot-Plate Design Using Circular Line-Heat Sources

ASTM C1044-16(2020) Standard Practice for Using a Guarded-Hot-Plate Apparatus or Thin-Heater Apparatus in the Single-Sided Mode

ASTM C1113/C1113M-09(2019) Standard Test Method for Thermal Conductivity of Refractories by Hot Wire (Platinum Resistance Thermometer Technique)

ASTM C1114-06(2019) Standard Test Method for Steady-State Thermal Transmission Properties by Means of the Thin-Heater Apparatus

ASTM D5470-17 Standard Test Method for Thermal Transmission Properties of Thermally Conductive Electrical Insulation Materials

ASTM D6744-06(2017)e1 Standard Test Method for Determination of the Thermal Conductivity of Anode Carbons by the Guarded Heat Flow Meter Technique

ASTM E1225-20 Standard Test Method for Thermal Conductivity of Solids Using the Guarded-Comparative-Longitudinal Heat Flow Technique

ASTM E1530-19 Standard Test Method for Evaluating the Resistance to Thermal Transmission by the Guarded Heat Flow Meter Technique

ASTM F433-02(2020) Standard Practice for Evaluating Thermal Conductivity of Gasket Materials

3.2.2 Non-Steady-State Measurement Method

ASTM C714-17 Standard Test Method for Thermal Diffusivity of Carbon and Graphite by Thermal Pulse Method

ASTM D5930-17 Standard Test Method for Thermal Conductivity of Plastics by Means of a Transient Line-Source Technique

ISO 13826:2013 Metallic and other inorganic coatings—Determination of thermal diffusivity of thermally sprayed ceramic coatings by laser flash method

ISO 18555:2016 Metallic and other inorganic coatings—Determination of thermal conductivity of thermal barrier coatings

ISO 18755:2005 Fine ceramics (advanced ceramics, advanced technical ceramics)—Determination of thermal diffusivity of monolithic ceramics by flash method

ISO 22007-2:2022 Plastics—Determination of thermal conductivity and thermal diffusivity—Part 2: Transient plane heat source (hot disc) method

3.3 Steady-State Method

Steady-state methods can by definition measure only thermal conductivity. They all depend on the same principle: measure the heat flow rate φ in a heat resistance flux R delimited by two isothermal surfaces at temperatures T_0 and T_1. We deduce the value of R with $R = T_1 - T_0$.

Thermal conductivity λ is deduced if the relationship $R(\lambda)$ is established. They have the disadvantage of relying only on one measurement reading; however, the measurement can be repeated a great number of times. In this case, the uncertainty due to the measurement noise is divided by $\sqrt{N}$.

They are not suitable for materials with high thermal conductivity (Figure 3.1).

As shown in scheme above, a uniform and constant heat flow rate φ is provided on one face of a sample with a thickness L by means of a planar heater, whereas the other face is in contact with an isothermal surface held a constant temperature at T_0. It is assumed that the heat transfer is 1D, and the temperature T_1 of the heating part is measured when the steady state is reached.

The thermal resistance between the two temperature measurement points is deduced from

$$R = (T_1 - T_0)/\varphi$$

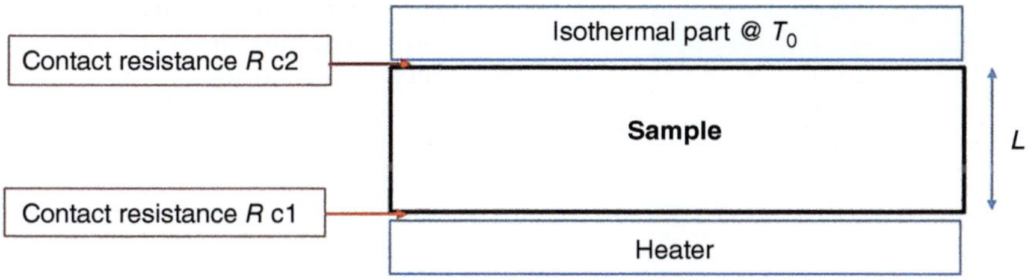

Figure 3.1 Schematic diagram for steady-state method. Source: Kang Zheng.

This resistance is the sum of three resistances:

- the thermal resistance $RL = L/\lambda S$ of the sample;
- the contact resistance R_{c1} between the heating element and the sample;
- the contact resistance R_{c2} between the sample and the surface maintained at T_0.

A contact resistance can reach a value of 2×10^{-4} m^2/K W; therefore, the thermal resistance of the sample must be at least 4×10^{-2} m^2/K W in order for us to be able to neglect the contact resistances.

We can deduce from the above: $L/\lambda > 0.04$ m^2/K W.

For a material of thermal conductivity $\lambda = 10$ W/m K, a sample of thickness equal to at least 40 cm would be required so that its thermal resistance is 100 times greater than the contact resistance, which is not realistic.

Therefore, it will be concluded that the steady-state methods only make it possible to measure a thermal resistance and that the value of the thermal conductivity can only be deduced therefrom if it is less than 1 W/m K.

Classical steady-state method uses stable heat transfer process, and λ value is measured under the equilibrium condition, where the heat transfer rate is equal to the heat dissipation rate. And steady-state method mainly includes guarded hot plate method and heat flow meter method. Such method possesses the advantages of clear principle, direct access to λ value, suitable for wide temperature measurement. However, such method is often time-consuming and requires rigorous measurement conditions, such as adiabatic conditions of measuring system, temperature control during measurement, and required specific shape and size of the samples.

3.3.1 Longitudinal Heat Flow Method

The instrument uses the sample DC longitudinal heat flow method, the cylindrical sample along the specimen longitudinal ends by the heat generated in the current conduction, to achieve stability, that is one-dimensional longitudinal heat flow and sample and the measurement is corrected to the environmental exchange. The testing principle is to place a sample of a certain thickness between the upper and lower plates, and apply a certain heat flow and pressure to the sample; to measure the heat flow through the sample by using a heat flow sensor, the thickness of the test sample,

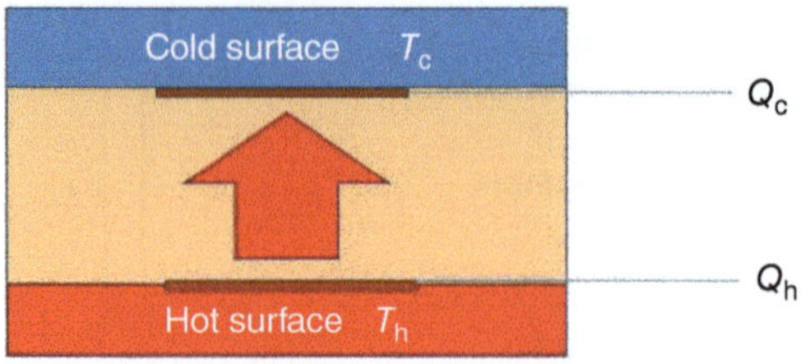

Figure 3.2 Schematic diagram for longitudinal heat flow method. Source: Kang Zheng.

and the temperature gradient between the hot plate and the cold plate; and then to obtain the corresponding thermal resistance data under different thicknesses. The thermal conductivity of the sample is calculated according to the following equation:

$$\lambda = \left(\frac{Q_h}{2} + \frac{Q_c}{2}\right)\frac{L}{\Delta T}$$

where λ is the thermal conductivity of sample, in W/m K; Q_h is the output of heat flow for upper thermal sensor, in W/m^2; Q_c is the output of heat flow for down thermal sensor, in W/m^2; L is the thickness of sample, in m; and ΔT is the difference of temperature between the upper and down surfaces of sample (Figure 3.2).

The advantages of this method are as follows: First, it can test the thermal resistance and thermal conductivity of the product. Second, it is especially suitable for simulating the use state of products under actual working conditions. However, there also exist some disadvantages. The thickness of samples needs to meet certain requirements. The contact thermal resistance will affect the test results. In order to reach steady state, the test will take a long time.

3.3.2 Guarded Heat Flow Meter Method

Guarded heat flow meter method is quasi-steady-state method, a variant of the guarded hot plate technique. Specimen is typically in the form of discs 50 mm in diameter and 1 to 20 mm thick. During the test, the specimen is held under a compressive load between two polished metal surfaces. The upper surface is the heat source maintained at the testing temperature. The lower surface constitutes the calibrated heat flux transducer that is attached to a liquid-cooled heat sink at a constant temperature. The temperature across the specimen is determined from temperature sensors on the metal surface on either side of the specimen. The time to reach the steady state prior to making measurement is typically about 2 hours for test at near-ambient condition. The λ value is calculated from the calibration factor, the specimen thickness, and the temperature across the specimen. More details of the test method and data analyses can be found in ASTM E-1530. The typical instrument and the corresponding schematic diagram for mechanism of the guarded heat flow meter method are shown in Figure 3.3.

3.3.3 Guarded Hot Plate Method

The mechanism of the hot plate method or the protective hot plate method heat conductivity meter is similar to that of the heat flow method heat conductivity

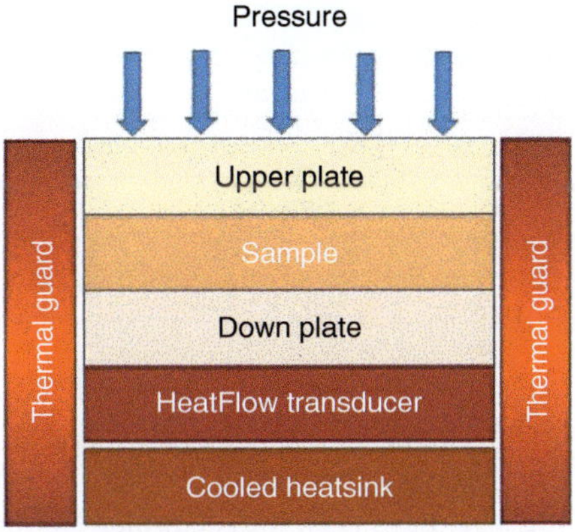

Figure 3.3 Schematic diagram for mechanism of the guarded heat flow meter method. Source: Kang Zheng.

meter using hot plate and cold plate. The heat source is located between two samples of the same material. The purpose of using two samples is to obtain the heat flow symmetrical in the upward and downward directions, and to ensure the complete absorption of energy from the heater by the test sample. During the measurement, the energy input to the hot plate is set accurately. Adjust the measured temperature and temperature gradient between the heat source and the auxiliary panel (cold plate) by adjusting the energy that input to the auxiliary heater. The protective heater and sample around the hot plate shall be placed in the way that the heat flow from the hot plate to the auxiliary heater (cold plate) is linear and one-dimensional. The radiator is behind the auxiliary heater, and the radiator and auxiliary heater are in good contact to ensure the removal of heat and improve control. The thermal conductivity of the material can be calculated by measuring the energy and temperature gradient applied to the hot plate and the thickness of the two samples.

Compared with the hot plate method, the advantage of the protective hot plate method is that the temperature range is wider (–180 to 650 °C) and the range is wider (up to 2 W/m K). In addition, the protection hot plate method uses the absolute method – there is no need to calibrate the measuring unit.

Guarded hot plate is a method presented whereby the thermal diffusivity of a solid is measured by observing the temperature excursion which results from a radial flow of heat. The radial heat flow is produced by the instantaneous deposition of energy on a disk region of one surface of a planar specimen. Experimental results which utilized this technique are presented for Armco iron, and they are compared to results based on an axial heat flow method with a radial heat flow correction. It is a reference method for measuring the thermal conductivity of materials with low thermal conductivity, developed and used for many years. It has been the subject of standards such as ISO 8302 and ASTM C-177, and numerous publications aimed mostly at proposing construction variants or estimating measurement uncertainties. Its schematic diagram is given in Figure 3.4.

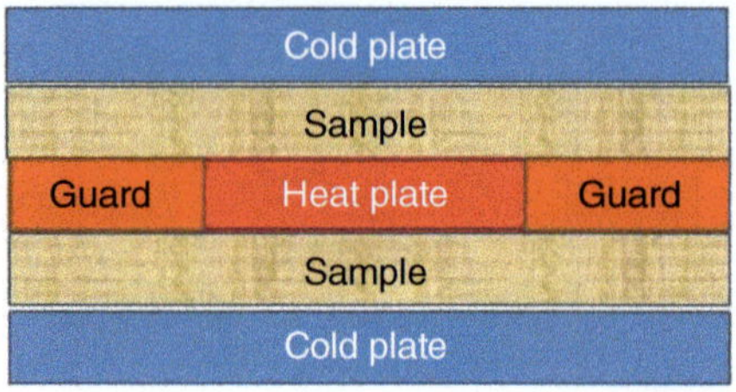

Figure 3.4 Schematic diagram for mechanism of the guarded heat plate method. Source: Kang Zheng.

3.4 Non-Steady-State Method

The non-steady-state method has been developed in recent years, mainly including laser flash [11], transient plane source [12], temperature wave analysis [13], and hot-wire [14, 15] methods. These methods are performed to analyze and investigate the highly thermally conductive materials, with the advantages of the short measuring time, high-accuracy, and low environmental requirements.

3.4.1 Laser Flash Method

The laser flash method is based on the measuring temperature change on the back face of a thin disk specimen, resulting from a short energy pulse on the front surface [16]. A short laser pulse irradiates one side of the specimen. The temperature change on the opposite specimen face is measured by an IR detector. A high-speed recorder collects data representing the temperature change. The diffusivity is calculated from the shape of the temperature–time curve (thermogram) and the thickness, *e*, of the specimen. The λ value is related to the thermal diffusivity (α), and $\lambda = \rho\alpha Cp$, where Cp is the hot melt and ρ is the density of the specimen. The λ value can be determined from α value. More details of the test method and data analyses can be found in ISO/DIS 22007-4 [11, 17]. The corresponding schematic diagram for mechanism of the laser flash method is shown in Figure 3.5.

3.4.2 Hot-Wire Method

Hot-wire method can be used to determine the λ value of the specimen as a function of temperature, which is applicable only to isotropic materials and mainly for solid polymers. A wire heater is placed in a test specimen or between two test specimens of the same materials. The temperature change can be measured either by the wire itself acting as a platinum resistance temperature detector or by a thermocouple placed in close proximity to the wire. The heater current is switched on, and the temperature in the thermocouple is measured as a function of time. The λ value is calculated from the differential equation of temperature fields of the specimen and wire heater. More details of the test method and data analyses can be found in ISO 8894-1 [14] and ISO 8894-2 [18]. And the corresponding schematic diagram for mechanism of the hot-wire method is shown in Figure 3.6.

Figure 3.5 Schematic diagram for the laser flash method. Source: Kang Zheng.

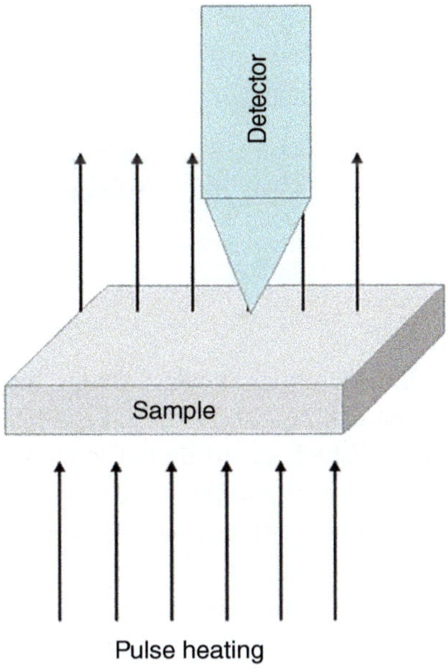

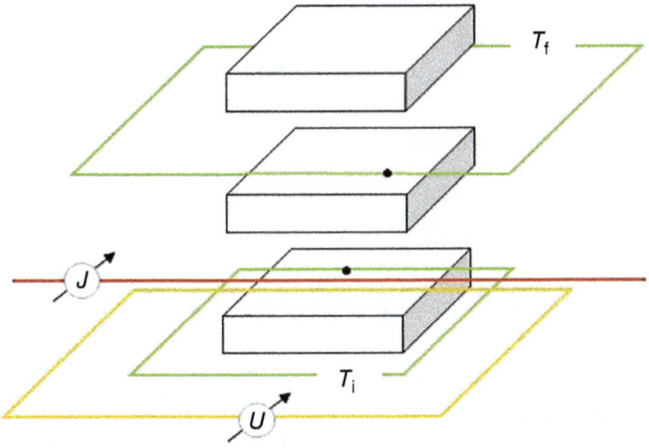

Figure 3.6 Schematic diagram for hot-wire method. Source: Kang Zheng.

3.4.3 Transient Planar Heat Source (TPS) Method

Transient planar heat source (TPS) is the most convenient and accurate method to study the thermal conductivity of materials, which is improved from the hot-wire method. This method uses a hot disk probe, which is also called the hot disk method. The transient plane source or hot disk method uses a thin bifilar spiral, which acts as both a heat source and a temperature sensor. Hot disk probe is made of thermal resistance material nickel, coated with insulating material (polyimide, mica, etc.) and has self-heating function. The element is sandwiched between two planar specimens of the same materials. A constant heating power is supplied to the element and its changing resistance as a function of time, so that voltage across the element

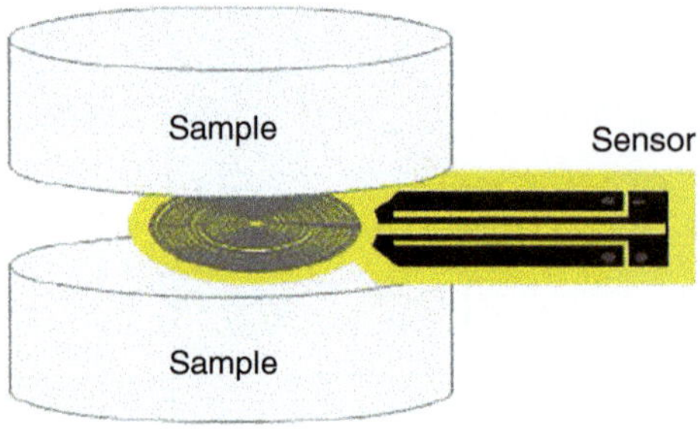

Figure 3.7 Schematic diagram of hot disk while test solid-state sample. Source: Kang Zheng.

changes. By changing voltage, the changing resistance value of the element with time can be obtained, and the temperature response curve of the element can be established according to the relationship between the changing resistance value of the hot wire and the temperature. The principle of this method is that the temperature probe with self-heating function is placed in the sample, and a constant heating power is applied to the probe during the test to increase its temperature. The relationship between the thermal resistance coefficient temperature and resistance of nickel is linear, that is, the heat loss can be known by understanding the change of resistance, thus reflecting the thermal conductivity of the sample. Then, measure the temperature rise relationship between the probe itself and the spherical surface at a certain distance from the probe with time, and obtain the thermal conductivity λ and thermal diffusion coefficient α of the sample by fitting the mathematical model (Figure 3.7).

The advantages of this method are as follows:

① It can simultaneously measure thermal conductivity, thermal diffusivity, and heat capacity per unit volume;
② Wide test range (0.005~500 W/m K), high accuracy (±3%), good repeatability (±1%), short measurement time (single measurement for 3–5 minutes), and simple operation;
③ There are many kinds of samples that can be tested, including liquid, powder, gel, polymer, and composite materials;
④ It is not affected by the contact thermal resistance, and the test result is closer to the thermal conductivity of the material itself.

The disadvantage is that this method is suitable for measuring the thermal conductivity of homogeneous materials but unsuitable for measuring anisotropic materials (such as graphite sheet).

3.5 Electrical Properties and Measurement Techniques

Electrical testing plays a very important role in failure analysis and is the starting point for every analysis. The initial electrical verification process provides a general understanding of how the device is failing electrically. This is generally in the form of a data log, the measured continuity, parametric, and functional outputs of the device from production test equipment. The results may drive the need for more detailed electrical characterization. For example, electrical leakage identified may lead to detailed I–V (current–voltage) characterization on a curve tracer or parametric analyzer. For functional failures, observation of functional failures may lead to a plotting of characterization as a function of temperature, power supply voltage or frequency, or scan testing to better understand the characteristics of the failure. This further electrical characterization may provide partial or even complete failure site identification. In highly structured devices such as memories, electrical testing may conclusively identify a failure site as single memory bit which covers less than a square micron. The initial electrical characterization can also provide indications of a package-related problem. In such cases, nondestructive analysis of the package may take place before any subsequent activity.

Electrical characterization plays a key role in the success of failure analysis. It provides the starting point for the complex process of narrowing the scope of analysis from an entire IC to a single bond wire, signal net, transistor, or defective component. Therefore, for electrical properties of electric packing materials, the most important parameters are electric conductivity, resistivity, and dielectric constant.

3.5.1 Electric Conductivity and Resistivity

Generally, the electrical conductivity is related to the density of charge carriers (n) and their mobility (μ), typically given by

$$\sigma = ne\mu$$

The mobility is given by

$$\mu = e\tau / m_{\mathrm{e}}$$

where μ is the mobility of the carriers, m_{e} is the effective mass, and τ is the mean scattering time between collisions for the carriers. In metals, there are many carriers and states available for conduction, typically $n \approx 10^{22}$ carriers cm^{-3}. For metals, the electrical conductivity is then very large on the order of 10^6 $(\Omega\,\mathrm{cm})^{-1}$.

For semiconductors, the carriers must be thermally excited across a gap for conduction to occur.

$$\sigma \approx \sigma_0 e^{(E_{\mathrm{G}}/k_{\mathrm{B}}T)}$$

The conductivity can occur through the contributions of both holes and electrons

$$\sigma = ne\mu_{\mathrm{e}} + pe\mu_{\mathrm{h}}$$

where n and μ_{e} are the electron concentration and mobility, respectively, and p and μ_{h} are the hole concentration and mobility, respectively. There are two primary ways to achieve a high conductivity in a semiconductor, either by having a very small gap to excite across ($E_{\mathrm{G}} < k_{\mathrm{B}}T$) or by having very high mobility carriers. Typical values of the electrical conductivity lie between 10^{-4} and 10^4 $(\Omega\,\mathrm{cm})^{-1}$; however, these are somewhat arbitrary boundaries.

The method of measuring the resistivity of conductor is to determine the resistance of samples with known geometric dimensions by applying a current through one pair of leads and measuring its voltage drop with another pair of leads, although the specific method used to measure resistivity depends on the size and shape of the sample. However, all methods require sensitive voltmeter and current source or micro-ohmmeter for measurement, because the resistance to be measured is generally very small.

3.5.1.1 Testing Resistivity of Bulk Material

Figure 3.8 shows a system for testing the resistivity of a bulk material, such as a metal bar or metallic strip. Connect the current source to both ends of the sample, and the voltmeter leads are placed at a certain distance. The resistivity is calculated according to the cross-sectional area of the sample and the distance between the voltmeter leads, as the equation below:

$$\rho = \frac{V}{I} \cdot \frac{A}{L}$$

where ρ is the resistivity in cm ohm,
V is the voltage measured by voltmeter,
I is the current supplied by current source,
A is the cross-sectional area of the sample in cm^2 ($w \times t$), and
L is the distance between voltmeter leads in cm.

To compensate for the influence of electrodynamic force, a voltage value under the positive test current is obtained first, and then another voltage datum is measured under the negative test current. Average the absolute values of these two voltage data and use them as V, I in equation $\rho = \frac{V}{I} \cdot \frac{A}{L}$. Most materials have large temperature coefficients, so it is important to keep the samples under a stable temperature.

3.5.1.2 Four-Probe Method

The four-probe method is used for very thin samples, such as epitaxial wafers and conductive coatings. Figure 3.9 shows the configuration diagram of four-point collinear probe used for resistivity measurement. The current is applied from two

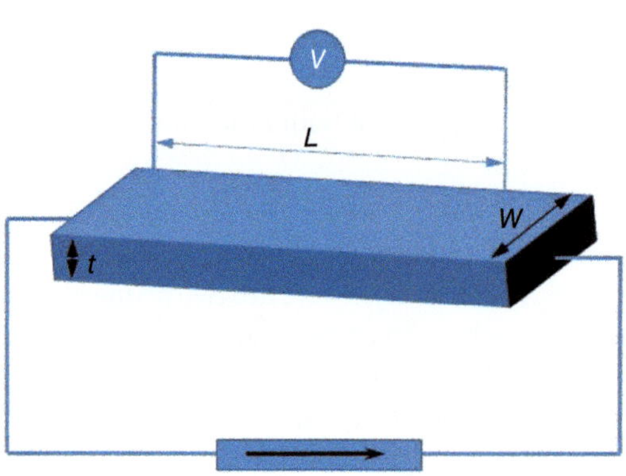

Figure 3.8 The system for testing the resistivity of a bulk material.

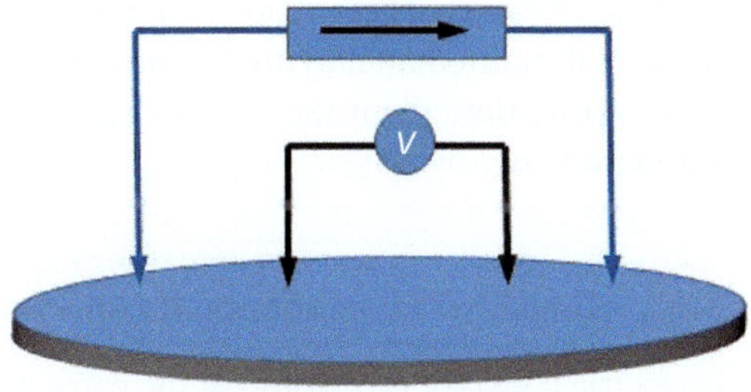

Figure 3.9 The configuration diagram of four-point collinear probe used for resistivity measurement. Source: Kang Zheng.

external probes, and the voltage drop is measured between two internal probes. The calculation formula for surface resistivity is

$$\sigma = \frac{\pi}{\ln 2} \cdot \frac{V}{I}$$

where σ is the surface resistivity in ohm/,

V is the voltage measured by voltmeter, and

I is the current supplied by current source.

It is worth noting that the unit of surface resistivity is expressed as ohm/ to distinguish it from the measured resistance (V/I). For very thin or very thick samples, a correction factor may be required to correct the resistivity calculation.

3.5.1.3 The Van der Pauw Method

Although the Van der Pauw method for resistivity measurement is mainly used in the semiconductor industry, it can also be used in other applications, such as determining the resistivity of superconductors or other thin materials. The Van der Pauw method is usually used for flat, uniform-thickness, arbitrarily shaped sample materials without any isolated holes. As shown in Figure 3.10, the contact point should be very small and placed on the periphery of the sample.

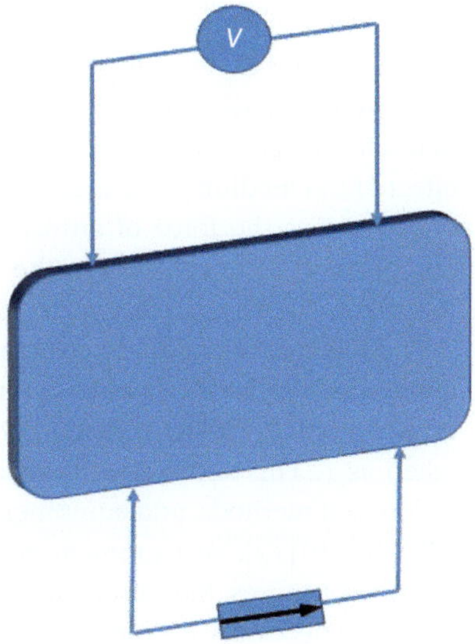

Figure 3.10 The configuration diagram of the Van der Pauw method used for resistivity measurement. Source: Kang Zheng.

After carrying out several measurements around the sample, the obtained data are combined mathematically to determine the average resistivity of the sample. Further information about the van der Pauw method in detail can be found in ASTM Standard F76.

3.5.2 Dielectric Constant and Its Characterization

Dielectric constant is the ratio of the capacitance formed by two plates with a material between them to the capacitance of the same plates with air as the dielectric. For low megahertz, frequencies are less than or equal to 1 GHz. For high megahertz, frequencies are greater than 1 GHz.

Insulating materials are used in two ways. First is to insulate and support components of an electric system from each other and from the ground, and the second is to function as the dielectric of a capacitor. Low dielectric constant values are preferred for high-frequency or power application fields to minimize electric power loss. High values of dielectric constant are recommended for capacitance applications of small sizes.

Dielectric constant (ε) and dissipation factor (D_F) dissipation factor are related to capacitance (C) and can be calculated from capacitance measurements. Capacitance, between a parallel-plate capacitor, is proportional to the dielectric constant of the insulator separating the conductors and the conductor area, A, and inversely proportional to the distance, d, between the conductors as shown in

$$C = \varepsilon A/d$$

If the gap between the plates is filled with a dielectric material, the capacitance increases by a constant factor characteristic of that material. The *dielectric constant*, ε, is defined as

$$C_m = \varepsilon C_v \text{ or } \varepsilon = C_m/C_v$$

where C_m and C_v are the capacitance of the material and a vacuum, respectively. Because dielectric material affects the force with which two oppositely charged plates attract each other, the dielectric constant may also be defined as the relative effect of the medium on this force of attraction, as shown in equation $F = QQ'/\varepsilon d^2$.

where F is the force of attraction between the two plates, Q is the charge on one plate, Q' is the charge on the second plate, ε is the dielectric constant, and d represents the distance between plates.

The higher the dielectric constant of the material between the plates, the more attenuated will be the force of attraction between the plates. The dielectric constant of vacuum is 1. For air, it is slightly above 1, which for all practical purposes is also taken as 1. This simplifies the measurement somewhat. Details of sample preparation, test methods, and equipment for measuring dielectric constants are given in ASTM D150 [19]. Measurements are made at a standard frequency of 1 MHz for comparative purposes. Measurements at higher frequencies extending into the gigahertz range are necessary for high-speed and high-frequency circuits.

The dissipation factor, D_F, is the ratio of the resistive (loss) component of current I_r to the capacitive component of current, I_c, and equals the tangent of the dielectric loss angle, δ, as follows:

$$D_F = \frac{I_r}{I_c} = \tan\delta$$

As for dielectric constant, test methods for dissipation factor may also be found in ASTM D150 [19].

In addition, the dielectric relaxations are present in both polymers and polymer matrix composites. For inorganic filler/polymeric composites, the relaxation phenomena include contributions from both the polymeric matrix and the distribution of filler. Dielectric data could provide valuable information regarding polymer/filler interactions, distribution of composite's morphology, improving our knowledge on their structure–properties relationship. The electrical performance of composites can be adjusted by selecting the appropriate type and amount of filler. Transition from insulating to conductive behavior can be studied in terms of the scaling law of percolation theory. Finally, the influence of temperature and frequency on conductivity data can be exploited in revealing the occurring charge transport mechanisms.

When the electromagnetic wave interacts with the material, it will bring out reflection, transmission, scattering, and other physical phenomena. By measuring and studying these phenomena, the relevant parameters of the material can be obtained.

Generally, the current methods for measuring dielectric constant mainly include concentrated circuit method, transmission line method, resonance method, and free space wave method. Among them, transmission line method, concentrated circuit method, and resonance method belong to measurement methods in laboratory. The measurement is usually carried out in the laboratory, and the corresponding sample collection technology is required.

For measuring the dielectric constant of solid materials, such as ceramic materials, the dielectric temperature spectrometer is required for measurement. The test fixture is designed according to the ASTM D150 standard method [19], by using the principle of parallel plate electrode, and the test electrodes are composed of upper, lower electrodes and protective electrode. The upper and lower electrodes have good concentricity and parallelism. The protective electrode can weaken the influence of surrounding air capacitance, so that the test data are more accurate and reliable. The sample shall be well prepared before measurement. In order to ensure the accuracy of the measurement results, the sample preparation needs to follow these points: (i) sample size: 5–40 mm in diameter (electrode diameter is 26.8 mm), less than 8 mm in thickness; (ii) the shape of the sample is prepared as a disc, and both sides are coated with electrodes; and (iii) the surface of sample must be flat and smooth to ensure good contact with the parallel electrode. Otherwise, the error of measured capacitance value may occur due to the existence of contact gap, which affects the test results. Finally, the high-temperature dielectric constant of the material is measured automatically by the dielectric temperature spectrometer. The software can simultaneously measure and output the data of frequency spectrum, voltage spectrum, bias voltage spectrum, temperature spectrum, and dielectric temperature spectrum.

3.6 Material Characterization Analysis Technology

After a die is separated and selected, it must be packaged. Different microchips require different packages. In most cases, the die and the case are packaged together by wire bonding. The bonding process is to use very thin metal wires, usually aluminum or gold wires, to connect the die and the leads of the case together. After final function testing, reliability testing, and other special testing if it is needed, the product can be transferred to customers.

The advantage of all of these techniques below is that they can be used to investigate larger material volumes and to give integral parameters for the measured structure.

3.6.1 Optical Microscope

Optical microscopy with magnifications of up to about 1000x is very useful for gaining an overview and as a first step in any morphology analysis.

Global techniques provide methodologies for the isolation of failures without a detailed understanding of IC operation. Ideally, global analysis techniques should be easy to use, nondestructive, sensitive, and should provide high spatial resolution. Photoemission microscopy, or light emission microscopy (LEM), is a relatively new failure analysis technique for detecting photonic radiation from a defect site, primarily due to carrier recombination mechanisms. Photon probing of ICs has been a powerful approach to failure analysis. Photon or light emission microscopy can be viewed as passive photon probing. Active photon probing takes advantage of a light beam's ability to interact with an IC and precisely localize the effects of this interaction (Figure 3.11).

The device has insulation defects in the design, production, and manufacturing process, or undergoes the external static penetration during the process mentioned above, which will cause the device leakage failure. When the device in the leakage failure mode is powered on, the flowing current is formed inside, and the electrons at the leakage position will migrate, leading to the conversion of electrical energy to optical energy, that is, electrical energy is released in the form of optical energy, thus forming 200–1700 nm infrared. Most defects in semiconductor or IC devices show weak infrared luminescence, which can quickly and accurately locate the failure point. The light emission microscope mainly uses an infrared detector to detect these released infrared rays through the infrared microscope, so as to accurately locate the leakage point of the device. It can analyze gate oxide leakage, p–n junction leakage, thermionic effect, CMOS latch-up effect, EOS/ESD damage, saturated MOS devices, and other failures. Such defect sites emit light during device operation which would otherwise be absent in a normal device. Such photoemissions, being very low-level, are not visible to the naked eye. Photoemission microscopy uses a powerful image intensification technology to amplify the light emitted by photo-emitting defect sites. The resulting radiation image is then overlaid with its corresponding die surface image, such that the emission spot coincides with the precise location of the defect. A CCD camera and a computer are used to accomplish this feat.

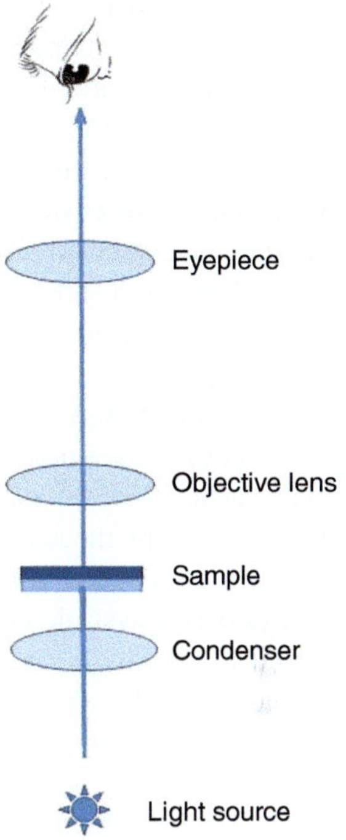

Figure 3.11 The configuration diagram of optical microscopy. Source: Kang Zheng.

3.6.2 X-ray Diffraction

Although neutron diffraction, electron diffraction, infrared spectroscopy, Mossbauer spectroscopy, and other methods can be used to analyze the material structure, X-ray diffraction is the most effective and widely used means, and X-ray diffraction is the first method used by humans to study the material microstructure. X-ray diffraction has been widely used in physics, chemistry, geosciences, materials science, and various engineering technology sciences. It has become an important experimental method and structural analysis means, and has the advantages of nondestructive samples.

X-ray is a kind of electromagnetic wave with very short wavelength (about 20–0.06 Å), which can penetrate materials with a certain thickness and can make fluorescent materials luminous, photographic emulsion sensitive, and gas ionization. When high-energy electron beams are used to bombard metal "target" materials to generate X-ray, it has a specific wavelength corresponding to the elements in the target, which is called characteristic X-ray. X-ray diffraction (XRD), as a nondestructive method, has been widely utilized to attain precise information about the physicochemical attributes of materials (i.e. amorphous/crystalline structures, crystalline lattice parameters, composition profile, etc.). Considering the angstrom range of XRD wavelengths, and their extreme energy for penetrating at an atomic level, this technique has been introduced as an efficient tool to investigate various properties of materials.

When a beam of monochromatic X-rays incident on a crystal, the distance between these regularly arranged atoms is the same order of magnitude as the wavelength of the incident X-rays, because the crystal is composed of crystal cells with atoms arranged in regular order. Therefore, the X-rays scattered by different atoms interfere with each other and produce strong X-ray diffraction in some special directions. The orientation and intensity of the spatial distribution of the diffracted rays are closely related to the crystal structure. This is the basic principle of X-ray diffraction.

X-ray diffraction technology has become the most basic and important structural testing method. Its main applications are as follows.

3.6.2.1 Phase Analysis

Phase analysis is the most widely used measurement of X-ray diffraction in metals, which can be divided into qualitative and quantitative analysis. The qualitative analysis compares the measured lattice plane spacing and diffraction intensity with the diffraction data of the standard phase to determine the phase in the material. The quantitative analysis determines the content of each phase in the material according to the intensity of the diffraction pattern. It has been widely used in studying the relationship between properties and the content of each phase and checking the composition ratio of materials.

3.6.2.2 Determination of Crystallinity

The crystallinity is defined as the percentage of the ratio of the weight of the crystalline part to the total sample weight. Amorphous alloys are widely used, such as soft magnetic materials, and the crystallinity directly affects the performance of materials. The measurement of crystallinity is particularly important. The crystallinity is determined by the diffraction pattern area of crystalline phase and the area of amorphous phase.

3.6.2.3 Precise Measurement of Lattice Parameters

Precise determination of lattice parameters is often used to determine the solid solubility curve of the phase diagram. The change of solubility often causes the change of lattice constant. When the solution limit is reached, the continuous increase in solute will cause the precipitation of new phase, and the change of lattice constant will no longer happen. This turning point is the dissolution limit. In addition, the number of atoms per unit cell can be obtained by precise measurement of lattice constant, so as to determine the type of solid solution. Useful physical constants such as density and expansion coefficient can also be calculated.

Basic composition of X-ray diffractometer: (i) **The high stability X-ray source**, which provides the X-ray needed for measurement. The choice of different anode target materials of the X-ray tube can change the wavelength for the X-ray, and adjusting the anode voltage can control the intensity of the X-ray source. (ii) **The sample and the adjustment system of sample position and orientation**. The sample must be a solid block of single crystal, powder, polycrystalline, or microcrystalline. (iii) **The X-ray detector**, which detects the diffraction intensity or the diffraction direction at the same time, and the polycrystalline diffraction pattern data can be obtained through the measuring and recording system or the computer processing system. (iv) **Diffraction pattern processing and analysis system**. Modern X-ray diffractometers are installed with computer systems equipped with special diffraction pattern processing and analysis software, which can automatically and intelligently finish the diffraction pattern processing and analysis.

3.6.3 Scanning Electron Microscope

Scanning electron microscope (SEM) is an observation method, of which its function is between transmission electron microscope and optical microscope. It scans the sample with a narrow focused high-energy electron beam, excites the surface of material, and collects, amplifies, and re-images various physical information to characterize the microscopic morphology of the material. The resolution of the new scanning electron microscope can reach 1 nm. The magnification can reach 300 thousand times or more and can be adjusted continuously. SEM provides a high magnification and a large depth of field, allowing examiners to view fiber ends, surfaces, or cross-sectional areas in detail. The SEM technique can be used for the evaluation of surface morphology like roughness and homogeneity of dispersion of nanomaterials in the polymer matrix. Easy sample preparation and a simple understanding of images make SEM the most common electron microscopy for this purpose. In addition, SEM combined with other analytical instruments can be used to observe the micromorphology and analyze the composition of the material in the micro-area at the same time. On the other hand, field-emissive SEM (FESEM) can provide a much higher resolution and magnification than SEM. Furthermore, a high-quality image with negligible electrical charging is possible to obtain from FESEM. It also eliminated the need of depositing conducting coatings onto the polymer nanocomposites, which is a common practice for SEM. SEM is widely used in the research of rock and soil, graphite, ceramics, and nanoscaled materials (Figure 3.12).

The scanning electron microscope is a multi-functional instrument with many superior performances and is the most widely used instrument. It can perform the following basic analysis: (i) Observation and analysis of morphology in three

Figure 3.12 The configuration diagram of SEM. Source: Kang Zheng.

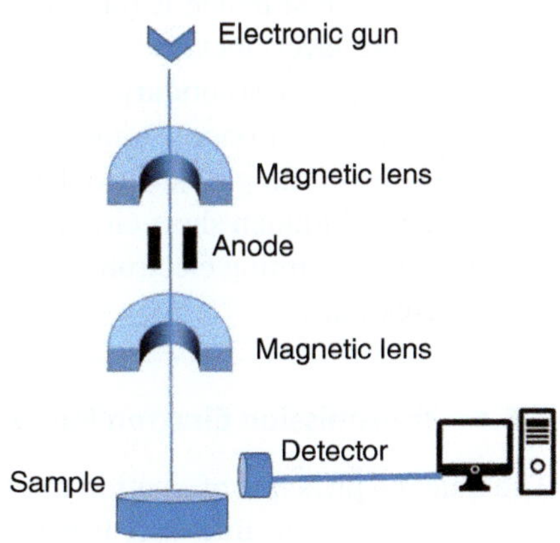

dimensions; (ii) It can observe the morphology and analyze the composition of the micro-area, simultaneously.

For the analysis of the fracture of material, owing to its important feature of large depth of field and stereoscopic imaging, the focal depth of scanning electron microscope is 10 times larger than that of transmission electron microscope and hundreds of times larger than that of optical microscope. Due to the large depth of field of the image, the scanning electron image obtained is full of stereoscopic feeling and three-dimensional shape, which can provide much more information than other microscopes. This feature is very valuable for structural analysis. The fracture morphology displayed by the scanning electron microscope presents the essence of material fracture from the perspective of deep level and high depth of field. It plays an irreplaceable role in teaching, scientific research, and production, and is a powerful means to analyze the causes of material fracture, analyze the causes of accidents, and determine the rationality of the process.

For observation of the original surface of the large sample, it can directly observe the sample with a diameter of 100 mm, a height of 50 mm, or larger size, without any restriction on the shape of the sample. The rough surface can also be observed, which avoids the trouble of sample preparation, and can truly obtain the difference in contrast of the sample (back reflection electron image) reflecting by different components.

For observation of thick specimen, when observing thick samples, SEM can obtain high resolution and the most real morphology. The resolution of SEM is between that of optical microscope and transmission electron microscope. However, when comparing the observation of thick block samples, since the coating method used in the transmission electron microscope leads to the resolution of the coating is only 10 nm, and the observation is not the sample itself, it is more advantageous to observe the thick block samples with the scanning electron microscope, which can obtain real information of the sample surface.

For applications in semiconductor device research, the imaging of scanning electron microscopy with electron-beam-induced current can be used to locate and study damage in integrated circuits. With the sample current imaging, the results can indicate the open and short circuits of the metal layer in the circuit. Therefore, the resistance contrast image is often used to check the metal wiring layer, polycrystalline wiring layer, metal-to-silicon test pattern, and the conductive form of the film resistance. The secondary electron potential contrast image reflects the potential on the sample surface. From this image, the height and distribution of potential on the surface of sample can be obtained, especially for the determination of hidden open-circuit or hidden short-circuit parts of devices.

To sum up, scanning electron microscope is a powerful means of semiconductor device research.

3.6.4 Transmission Electron Microscope

The imaging principle of electron microscope is basically the same as that of optical microscope, but the difference is that the former uses electron beam as light source and electromagnetic field as lens. In addition, due to the weak penetration of the electron beam, the specimen used for electron microscopy must be made into ultrathin sections with a thickness of about 50 nm. This kind of slicing needs to be made with ultramicrotome. The magnification of the electron microscope can reach nearly one million times, and it is composed of five parts: lighting system, imaging system, vacuum system, recording system, and power supply system.

Since the de Broglie wavelength of the electron is very short, the resolution of the transmission electron microscope is much higher than that of the optical microscope, which can reach 0.1–0.2 nm, and the magnification is tens of thousands to millions of times. Therefore, the transmission electron microscope can be used to observe the fine structure of the sample, or even the structure of only a column of atoms, which is ten thousand times smaller than the smallest structure observed by the optical microscope. TEM is an important analytical method in many scientific fields related to physics and biology, such as virology, material science, nanotechnology, and semiconductor research.

Transmission electron microscopy (TEM), with its various imaging modes and analytical abilities, is now an indispensable tool for chemical and structural characterization at the nanoscale of all types of materials. Advanced packaging materials do not differ fundamentally from other materials except that they are more heterogeneous, with a more complex structure. In addition, many of them contain nanoparticles or have a nanoscale structuration, which plays a significant role in their physical properties or is rich in information concerning their manufacture. TEM techniques are thus well suited to investigate them, especially because the developments of these last decades afford both a more efficient sample preparation and faster data recording (Figure 3.13).

Transmission electron microscopy is an excellent tool for investigating the microscopic nature of defects, such as dislocations in Si/SiGe heterostructures.

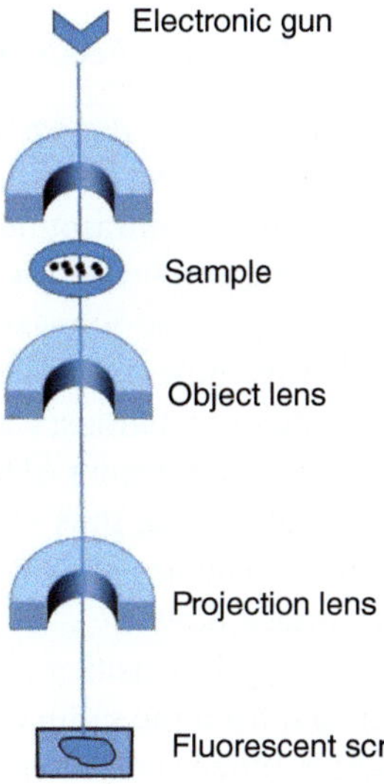

Figure 3.13 The configuration diagram of TEM. Source: Kang Zheng.

The physics of electron-beam diffraction is similar to that of X-ray diffraction described previously. However, unlike X-ray rocking curves, cross-sectional transmission electron microscopy provides depth resolution of defects and can easily resolve dense arrays of dislocations at interfaces. The limitations of transmission electron microscopy are that samples have to be mechanically and/or chemically thinned first to ≈0.15 μm to allow the electron beam to traverse the sample, and these destructive preparation procedures usually provide only small areas suitable for examination. The sensitivity of cross-sectional transmission electron microscopy for dislocation observation is at best ~105–106 cm^{-2}, which is similar to careful X-ray rocking curve measurements. Plan-view transmission electron microscopy can have slightly better resolution of defect densities; however, depth information on these defects is not available in this case. For interfaces with high dislocation densities, TEM can easily observe these arrays, whereas X-ray topography may not be able to.

One thing that is worth reemphasizing is that in situ TEM is a very powerful method for studying the defects and the strain-relaxation process, when the epi-SiGe layer can be heated up or ion-implanted, and the kinetics of the defect generation of the strain-relaxation process can be videotaped.

TEM can provide atomic-level micro-details and precision equipment handling data collection for device critical dimension control and failure analysis. New semiconductor device manufacturing requires high-accuracy imaging and fault analysis to produce improved and optimized manufacturing process. These complex analysis processes usually require the TEM products used to have the automation function of precision equipment handling rate, extremely easy operating user interface, image distortion rate of precision equipment handling, and other rapid characterization capabilities, such as high-speed EDS energy spectrum acquisition. All these have promoted TEM to become one of the indispensable tools for semiconductor device R&D, yield improvement, and failure analysis.

3.6.5 Scanning Acoustic Microscope

Widely used for nondestructive testing, ultrasonic waves are employed in the so-called scanning acoustic microscopy (SAM) technique based on the reflection or transmission of ultrasound in the sample, which enables the imaging of sample surface elastic properties at submicrometer scale with resolution that highly depends on the ultrasonic wavelength in the investigated material. It yields information on density differences, delamination, microvoids, cracks, and other defects in components, materials, wafers, and other samples, with sizes of less than 1 μm. The oscillating component of the cantilever deflection can be acquired at each point of the scanned area, thus obtaining an image which is related to the surface elastic modulus. The use of SAW energy is confined in a volume of material underneath the surface down to a depth of a few times the wavelength. Therefore, acoustic microscopy is sensitive to the mechanical properties of the material in a volume included from the sample surface to a depth of a few times the wavelength into its interior. When short wave is transmitted into sample, it passes through the interface between two different materials. Due to the different impedances of different materials, the degrees of absorption and reflection are different. The collected changes of reflected energy information or phase information can be used to check the defects such as delamination, cracks, or cavities in the sample interior. Advanced microscopic imaging technology is an important means for many industries to check and find defects in various kinds of samples. The advantages of this nondestructive testing technology are particularly prominent when inspecting materials and maintaining complete samples. SAM can be used as a failure analysis technique for detecting disbonds or delaminations between package interfaces, for example, interfaces between the plastic resin package material, the die, the die paddle, the lead frame, the die attach material, etc. It basically consists of sending a sound wave through the package and interpreting the interaction of the sound wave with the package. A typical scanning acoustic microscope may employ either pulse echo or transmission inspection to scan for disbonds or delaminations. The sound wave may be generated by a piezoelectric crystal, or transducer, that has been cut to provide a specific frequency. It is activated by a high-voltage pulse from a transmitter, which is also known as the pulser. The activation would cause the transducer to vibrate at the specified frequency, which transmits an ultrasonic wave through the package. By using the nature of the pulse echo, the piezoelectric transducer is excited to emit multiple beams which are transmitted to the measured sample through the coupling liquid medium. Pulse echo inspection consists of interpreting echoes sent back by the package, while transmission inspection consists of interpreting the sound wave at the other end of the package, after it has passed through the latter. This wave travels to the specimen through a medium or couplant, which is usually deionized water since sound waves could not travel through air at the frequencies used. The wave travels through the specimen's material at the material's velocity, with a portion of it being reflected back every time it hits an interface within the material. When passing through materials with different impedances, wave phase, energy changes, and other phenomena will occur. After a series of data acquisition and calculation, gray value pictures are formed, which can be used to analyze the internal conditions of the sample. SAM offers a versatile nondestructive method for determining the elastic moduli at different scales. The traditional SAM, based on the reflection coefficient, and $V(z)$ curve measurements using spherical lens are, however, not ideally suited for anisotropy measurements of elastic constants. The acoustic field generated by a spherical lens is symmetric. When the symmetric field interacts with the unknown substrate placed perpendicular to the lens axis, both the reflected wave and the Rayleigh wave contribution to the returned signal are averaged over all directions. Thus, the directional properties due to material anisotropy are smeared out. For our measurements, the DEJ slab was placed in a distilled (deionized) temperature-controlled H_2O bath and positioned relative to the acoustic lens using a positioning system. The positioning setup permits three-axis translational motion, X–Y motion for interrogating different regions of the specimen and Z-positioning to enable focusing and defocusing on the specimen. The positioning system also includes a tilt stage to allow for θ-rotation of the specimen relative to the transducer axis.

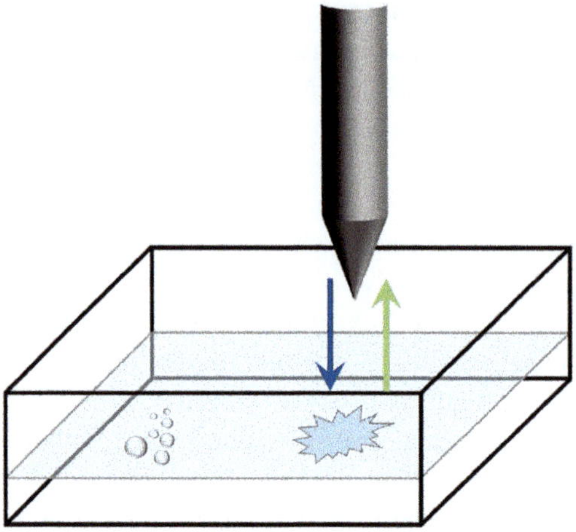

Figure 3.14 The configuration diagram of C-mode scan furnishes a two-dimensional (area) description at a particular depth. Source: Kang Zheng.

In the pulse echo method, the same transducer is used as sender and receiver of the sound waves. Pulses are repeated using repetition rates at which the echoes from one pulse will not interfere with those of another, for example, 10–20 KHz. The echoes received by the transducer are converted to voltages, amplified, digitized, and presented to the user as an image.

In the through-transmission technique, separate transducers are used to send and receive sound waves, both of which are on opposite sides of the specimen. The absence and presence of signals mean bad and good bonding, respectively.

Scanning acoustic microscopy has several modes. The A-scan mode is the real-time oscilloscope waveform of the acoustic signals based on the reflected echoes, or acoustic data collected at a single X–Y portion or point.

The B-scan mode is the cross-sectional display showing the ultrasonic reflection of the various interfaces along the depth of the package, or acoustic data collected along the X–Z plane at depth A. A B-mode scan furnishes a two-dimensional (cross-sectional) description along a test line (Y).

The C-scan mode is the display of the image of reflected echoes at the focused plane of interest, or acoustical data collected along an X–Y plane at depth Z. A C-mode scan furnishes a two-dimensional (area) description at a particular depth (Z) (see Figure 3.14). Usually, the B-scan images are based on the C-scan image for precise determination of the depth of flaws detected.

As a kind of nondestructive testing analysis, it can detect materials without damaging the electrical energy of materials and maintaining the structural integrity. It is widely used in material inspection (IQC), failure analysis (FA), destructive physical analysis (DPA), reliability analysis, secondary screening of components, quality control (QC), quality assurance and reliability (QA/REL), research and development (R&D), and other fields.

3.6.6 Atomic Force Microscope

The atomic force microscope (AFM) is a very powerful microscope invented by Binnig, Quate, and Gerber in 1986. Besides imaging, it is also one of the foremost tools for the manipulation of matter at the nanoscale (Figure 3.15).

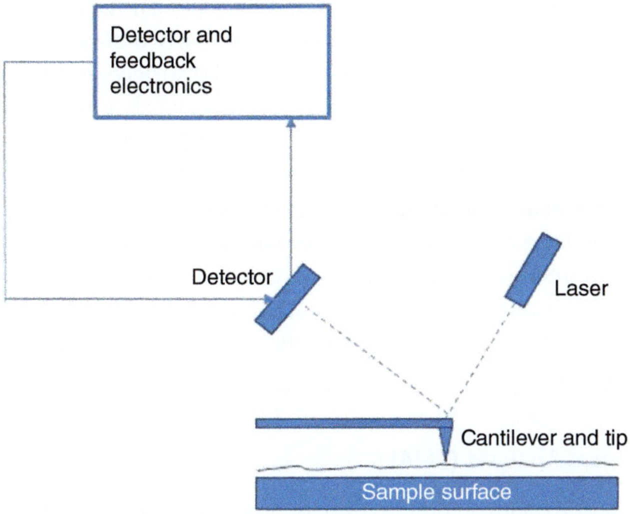

Figure 3.15 The configuration diagram of AFM. Source: Kang Zheng.

The AFM consists of a cantilever with a sharp tip at its end. The tip is brought into close proximity to a sample surface. The force between the tip and the sample leads to a deflection of the cantilever according to Hooke's law. Typically, the deflection is measured using a laser spot reflected from the top of the cantilever.

The primary mode of operation for AFM is contact mode. If the tip were scanned at constant height, there would be a risk that the tip would collide with the surface, causing damage. Hence, in most cases, a feedback mechanism is employed to adjust the tip-to-sample distance to keep the force between the tip and the sample constant. This can be achieved by mounting the sample on a piezoelectric crystal. The tip is then scanned across the sample surface, and the vertical displacement s necessary to maintain a constant force on the tip is recorded. The resulting map of $s(x,y)$ represents the topography of the sample.

Over decades, several modes of operation have been developed for the AFM. The primary modes of operation are contact mode and dynamic mode. In the contact mode operation, the force between the tip and the surface is kept constant during scanning by maintaining a constant deflection. In the dynamic mode, the cantilever is externally oscillated close to its resonance frequency. The oscillation gets modified by the tip-sample interaction forces; these changes in oscillation with respect to the external reference oscillation provide information about the sample's characteristics. The dynamic mode generates lower lateral forces on the sample and is widely used to image biological samples.

Schemes for dynamic mode operation include frequency modulation and the more common amplitude modulation. In frequency modulation, changes in the frequency of modulation provide information about the characteristics of sample. In amplitude modulation (also known as intermittent contact or tapping mode), changes in the oscillation amplitude yield topographic information about the sample. Additionally, changes in the phase of oscillation under tapping mode can be used to discriminate between different types of materials on the surface.

The AFM has several advantages over the electron microscope. Unlike the electron microscope which provides a two-dimensional projection or a two-dimensional image of a sample, the AFM provides a true three-dimensional surface profile. In addition, samples observed by an AFM do not require any special treatment that may actually destroy the sample and prevent its reuse. Different from that the electron microscope needs an expensive vacuum environment for proper operation, the AFM can work perfectly well in an ambient or even liquid environment.

The main disadvantage that the AFM has compared to the electron microscope is the image size. The electron microscope can show an area on the order of millimeters by millimeters and a depth of field on the order of millimeters. The AFM can only show a maximum height on the order of micrometers and a maximum area of around 100 by 100 micrometers.

3.6.7 Thermal Mechanical Analysis (TMA)

Electronic components require close combination of many different materials, which requires that they have similar thermal expansion coefficients to cope with the temperature fluctuations during use. The expansion coefficient and glass transition temperature T_g can be measured by the thermal expansion curve. The glass transition temperature T_g is the inflection point of the curve of expansion versus temperature. When the temperature is higher than the glass transition temperature, the coefficient of thermal expansion will increase by 3–5 times. The coefficient of thermal expansion (CTE) of chosen materials determines whether the materials are suitable for specific electronic component packaging technology.

Measurement of the thermal expansion of solids and liquids is a classical physical technique named dilatometry. When special emphasis that such dimensional changes is a function of temperature, during a controlled temperature program, the technique is labeled thermos-dilatometry and is one of the special techniques derived from thermos-mechanometry. For solids, we can distinguish between linear and volume expansion and most commonly measure the length of a solid sample as a function of temperature.

Most solids expand on heating. Exceptions include vitreous silica and ZnS at low temperatures. The change in a linear dimension, L, with T is given by

$$L_2 = L_1 \left(1 + \int_{T_1}^{T_2} \alpha \mathrm{d}T\right)$$

where L_1 is the length at T_1, L_2 is the length at T_2, and α is the coefficient of linear expansion. Over a small temperature interval, ΔT is approximately constant for many materials, so the dilation

$$L_2 - L_1 = \Delta L = L_1 \alpha \Delta T$$

The value of α is related to the structure and type of bonding in the solid. In general, strong bonding results in low values of α so the order of expansion coefficients is usually following the relationship: covalent and ionic materials < metals < molecularly bonded materials. Some substances may vary with direction (anisotropic behavior).

Among the materials that may show anisotropy are single crystals, drawn fibers, drawn or blown films, fiber-reinforced composites, liquid crystals, and some metallic glasses. In some anisotropic solids, the coefficient of expansion in direction of one of the principal axes may be negative.

Measurements of thermal expansion may be classed as either absolute or difference measurements. The pointer and scale are outside the furnace and the movements of the sample and of the reference are transmitted by identical push rods. Measurements may be made in either the horizontal or vertical position. In horizontal instruments, arrangement has to be made for suitable spring loading, so that any contraction can be detected. In vertical instruments, the weight of the push rod and measuring system has to be counterbalanced to avoid compression or deformation of the sample.

Another important method to evaluate the deformation of a material is thermomechanical analysis (TMA). TMA, as defined by ASTM E473-85, is a method for measuring the deformation of a material under a constant very small load as a function of temperature, while the material is under a controlled temperature program. The measuring system consists of a linear voltage differential transformer (LVDT) connected to the appropriate probe (Figure 3.16). Various probes are available, and the measurements can be done in either compression, expansion, penetration, flexure, or in tension mode. It is this variety of probes which allows for the measurement of samples of different configurations. Any displacement of the probe generates a voltage that is then recorded. The dimensional change of a sample with an applied force is measured as a function of time or temperature. The plot of expansion (or contraction) versus temperature (or time) can then be used to obtain T_g, the coefficient of thermal expansion (CTE), softening temperature, and Young's modulus [20–23].

The change in linear dimension as a function of temperature can be described by the following:

$$L_2 = L_1 \left(1 + \int_{T_1}^{T_2} \alpha_1 \mathrm{d}T\right)$$

where α_1 is the coefficient of linear expansion, and L_1 and L_2 are the lengths of the specimen at temperatures (or time) T_1 and T_2, respectively. If the difference between T_2 and T_1 is relatively small, then the equation can be represented by

$$L_2 - L_1 = L_1 \alpha_1 (T_2 - T_1)$$

or it can be rewritten as

$$\alpha_1 = \frac{1}{L_1} \frac{\Delta L}{\Delta T}$$

Therefore, the slope of the curve of length versus temperature yields $\alpha_1 L_1$, and the coefficient of linear thermal expansion is obtained by dividing by L_1.

There are some drawbacks to thermal mechanical analysis. For example, proper calibration is required to obtain reliable and reproducible data. Other sources of errors include slippage of the probe on the specimen and specimens undergoing creep in addition to length changes.

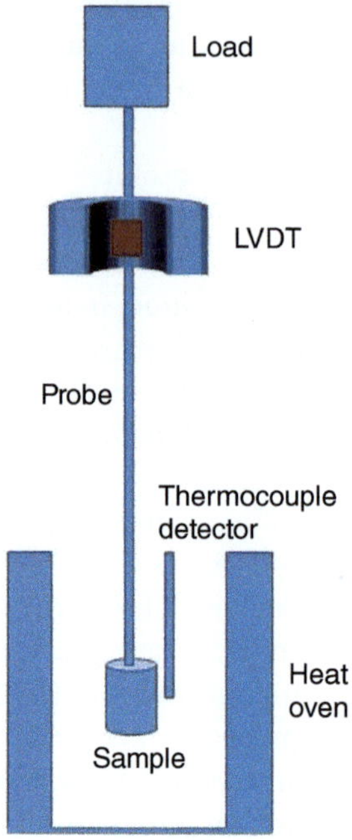

Figure 3.16 The configuration diagram of TMA. Source: Kang Zheng.

3.6.8 Dynamic Mechanical Analysis (DMA)

Dynamic thermal mechanical analysis (DMA) is used to measure the relationship between the mechanical properties of viscoelastic materials and time, temperature, or frequency. The sample is deformed when a controlled periodic (sinusoidal) mechanical stress is applied. Dynamic thermal mechanical analysis measurements are ideal solutions for obtaining information about glass transition and melting, secondary transformation, frequency effect, optimization of transformation process, characterization of nonlinear properties of elastomer, fatigue, characterization of material aging, immersion test, long-term creep prediction, other optimal material characterization schemes, etc.

The following equations describe the stress–strain relationship (Figure 3.17) as measured by DMA:

$$\sigma = \sigma_0 \sin(\omega t) \cos\delta + \sigma_0 \cos(\omega t) \sin\delta$$

$$\varepsilon = \varepsilon_0 \sin(\omega t)$$

where σ is the stress, ω is the angular frequency, t is the time, δ is the phase angle, and ε is the strain.

The *real* component, $\sigma_0 \cos\delta$, occurs when stress is in-phase with strain. The *imaginary* component is 90° out-of-phase with strain and corresponds to $\sigma_0 \sin\delta$. The stress–strain components can be resolved, and real and imaginary components of modulus are obtained:

$$\sigma = \varepsilon_0 E' \sin(\omega t) + \varepsilon_0 E'' \cos(\omega t)$$

Figure 3.17 Stress–strain relationship obtained by DMA. tan $\delta = E''/E'$. Source: Kang Zheng.

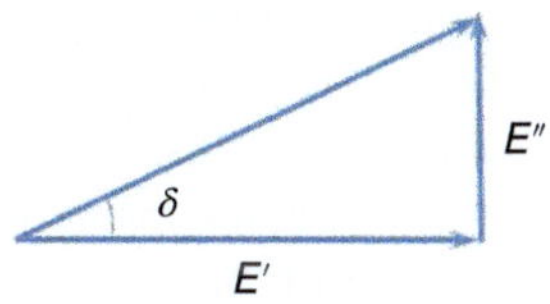

Figure 3.18 The configuration diagram of DMA. Source: Kang Zheng.

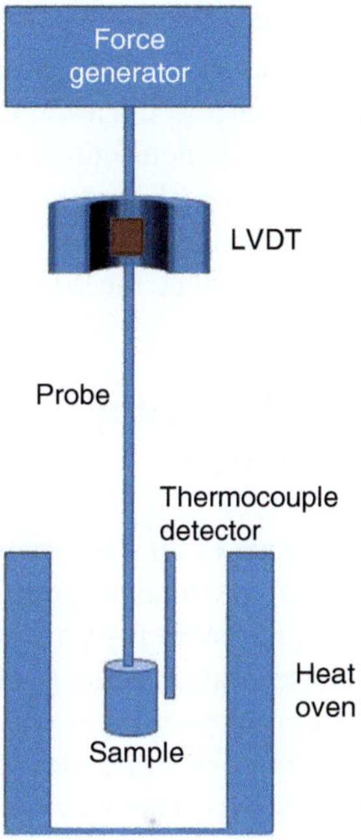

where $E' = (\sigma_0 \cos \delta)/\varepsilon_0$ and is a measure of recoverable strain energy in a deformed body and is known as the *storage* modulus. $E'' = (\sigma_0 \sin \delta)/\varepsilon_0$, where E'' is the loss modulus and is associated with the loss of energy as heat due to the deformation of the material. The loss tangent or damping factor, tan δ, is defined as the ratio of E''/E' (Figure 3.18).

In a DMA experiment, E' and/or E'' and/or tan δ are plotted as a function of time or temperature. T_g is usually obtained from the most intense peak observed for either the E'' or tan δ (and significant inflection for E') curve. Amorphous polymers have a more intense peak than semi-crystalline polymers because the former are less rigid.

The glass transition temperature, determined by DMA, is dependent on the heating rate and frequency. Therefore, T_g values obtained by this dynamic technique are generally different from that obtained by static techniques, such as differential scanning calorimetry (DSC). Moreover, the T_g temperature of a polymer can also be increased by subjecting the material to high-frequency and high-amplitude oscillations. Thus, when studying dynamic mechanical properties, low frequencies and low strain amplitudes should be used. Low strain amplitude is associated with the linear region of a stress–strain curve, but if a large stress or strain amplitude is applied to a viscoelastic material, high internal heat due to molecular vibration is generated. This results in a nonlinear viscoelastic response that is quite complex to analyze. Also, in nonlinear viscoelastic regions, the material is permanently modified. For example, microscopic crack formation or failure due to fatigue can result. This will be useful for analysis of the behavior of electronic package material during heating process.

The clamp used in DMA test will affect modulus results; therefore, absolute modulus values are obtained with significant difficulty using DMA. The comparison among various materials is feasible only when procedures of measurement were taken care of in laboratory and results can be reproducible. Although DMA is weak with respect to the accuracy of absolute modulus, the transition temperatures can routinely be determined with great accuracy. The method used to obtain T_g (i.e. E'' or tan δ peak temperature) affects the value, and, therefore, the parameter must be specified. As long as the same parameter is used throughout a study, the trend observed will be the same regardless of the parameter used [24–26].

3.7 Reliability Analysis and Environmental Performance Evaluation

According to the definition of general standards, reliability refers to the ability of a product to complete specified functions under certain conditions and within specified time. With the progress of technology, semiconductor devices have developed from separation to integration, from large-scale integration to ultra-large-scale integration. At present, nearly 100 million devices can be integrated on a chip. Due to the minimizing of device size, the number of devices per unit area is increasing and the function of circuit is improving, which makes the circuit more complex, and the reliability problem is more important than it has ever been.

The reliability of electronic packaging is evaluated by the ability of electronic packaging system to resist the functional deterioration of device. In the design of microelectronic packaging, packaging reliability must be a key consideration. With the increase in packaging complexity of electronic seal, it is a disaster that the failure of reliability in design was found after the packaging proceeded. Reliability must be an important part of electronic development design, and not an additional work after the device is developed.

The reliability experiment is to apply a certain load to the sample for the purpose of evaluating and analyzing the product, including electrical load, thermal mechanical load, or their combination. The tests are to check whether various performance parameters of the sample still meet the technical requirements, so as to judge whether the sample is invalid or reliable. Common mechanical tests include mechanical vibration test, mechanical impact test, constant acceleration test, lead strength test, and bonding strength test. Commonly used environmental tests include temperature cycle test, thermal shock test, high-temperature storage test, damp heat test, high pressure steam test, and sealing test. In addition, there are reliability screening tests, which can eliminate products with quality problems that may cause early failure via various methods.

3.7.1 Failure Modes and Mechanisms

The reliability research and verification of devices include reliability experiment and analysis. On the one hand, the aim is to evaluate and identify the reliability level of integrated circuit devices and provide reference data for the reliability design of the whole machine and, on the other hand, is to improve the reliability of products. This requires necessary analysis of the failed products, finding out the failure mode, analyzing the failure mechanism, studying the causes of failure, and formulating corrective and improvement measures. In addition, product failure analysis can help us to determine which performance filtering should be effective for improving the reliability of products, and which performance filtering conditions are most appropriate. The main failure modes of integrated circuits are shown in Table 3.1. Common failure mechanisms include the failure caused by chip damage, by wire bonding, by surface contamination, and by external wire breakage. The instruments used for failure analysis and damage detection mainly include ultrasonic microscope, X-ray microscope, and other nondestructive testing methods, as well as scanning electron microscope, transmission electron microscope, electron probe, ion probe, Auger electron spectroscopy analyzer, infrared thermal analysis, etc.

3.7.1.1 Residual Stress

Almost all manufacturing processes create residual stresses. Further, stresses can also develop during the service life of the manufactured component. These stresses develop as an elastic response to incompatible local strains within the component, for example, due to non-uniform plastic deformations. The surrounding material

Table 3.1 Main failure modes of electronic packaging.

Electrical signal failure mode	Physical failure mechanism
Open circuit	Lead wire bonding failure
	Internal micro-cavity
	Contact spring creep
	Corrosion
	Fatigue crack
	Mechanical damage
	Separable contact fracture
Short circuit	Corrosion
	Electrolytic dendritic crystal formation
	Lead transfer
	Dielectric breakdown
	Overheating
The interference in circuit	Unstable contact
	Unstable temperature distribution
	Crack initiation

must then deform elastically to preserve dimensional continuity, thereby creating residual stresses. The mechanisms for creating residual stresses include

1. Non-uniform plastic deformation. Examples occur in manufacturing processes that change the shape of a material including forging, rolling, bending, drawing, and extrusion, and in service during surface deformation, as in ball bearings and railway rails.
2. Surface modification. Examples occur in manufacture during machining, grinding, plating, peening, and carburizing, and in service by corrosion or oxidation.
3. Material phase and/or density changes, often in the presence of large thermal gradients. Examples occur in manufacture during welding, casting, quenching, phase transformation in metals and ceramics, precipitation hardening in alloys, and polymerization in plastics, as well as in service from radiation damage in nuclear reactor components and moisture changes in wood.

Residual stresses can also be harmful and significantly reduce material strength and cause premature fracture.

3.7.1.2 Stress Void

Stress voiding refers to the formation of voids within IC conductors as a result of the tensile mechanical stresses that are present following thin-film deposition and patterning. It became an important technological issue as feature sizes approached 1 mm in the late 1980s. The growth of stress voids is driven by two competing mechanisms: thermal stresses arising from mismatch in the coefficient of thermal expansion between the conductor and the Si substrate and vacancy diffusion that allows void growth. The interplay of these mechanisms leads to the rate of void growth showing a peak typically in the temperature range 150–250 °C. Al interconnects exhibit stress voids at grain boundary intersections with the film surface, and so refractory metal cladding (as described earlier for electro-migration improvement) also serves to provide electrical redundancy to limit the electrical impact of the voids.

3.7.1.3 Adherence Strength

If the adhesion between the plastic packaging material and the chip, the chip base, and the lead frame is too poor, it will leave a serious hidden danger to the product, leading to defects or even failure, for example, the "popcorn" effect, delamination, package cracking, chip fracture, and metallization deformation on the chip during the mounting process. By adjusting the reaction additives, polymer adhesion, and polymer reaction rate, the specific design requirements of the device can be achieved.

In the process of wire bonding packaging, a small amount of copper rust on the pin will seriously affect the adhesion strength between the plastic package and the pin, causing potential danger. The adhesive strength can be improved to a certain extent, and cracking can be reduced by drilling holes in the plastic package.

3.7.1.4 Moisture

Moisture in the air will have a continuous impact on the package, gradually changing the performance of the device (corrosion, T_g drop, expansion mismatch, etc.).

Therefore, accurate measurement of moisture content and diffusion of plastic packaging materials is very necessary for packaging design and material selection. The biggest problem after moisture absorption is moisture absorption expansion, which can even reach more than three times of thermal expansion, seriously affecting the reliability of packaging. The research shows that the reason for the expansion may be that water molecules enter the polymer material and form hydrogen bonds, which lead to the expansion of the polymer chain and the expansion of the plastic package. The hygroscopic expansion characteristics of the material were characterized by the coefficient of hygroscopic expansion (CHE).

3.7.2 Reliability Certification

The influence of package failure on IC's reliability under high temperature has been important. The failure of wire bonding will result in poor contact, wire shifting, or falling off. It degrades the device performance. Popcorn effect leads to delamination between its plastic mold compound/lead frame interface. Thermal expansion coefficient mismatch of the plastic mold compound and the die will also cause IC failure. Thermal stress induced by the internal and external reflowing temperature difference when assembled and high-temperature reflowing is the root cause of the BGA samples. The thermal stress can make the solder bump on the flip chip melt again and bridge the adjacent solder bumps to result in device short-circuit failure. It can also induce crack or delamination in the underfill, crack in solder bump, or solder bump fallen off to result in device open-circuit failure. Generally speaking, the evaluation of performance for electronic packaging materials is majorly focused on the following four aspects:

1. Process performance, such as viscosity, fluidity, gel time, (post) curing time, and temperature;
2. Mechanical properties under wet and thermal conditions, such as CTE, bending modulus and strength, thermal conductivity, and moisture diffusion coefficient;
3. Electrical properties, such as dielectric constant, breakdown strength, and loss factor;
4. Chemical properties, such as flammability and amount of ionic impurities;
5. Among them, the process performance is evaluated from the perspective of process, and the other three properties are evaluated from the perspective of performance and function.

3.7.2.1 Viscosity of Plastic Packaging Material

The methods to measure the viscosity of plastic packaging materials include punching shear, hard die shear, 180 °C peeling, and lead frame bump pull-off test. The method of punching and shearing experiment is commonly used in industry.

3.7.2.2 The Moisture Test

The moisture test is generally conducted by means of "accelerated experiment." For example, the general method of moisture content test is to soak in boiling water for

24 hours. The diffusion coefficient is usually measured at 85°C/85% RH for one week (168 hours).

3.7.2.3 Hygroscopic Strain and Humidity Measurement

Hygroscopic strain and humidity can be measured by simultaneous thermomechanical analysis and thermogravimetric analysis during desorption. Thermomechanical analysis can be used to measure the linear deformation of wet samples in the process of moisture release, and thermogravimetric analysis can be used to measure the loss of moisture content.

3.7.2.4 Temperature Adaptability

The general working temperature in the high-reliability field is −55 to 125 °C, so the plastic-encapsulated devices used in the high-reliability field must at least meet the temperature range of −55 to 125 °C. The materials of plastic-encapsulated devices will expand and contract at high and low temperatures. Since the plastic-encapsulated materials are in contact with other materials of the devices, the degrees of expansion and contraction between the two materials are different. The shear force induced by expansion and contraction between the two materials will be generated at the junction, leading to material cracking. When the device is cycled repeatedly under cold and hot conditions, the shear stress of the material changes repeatedly from compression to tension and finally cracks the metal layer or polysilicon under the passivation layer, cracks the chip, and deforms the metallized layer.

3.7.2.5 Tightness

In highly reliable applications, plastic-encapsulated devices are very sensitive to moisture and humidity due to the defects of plastic-encapsulated materials, with a high failure rate. With the effect of temperature and humidity, plastic-encapsulated devices are prone to delamination and cracking, leading to bond damage, bond displacement, etc. Four kinds of delamination can be detected by acoustic scanning test: delamination between chip and package, pin bonding area delamination, delamination between chip and coating, and delamination between the frame and the plastic package.

3.7.2.6 Defects in Manufacturing Process Control

The defects of plastic-encapsulated semiconductor devices in the production process include passivation layer defects, impurities, or superfluous materials introduced in packaging, which may cause chip peeling, increasing device leakage current, and open circuit or short circuit. These defects will cause failure related to time and stress, such as initial fatal failure or early life failure for defective devices under service conditions after aging test to expose them.

With the increasing application of high reliability of plastic packaging devices, the demand for the development and production of plastic packaging products continues to increase. The screening, identification, inspection, and quality control measures for high-reliability plastic packaging devices are used to eliminate the nonconforming products with potential defects in the production stage, so as to ensure the high reliability of the products.

3.7.2.7 Quality Control Procedure for High-Reliability Plastic Packaging Devices

For example, the specially treated copper frame, epoxy resin plastic sealant with low moisture content and high adhesion, silver paste with low moisture content and high shear strength, and high-purity gold wire with certain hardness and ductility of these specially treated materials can weaken the factors that affect the adhesive strength as much as possible. Special pre-curing (non-oxidation) shall be carried out after core installation to prevent oxidation of the frame and remove the moisture adsorbed in the silver paste. After pressure welding, the lead frame shall be cleaned by plasma dry method to ensure that the frame is free of oxidation and contamination, and the plastic sealing pressure and heat preservation time shall be appropriately increased. After high-temperature storage, strongly accelerated steady-state damp heat, and reflow soldering test, the package sample was confirmed to be free of delamination by ultrasonic testing. Through failure mechanism analysis, selecting appropriate packaging materials and improving process technology can effectively control the number of defects in the design and processing stages, and greatly reduce the possibility of early failure and service life failure of plastic packaging devices.

3.7.2.8 Selection of High-Reliability Plastic Packaging Devices

Filtering is divided into non-stress filtering and stress filtering. The non-stress one is used to eliminate obvious defects in design and process, as well as some early failure products, mainly including external visual inspection, X-ray, ultrasonic scanning, and electrical test. The stress one should apply sufficient load to the device to accelerate the failure of the defective device and eliminate the potential defective device. Temperature cycling and high-temperature aging belong to stress filtering. It shall not introduce new failure mechanism and shall be designed and arranged scientifically. Typical process mainly includes external visual inspection, temperature circulation, X-ray, ultrasonic scanning, high-temperature aging, and other items, and PDA is required to be less than 5%.

High working temperature, poor heat-spreading of the die, plastic, or underfill sensitive to moisture are all the causes of the IC's failure. All these will directly affect the using life and reliability of ICs. During being used, the process design and using condition should all be strictly controlled to reduce the package-related failure and the failure rate.

3.8 Conclusion

With the rapid development of electronic technology, the market for thermal conductive polymer materials is growing. Thermal conductivity is very important for the application of materials, so these properties must be accurately measured. Since there are many methods for measuring the thermal conductivity, the approximate range of the thermal conductivity of the material and the approximate range of the service temperature must be considered in advance to select the correct measurement method.

Supported by new device structures, advanced manufacturing processes, wide bandgap materials and devices, and other research achievements, based on application needs and collaborative innovation throughout the industry chain, semiconductor devices are rapidly iterating toward higher current density, higher operating temperature, higher operating frequency, lower loss, and lower cost. The first-generation semiconductor materials and devices are fully mature and occupy the market. The second-generation materials and devices have unique performance and special application advantages, while the third-generation materials and devices are focusing on breakthroughs and have huge potential, forming a power semiconductor device family represented by IGCTs, MOSFETs, and IGBTs, meeting the needs of various power generations, transmission, control, conversion, and other different application scenarios.

With the rapid development of electronic manufacturing technology, the needs of application pay more attention to the diversification and integration of functions. Advanced processes are used to integrate modules manufactured by different process routes through packaging technology to achieve strong comprehensive performance and excellent integration. Its potential failure mechanism and mode are also more complex, showing the characteristics of multi-scale, multi-accumulation composite failure. Therefore, on the basis of further improving the material testing and evaluation technology described above, it is more important to develop online monitoring technology for electronic devices, modules, and systems.

References

1 Shen, X., Wang, Z.Y., Wu, Y. et al. (2016). Effect of functionalization on thermal conductivities of graphene/epoxy composites. *Carbon* 108: 412–422.

2 Jiang, H., Yi, Z., Chen, P. et al. (2017). Modified thermal resistance networks model for transverse thermal conductivity of unidirectional fiber composite. *Composites Communications* 6: 52–58.

3 Ngo, I.L., Jeon, S., and Byon, C. (2016). Thermal conductivity of transparent and flexible polymers containing fillers: a literature review. *International Journal of Heat and Mass Transfer* 98: 219–226.

4 Zhang, L., Deng, H., and Fu, Q. (2018). Recent progress on thermal conductive and electrical insulating polymer composites. *Composites Communications* 8: 74–82.

5 Li, Q., Chen, L., Gadinski, M.R. et al. (2016). Corrigendum: flexible high-temperature dielectric materials from polymer nanocomposites. *Nature* 536: 112.

6 Gu, J.W., Liang, C.B., Zhao, X.M. et al. (2017). Highly thermally conductive flameretardant epoxy nanocomposites with reduced ignitability and excellent electrical conductivities. *Composites Science and Technology* 139: 83–89.

7 Song, S.H., Park, K.H., Kim, B.H. et al. (2013). Enhanced thermal conductivity of epoxygraphene composites by using nonoxidized graphene flakes with non-covalent functionalization. *Advanced Materials* 25: 732–737.

8 Zhou, H., Deng, H., Zhang, L., and Fu, Q. (2017). Significant enhancement of thermal conductivity in polymer composite via constructing macroscopic segregated filler networks. *ACS Applied Materials & Interfaces* 9: 29071–29081.

9 Tseng, I.H., Chang, J.C., Huang, S.L., and Tsai, M.H. (2013). Enhanced thermal conductivity and dimensional stability of flexible polyimide nanocomposite film by addition of functionalized graphene oxide. *Polymer International* 62: 827–835.

10 Zha, J.W., Li, W.K., Zhu, Y.H. et al. (2014). Effect of micro-Si_3N_4-nano-Al_2O_3 cofilled particles on thermal conductivity, dielectric and mechanical properties of sillicone rubber composites. *IEEE Transactions on Dielectrics and Electrical Insulation* 21: 1989–1996.

11 ISO/DIS 22007-4 (2012). Plastics—determination of thermal conductivity and thermal diffusivity—part 4: laser flash method.

12 ISO/DIS 22007-2 (2012). Plastics—determination of thermal conductivity and thermal diffusivity—part 2: transient plane heat source (hot disc) method.

13 ISO 22007-1 (2012). Plastics—determination of thermal conductivity and thermal diffusivity—general principles.

14 ISO 8894-1 (2010). Refractory materials—determination of thermal conductivity—part 1: hot-wire method (cross-array and resistance thermometer).

15 Hammerschmidt, U., Hameury, J., Strnad, R. et al. (2015). Critical review of industrial techniques for thermalconductivity measurements of thermal insulation materials. *International Journal of Thermophysics* 36: 1530–1544.

16 Hay, B., Filtz, J.R., Hameury, J., and Rongione, L. (2005). Uncertainty of thermal diffusivity measurements by laser flash method. *International Journal of Thermophysics* 26: 1883–1898.

17 Parker, W., Jenkins, R., Butler, C., and Abbott, G. (1961). Flash method of determining thermal diffusivity, heat capacity, and thermal conductivity. *Journal of Applied Physics* 32: 1679–1684.

18 ISO 8894-2:2007 Refractory materials — Determination of thermal conductivity — Part 2: Hot-wire method (parallel).

19 ASTM D150-22 Standard test methods for AC loss characteristics and permittivity (dielectric constant) of solid electrical insulation.

20 ASTM E831-19 Standard Test Method for Linear Thermal Expansion of Solid Materials by Thermomechanical Analysis.

21 ASTM E228-17 Standard Test Method for Linear Thermal Expansion of Solid Materials With a Push-Rod Dilatometer.

22 ASTM E1545-22 Standard Test Method for Assignment of the Glass Transition Temperature by Thermomechanical Analysis.

23 ASTM E1824-19 Standard Test Method for Assignment of a Glass Transition Temperature Using Thermomechanical Analysis: Tension Method.

24 ASTM E2425-21 Standard Test Method for Loss Modulus Conformance of Dynamic Mechanical Analyzers.
25 E2254-18 Standard Test Method for Storage Modulus Calibration of Dynamic Mechanical Analyzers.
26 E1640-18 Standard Test Method for Assignment of the Glass Transition Temperature By Dynamic Mechanical Analysis.

4

Construction of Thermal Conductivity Network and Performance Optimization of Polymer Substrate

Hua Wang[1], *Xingyou Tian*[1], *Haiping Hong*[2], *Hao Li*[1], *Yanyan Liu*[1], *Xiaoxiao Li*[1], *Yusheng Da*[1], *Qiang Liu*[1], *Bin Yao*[2], *Ding Lou*[2], *Mingyang Mao*[2], *and Zhong Hu*[3]

[1] *Chinese Academy of Sciences, Institute of Solid State Physics, Hefei Institutes of Physical Science, 350 Shushanhu Road, Hefei 230031, Anhui, P. R. China*
[2] *South Dakota School of Mines and Technology, Department of Electrical Engineering, Rapid City, South Dakota 57701, USA*
[3] *South Dakota State University, Department of Mechanical Engineering, Brooking, South Dakota 57007, USA*

4.1 Synthesis and Surface Modification of High Thermal Conductive Filler and the Synthesis of Substrates

4.1.1 Synthesis of Hexagonal Boron Nitride Nanosheets by Halide-Assisted Hydrothermal Method at Low Temperature

Inspired by the synthesis of hBN quantum dots and halide-assisted effect, the HAHTA synthesis method for hBNNS synthesis is proposed here (Figure 4.1). In addition, in the HAHTA process, the halide etching also contributes to the in situ surface modification of boron nitride nanosheets, because the etched hBNNS surface has a low energy barrier for further functionalization. For the mechanism of halide auxiliary effect, we have carried out the following mechanism: the edge of hBNQDs formed in the initial stage of HAHTA is activated due to halide etching, which can reduce the energy barrier for further growth of hBNNS, and at the same time, the continuous hydrothermal reaction provides a good growth environment, so that the product eventually evolves from zero-dimensional hBNQDs to two-dimensional hBNNS.

4.1.2 Modification and Compounding of Inorganic Thermal Conductive Silicon Carbide Filler

Multilayered graphene (MG) and SiC nanowire composite fillers with three-dimensional nanostructures were prepared in a high-frequency heating furnace by a simple carbothermal reduction method [2]. Not only SiC nanowires but also SiC particles are grown on the surface of MG. Interestingly, SiC particles grow along the edge of MG, which provides favorable conditions for MG to have good

Thermal Management Materials for Electronic Packaging: Preparation, Characterization, and Devices, First Edition. Edited by Xingyou Tian.

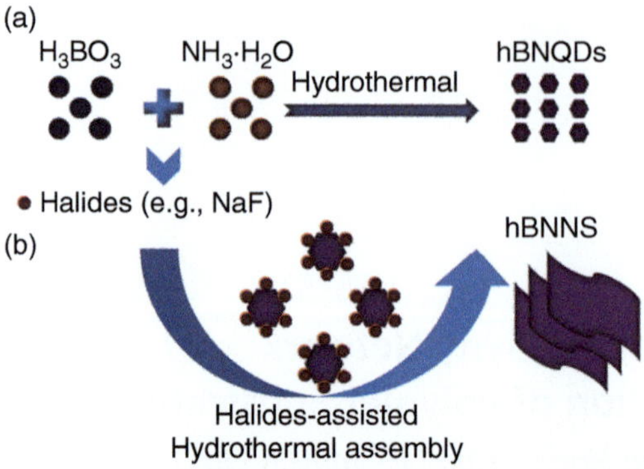

Figure 4.1 Schematic diagram of halide-assisted hydrothermal synthesis of hBNNS (HAHTA). Source: Reproduced with permission from Ref. [1]; John Wiley & Sons.

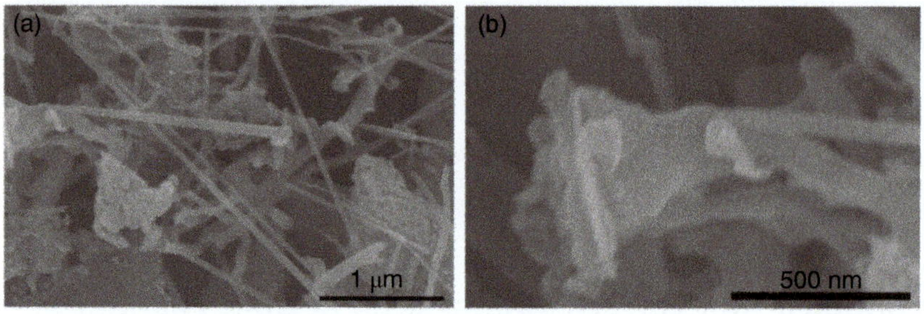

Figure 4.2 Characterization of MG SiC_{nw}: (a) SEM image, (b) SEM magnification. Source: Reproduced with permission from Ref. [3]; © 2017 IOP Publishing Ltd.

electrical insulation. The existence of SiC nanowires also hinders the electron transmission between MG layers and constructs an effective phonon channel between MG layers. The typical SEM image of MG SiC_{nw} is shown in Figure 4.2. The results show that the surface of MG contains a large number of SiC small particles and many microscale SiC nanowires orienting randomly. At the same time, it can be observed that the SiC nanowires are grown on the SiC small particles on the surface of MG, as shown in Figure 4.2b, which proves that the SiC nanowires and MG can form a specific three-dimensional structure through the tight connection and assembly of SiC small particles.

We propose Fe-assisted gas–liquid–solid mechanism to explain the growth of SiC nanowires. The growth process of SiC nanowires on MG is shown in Figure 4.3. The formation process is divided into three parts including nucleation, SiC nanoparticle growth, and SiC nanowire growth.

4.1.3 Preparation and Characterization of Intrinsic Polymer with High Thermal Conductivity

TMBPDGE-DDM is used as a polymer matrix because its thermal conductivity is 37.3% higher than that of amorphous epoxy resin. Second, we choose hexagonal boron nitride (h-BN) as the thermal conductive filler, which interacts with the curing agent (DDM). We prove the electrostatic interaction between DDM and h-BN FTIR by Raman and other techniques. The results are shown in Figure 4.4. Compared

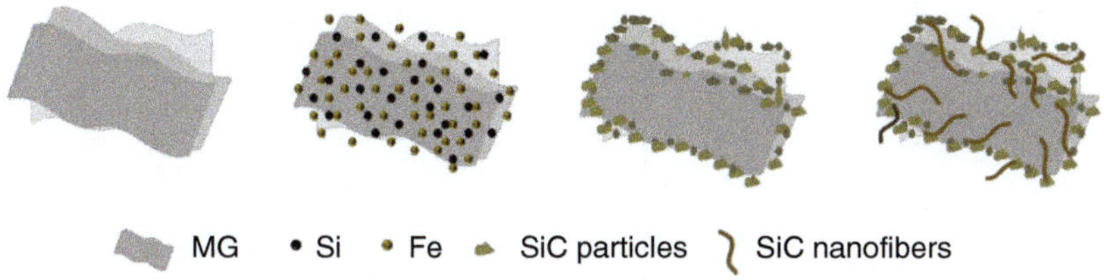

Figure 4.3 Schematic diagram of the growth process of MG SiC_{nw}. Source: Reproduced with permission from Ref. [3]; © 2017 IOP Publishing Ltd.

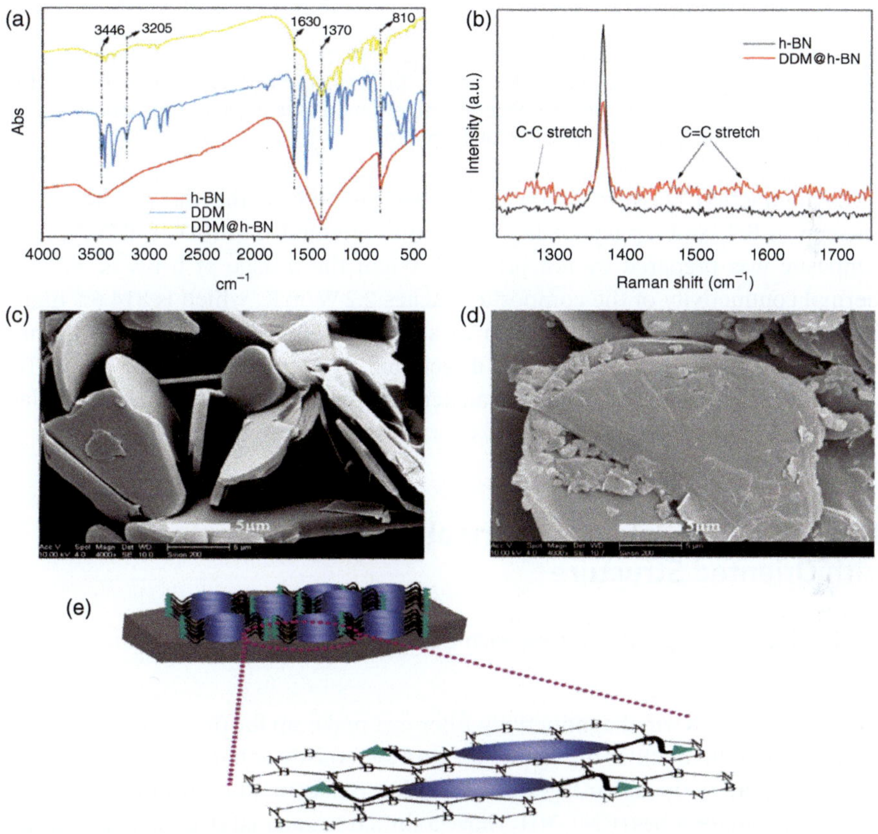

Figure 4.4 (a) FTIR spectra of h-BN, DDM, and DDM @ h-BN; (b) Raman spectra of h-BN and DDM @ h-BN; (c) SEM of h-BN; (d) DDM @ h-BN; and (e) Interaction between DDM and h-BN. Source: Reproduced with permission from Ref. [4]; © 2019 IOP Publishing Ltd.

with pure boron nitride in FTIR spectrum, DDM@h-BN generates benzene ring ($1630\,cm^{-1}$) peak and the primary amine peak (3446 and $3205\,cm^{-1}$). The characteristic peak of Raman spectrum h-BN appears at $1364\,cm^{-1}$, DDM@h-BN appears benzene ring peak, which is about $1470\,cm^{-1}$, $1570\,cm^{-1}$, and $1280\,cm^{-1}$. In addition, by comparing the pure h-BN and DDM@h-BN, we can see that the surface of DDM@h-BN is rougher than that of pure h-BN, and DDM attaches to BN sheet.

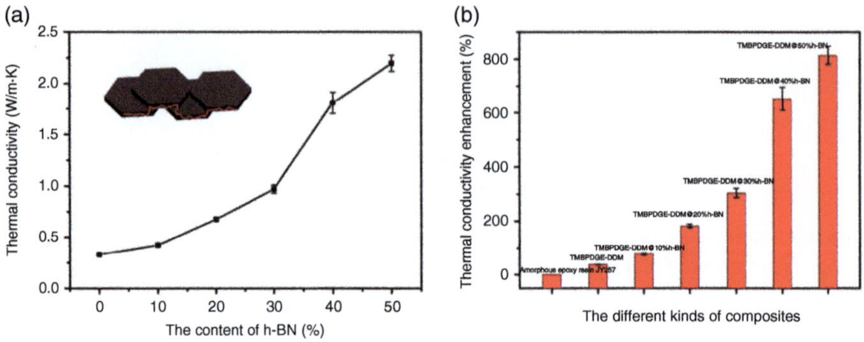

Figure 4.5 Thermal conductivity of TMBPDGE. (a) TMBPDGE-DDM@h-BN relationship between composite materials and h-BN content. (b) Compared with JY257 epoxy resin, the thermal conductivity enhancement effect of different materials. Source: Reproduced with permission from Ref. [4]; © 2019 IOP Publishing Ltd.

The interaction between DDM and h-BN can reduce the thermal resistance between h-BN and epoxy resin. Horizontal-oriented TMBPDGE-DDM@h-BN composite was prepared by hot pressing. When the dosage of h-BN is 50%, the thermal conductivity of the composite reaches 2.2 W/m K, which is 814.6% higher than that of amorphous epoxy resin as shown in Figure 4.5. The high thermal conductivity can be attributed to the directional arrangement of TMBPDGE-DDM and h-BN, which increases the heat transfer path and reduces the surface-to-surface thermal resistance between h-BN and epoxy resin.

4.2 Study on Polymer Thermal Conductive Composites with Oriented Structure

4.2.1 Epoxy Composites Filled with Boron Nitride and Amino Carbon Nanotubes

The binary high thermal conductivity filler can make up for the shortage of single thermal conductivity filler filling and create a synergistic enhancement effect while giving full play to the advantages of each filler. Hexagonal boron nitride and aminated carbon nanotubes (CNT-NH_2) have a similar hexagonal structure, and the π–π interaction force between them provides the possibility for the hybridization reaction between them [5–7]. The amino group on the CNT-NH_2 and the boron atom in the h-BN can form Lewis acid–base pairs, and Lewis acid–base forces likewise facilitate the preparation of the hybridized packing h-BN@CNT-NH_2.

As shown in Figure 4.6a, h-BN@CNT-NH_2 was prepared by the co-hybrid treatment of h-BN and CNT-NH_2, and then h-BN@CNT-NH_2 was added to the polymer matrix to prepare the composite CER/h-BN@CNT-NH_2. The h-BN and the hybrid h-BN@CNT-NH_2 are horizontally oriented in the polymer matrix by means of the gravitational field and the interaction between the high aspect ratio fillers. Through the characterization and analysis of the structure and properties

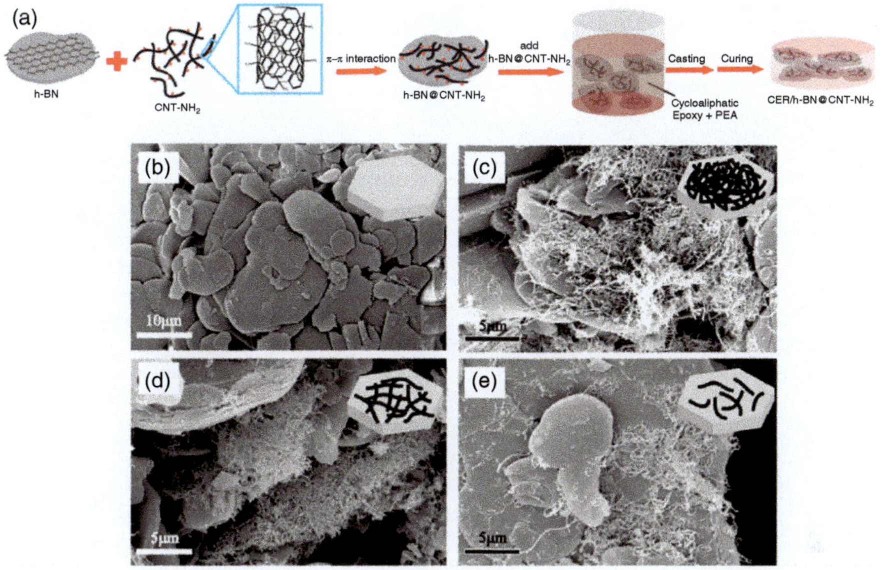

Figure 4.6 (a) Preparation of CER/h-BN@CNT-NH_2 composites. SEM: (b) h-BN; the mass ratio of CNT-NH_2 and h-BN is (c) 2 : 15; (d) 2 : 25; (e) 2 : 35. Source: Reproduced with permission from Ref. [8]; © 2018 Elsevier.

of CER/CNT-NH_2/h-BN, CER/h-BN@CNT-NH_2, and CER/h-BN composites, we believe that the composites have excellent thermal conductivity, electrical insulation, and thermal stability, which has important potential application value in the field of electronic packaging.

As shown in Figure 4.6b, the surface of h-BN is smooth and flat, while the surface of the hybrid filler h-BN@CNT-NH_2 is covered with CNT-NH_2 in Figure 4.6c–d. The mass ratio of CNT-NH_2 to h-BN determines the different surface morphologies of the hybrid filler h-BN@CNT-NH_5. Too much CNT-NH_2 will pile on top of each other on the surface of the h-BN.

The h-BN as well as h-BN@CNT-NH_2 will be oriented horizontally in the polymer matrix under the effect of gravitational field and the interaction between the high aspect ratio fillers. As shown in Figure 4.7a–i, the smooth part of the SEM photograph of the cross-section of the composite represents the polymer matrix, whose volume decreases with increasing filler content. The orientation of the filler is essentially stacked along the horizontal orientation. Interestingly, it is absent in the cross-sectional sections of the composite CER/h-BN@CNT-NH_2, which can be attributed to the fact that the —NH_2 adsorbed on the surface of the filler h-BN@CNT-NH_2 in the composite CER/h-BN@CNT-NH_2 participates in the curing of the alicyclic epoxy resin, thus promoting the curing of the filler in the composite CER/h-BN@CNT-NH_2.

Figure 4.7j,k shows the variation trend of horizontal thermal conductivity of composite materials with packing content. In the composite CER/h-BN@CNT-NH_2, CNT-NH_2 is confined to the surface of h-BN, and the amount of CNT-NH_2 on the surface of h-BN@CNT-NH_2 decreases with the increase in h-BN@CNT-NH_2

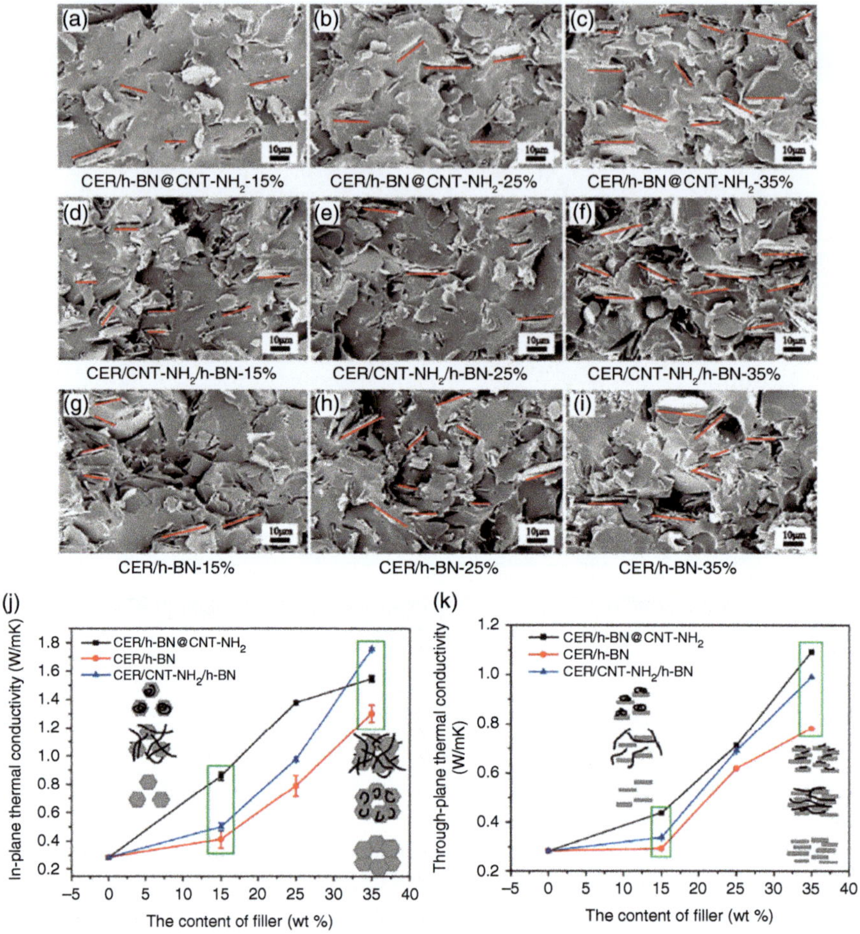

Figure 4.7 Cross-sectional SEM of (a–c) CER/h-BN@CNT-NH_2; (d–f) CER/CNT-NH_2/h-BN; (g–i) CER/h-BN; Thermal conductivity of (j) CER/CNT-NH_2/h-BN, (k) CER/h-BN@CNT-NH_2, and CER/h-BN. Source: Reproduced with permission from Ref. [8]; © 2018 Elsevier.

content of hybrid filler. As shown in Figure 4.7k, with the increase in the packing, the thermal conductivity of the composite in the vertical direction also increases due to the close packing between the packing. We used infrared thermal imager to monitor the actual heat transfer effect of the composite in real time. The adsorption of CNT-NH_2 on the h-BN surface of the hybrid filler h-BN@CNT-NH_2 in the CER/h-BN@CNT-NH_2 composite enhances the heat transfer of the composite in the vertical direction. Therefore, the surface speed of composite CER/h-BN@CNT-NH_2 reaches 50 °C that is faster than that of composite CER/CNT-NH_2/h-BN.

In the composite CER/h-BN@CNT-NH_2, the packing h-BN@CNT-NH_2 with rough surface has obvious enhancement effect on the tensile modulus of the composite. The network structure formed by randomly distributed fillers CNT-NH_2 in the CER/CNT-NH_2/h-BN composite has an obvious enhancement effect on the Young's modulus of the composite, as shown in Figure 4.8.

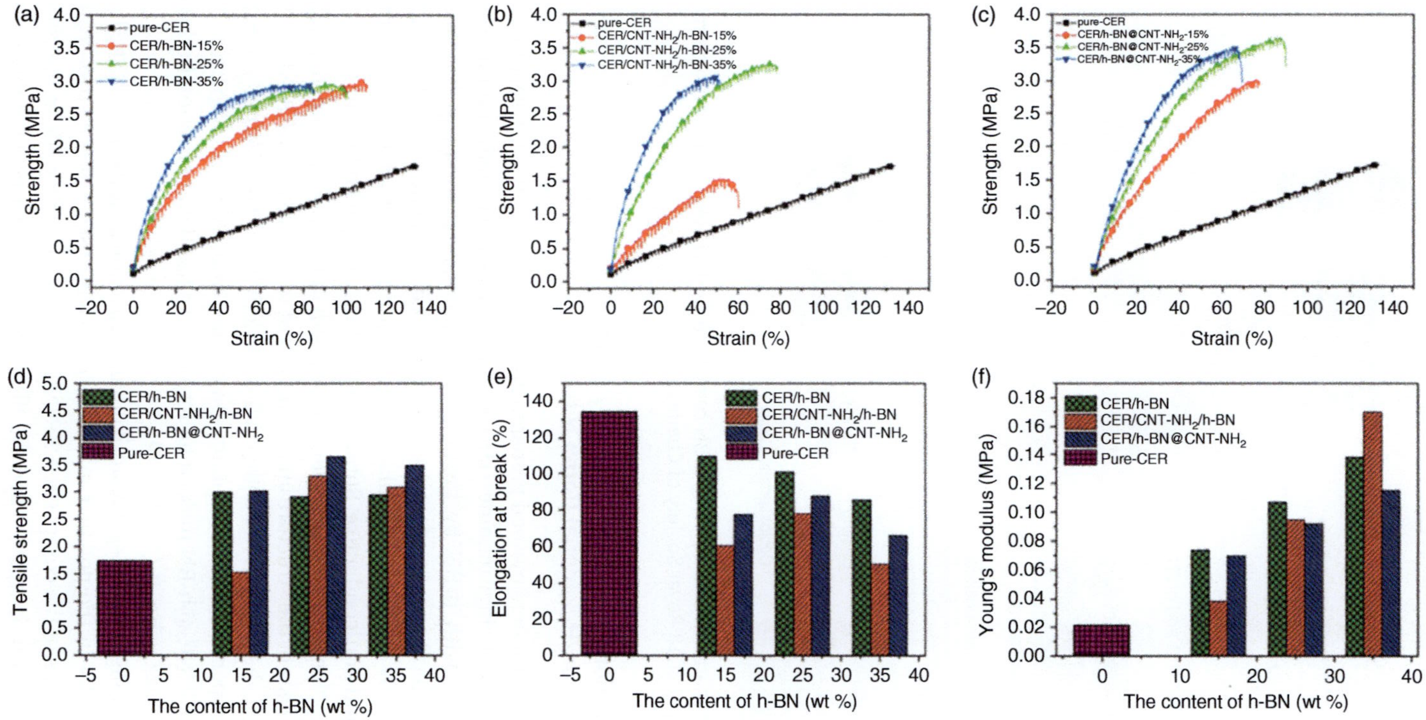

Figure 4.8 CER/CNT-NH_2/h-BN, CER/h-BN@CNT-NH_2, and CER/h-BN filled composites of (a), (b), and (c) stress–strain curves; (d) tensile modulus; (e) elongation at break; (f) Young's modulus. Source: Reproduced with permission from Ref. [8]; © 2018 Elsevier.

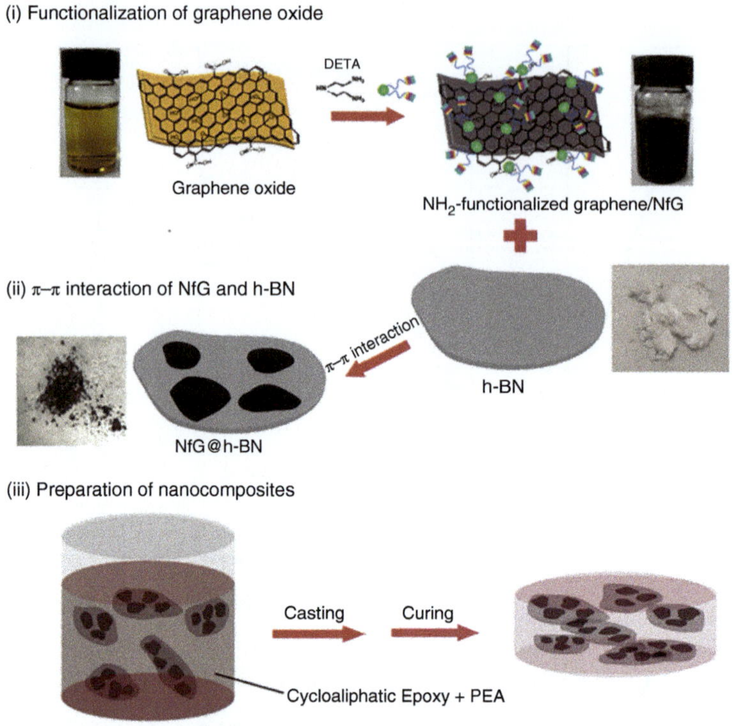

Figure 4.9 Preparation of CER/NfG@h-BN composite. Source: Reproduced with permission from Ref. [10]; © 2018 Royal Society of Chemistry.

4.2.2 Reduction of Graphene Oxide by Amino Functionalization/Hexagonal Boron Nitride

The modified Hummers method prepared graphite oxide (GO) was treated with the alkaline reagent divinyltriamine (DETA) for simultaneous reduction/amination and hybridized with h-BN by means of π–π interaction forces to produce the amino groups on the amino-functionalized reduced graphene oxide sheets@h-BN (NfG@h-BN). By structural modulation, the hybrid filler NfG @h-BN is thermally conductive and non-conductive. Subsequently, the hybridized filler NfG@h-BN was added to the polymer matrix and oriented as shown in Figure 4.9. After characterization and analysis of the structure and properties of the composite CER/NfG@h-BN, we concluded that the excellent thermal conductivity (~3.409 W/m K), electrical insulation properties, thermal stability properties, and adhesive properties of the composite CER/NfG@h-BN make it an important potential application in the field of electronic packaging.

As shown in Figure 4.10a–g, the horizontal orientation arrangement of the hybrid filler NfG@h-BN in the polymer matrix can be clearly seen in the cross-sectional SEM photographs of the composite CER/NfG@h-BN, and the voids in the composite structure start to increase when the filler content increases, probably because, during the preparation of the composite cross-section, we used the method of liquid

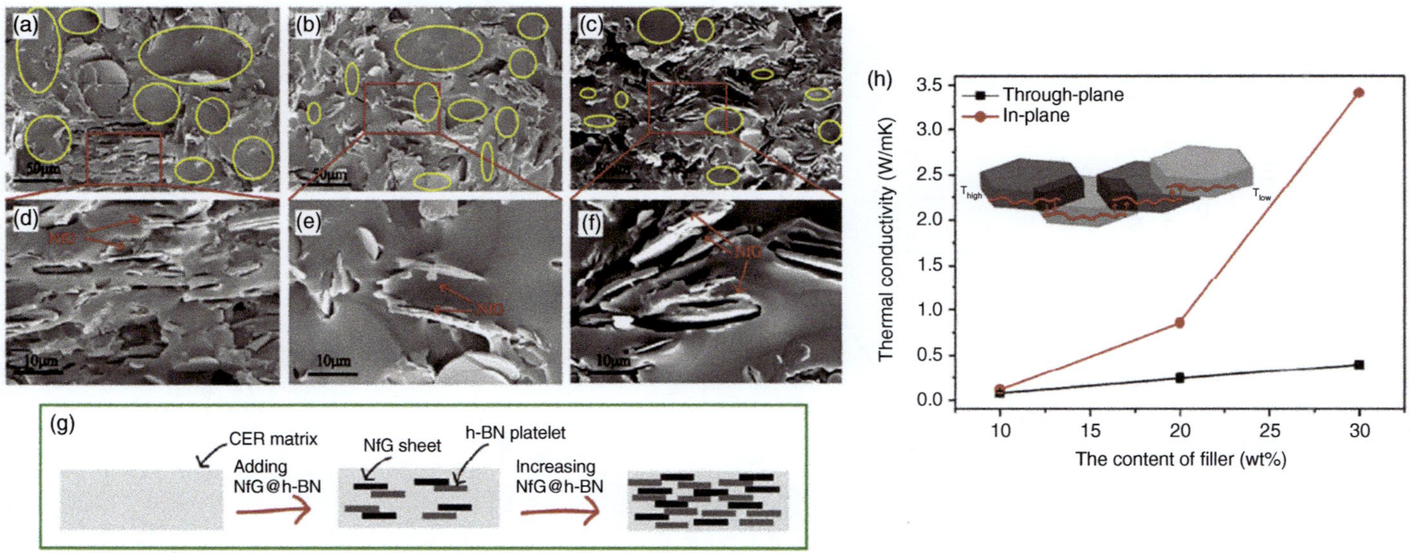

Figure 4.10 Cross-section of (a) and (d) CER/NfG@h-BN-10 wt%, (b) and (e) CER/NfG@h-BN-20 wt%, (c) and (f) CER/NfG@h-BN-30 wt%; (g) schematic diagram of the stacking form of filler NfG@h-BN in the matrix; (h) variation of thermal conductivity of composites with filler NfG@h-BN composition. Source: Reproduced with permission from Ref. [10]; © 2018 Royal Society of Chemistry.

nitrogen embrittlement fracture, the low-temperature treatment of liquid nitrogen makes the polymer matrix crumple, and the amino groups on the NfG will participate in the curing process of the alicyclic epoxy resin and are also peeled off from the h-BN when the polymer matrix crumples, as shown in Figure 4.10f. The increased content of filler NfG@h-BN added in the composite CER/NfG@h-BN not only provides more thermal conductivity pathways in the horizontal direction, but also the filler is stacked more tightly in the vertical direction.

The horizontal orientation of the thermally conductive filler NfG@h-BN in the composite CER/NfG@h-BN forms a horizontal thermal conductivity path inside the composite, so the thermal conductivity of the composite is higher in the horizontal

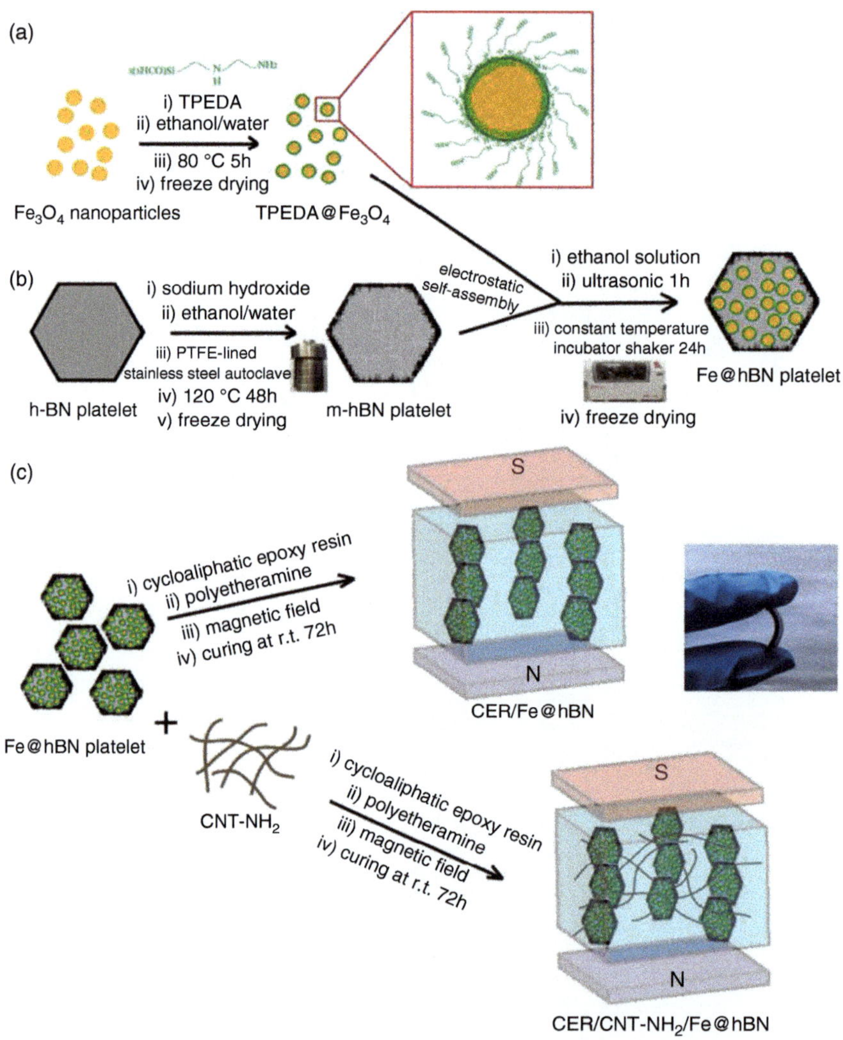

Figure 4.11 CER/$CNT-NH_2$/Fe@hBN preparation process. Source: Reproduced with permission from Ref. [9]; © 2018 American Chemical Society.

direction than in the vertical direction, and the horizontal thermal conductivity of the composite can reach 3.409 W/m K, when the filler content reaches 30 wt% as shown in Figure 4.10h. The actual heat transfer and heat dissipation effect of the composite CER/NfG@h-BN was monitored in real time by infrared thermography. Both in the warming and cooling processes, the increase in the content of the thermally conductive filler NfG@h-BN in the composite CER/NfG@h-BN accelerated the conduction of the composite CER/NfG@h-BN for heat, and the warming was faster and the cooling was more rapid.

4.2.3 The Interconnection Thermal Conductive Network of Three-Dimensional Staggered Boron Nitride Sheet/Amino-Functionalized Carbon Nanotubes

For the vertical orientation of high aspect ratio thermally conductive fillers, there are two common methods: the first one is to make thicker blocks after the horizontal orientation of high aspect ratio fillers, and then cut the blocks into shape by standing them up on their sides [11]. The second one is to use an applied magnetic field to vertically orient the materials with magnetic response properties [7, 8]. In this experiment, as shown in Figure 4.11, by filling the polymer matrix with vertically oriented hybrid filler Fe@hBN and randomly distributed CNT-NH_2, we successfully prepared three-dimensional interlaced boron nitride sheets with oriented and

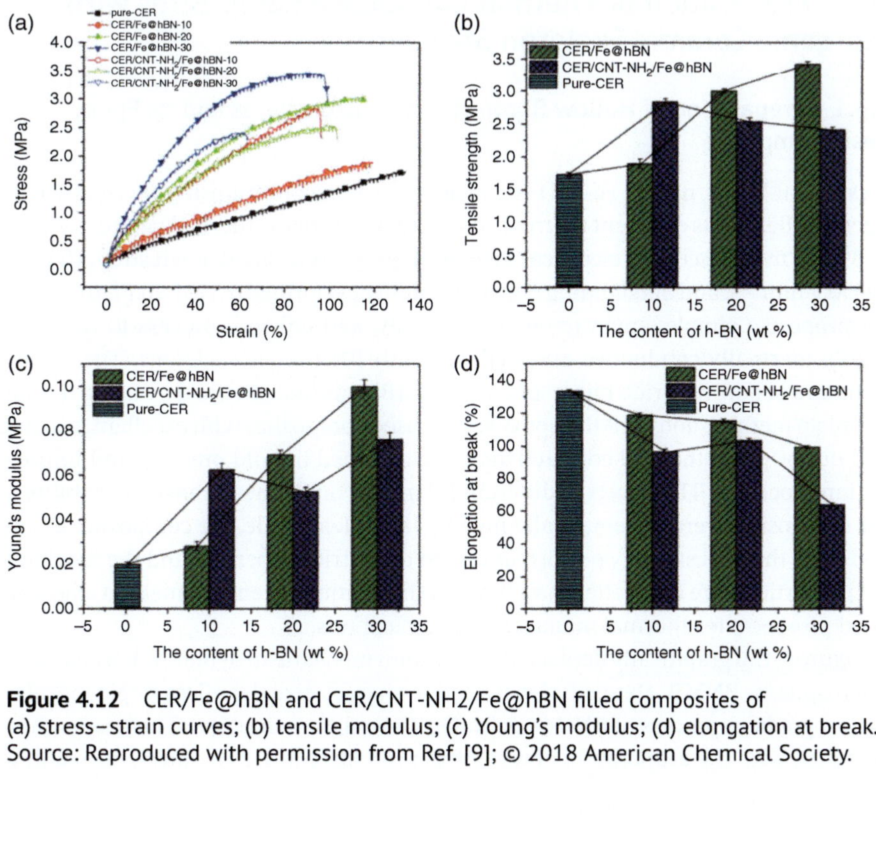

Figure 4.12 CER/Fe@hBN and CER/CNT-NH2/Fe@hBN filled composites of (a) stress–strain curves; (b) tensile modulus; (c) Young's modulus; (d) elongation at break. Source: Reproduced with permission from Ref. [9]; © 2018 American Chemical Society.

interconnected carbon nanotube network structures [11, 12]. By regulating the amount of fillers added, we achieved simultaneous orientation of the composite CER/CNT-NH_2/Fe@hBN in the vertical and horizontal directions, and the vertical and horizontal thermal conductivities reached 0.76 W/m K and 0.85 W/m K, respectively. At the same time, the composite CER/CNT-NH_2/Fe@hBN has excellent electrical insulation properties ($<1.2\times10^{-12}$ S/cm) and mechanical properties, and the results of simulating the thermal effects of the composite CER/Fe@hBN and CER/CNT-NH_2/Fe@hBN on electronic components in actual operation show that the composite has excellent heat dissipation effect and has potential value for application in the field of electronic packaging.

The stress–strain curves of the composites reveal the effects of the hybrid filler Fe@hBN and the addition of CNT-NH_2 on the mechanical properties of the composites. The tensile modulus of the composite CER/CNT-NH_2/Fe@hBN decreased instead, and there was a decrease in Young's modulus, as shown in Figure 4.12. The phenomenon of increased elongation at break can be mainly attributed to the intercalation of CNT-NH_2 in the structure of composite CER/CNT-NH_2/Fe@hBN, which to some extent affects the encapsulation and support of the polymer matrix on the hybrid filler Fe@hBN, making the structure of composite CER/CNT-NH_2/Fe@hBN loose.

4.3 Preparation of Thermal Conductive Composites with Inorganic Ceramic Skeleton Structure

4.3.1 Preparation of Hollow Boron Nitride Microspheres and Its Epoxy Resin Composite

Hexagonal boron nitride (h-BN) is considered the most promising thermally conductive filler for its excellent thermal conductivity (in-plane thermal conductivity up to 390 W/m K), excellent electrical insulation properties (breakdown strength up to 794 MV mm^{-1}), and outstanding thermal stability performance [13]. In Figure 4.13a, we propose a simple, environmentally friendly, and efficient process to construct highly thermally conductive and high-strength BN porous skeletons. The hollow structure of boron nitride microspheres (BNMB) was first obtained by using the salt template method, and then the epoxy resin-based composites with excellent in-plane and out-of-plane thermal conductivity were obtained by cold pressing and impregnation processes. The effects of different filler sizes on the heat transfer capability of the composites were systematically investigated. Meanwhile, the composites exhibited good thermal stability performance and dielectric properties, and the mechanical properties were also better than those of the common blending method, showing good prospects for thermal management applications.

Figure 4.13b graphically depicts the formation mechanism of hollow boron nitride microspheres BNMB. Here, SEM was used to characterize the microscopic morphological gradient process of BN-15 μm/NaCl/PVA mixture in stages. The results show that first, with the evaporation of water from the BN/NaCl/PVA slurry, the solution

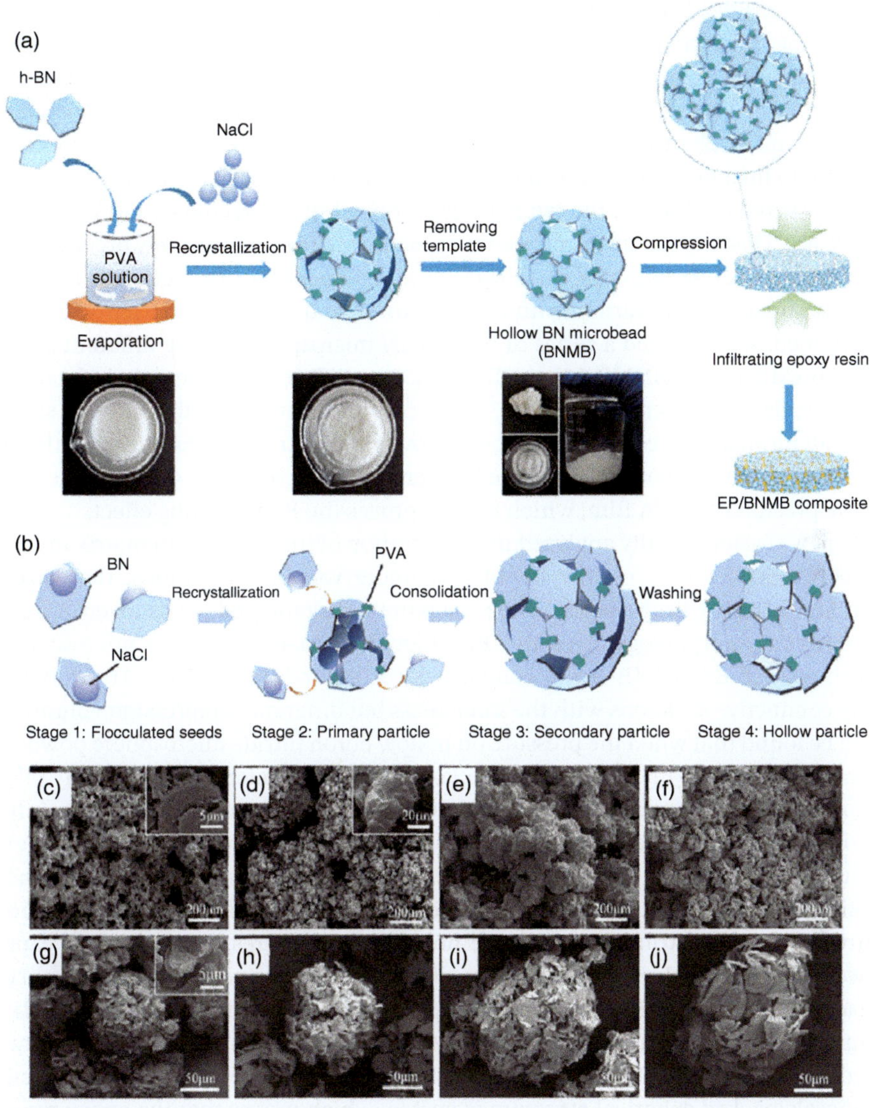

Figure 4.13 (a) Preparation of hollow h-BN and its epoxy resin composite; (b) Schematic diagram of BNMB formation of hollow h-BN microspheres; (c–f) the four stages SEM of BN-15 μm/NaCl/PVA mixture; (g–j) SEM of hollow h-BN microspheres formed by different sizes of h-BN. Source: Reproduced with permission from Ref. [14]; © 2019 Elsevier.

is gradually supersaturated with NaCl and starts to crystallize and precipitate on the surface of BN particles, at which time the homogeneous mixture starts to flocculate at this time, as shown in Figure 4.13c, the BN sheet layer acts as a crystalline nucleus, and PVA acts as an adhesive to closely connect the BN sheet and NaCl. With further evaporation of the solvent, the NaCl grains, driven by the reduction of free energy, start to agglomerate and grow, driving the BN sheet layer to pile up.

As shown in Figure 4.13d, at this time the flocculated particles start to gradually agglomerate into irregular spheres, which are called primary particles. When the water evaporates out, the third stage in Figure 4.13e, more and more BN/NaCl seeds are assembled layer by layer on the surface of primary particles to form regular BN/NaCl microspheres, as shown in Figure 4.13f, and the flocs in the mixture can be found to further agglomerate into uniform spheres. When the flocs were poured into the stirred cold water, the temperature dropped, PVA precipitated to bond BN tightly, NaCl lost the supersaturated environment, and the self-assembly process was terminated. After several water washes, NaCl is re-dissolved in water and the solid BN/NaCl microspheres turn into hollow structures, and hollow BNMB particles can be obtained. We prepared a series of hollow BN microspheres using different sizes of BN: BNMB-10 μm, BNMB-15 μm, BNMB-25 μm, BNMB-35 μm, and their detailed morphologies are shown in Figure 4.13g–j, and it can be found that the different sizes of BN all formed approximately regular hollow boron nitride microspheres with an average diameter of about 100 μm. The local enlargement of Figure 4.13g can clearly see the precipitated PVA film, which further proves the PVA bonding effect.

Then, we systematically analyzed the penetration of the epoxy resin matrix inside the porous skeleton. It can be found that under vacuum conditions, the epoxy resin matrix fully fills the pores inside the porous skeleton and forms a very dense bicontinuous morphology with the BN. Figure 4.14a–e shows the microscopic morphology of EP/BNMB-10 μm composites corresponding to BNMB-10 μm thermally conductive skeletons with the same mass but different compression volumes. It can be found that when the pressure on hollow boron nitride microsphere powder is small, such as EP/BNMB-10 μm-6V, a large number of microspheres simply pile up together, and the microspheres are loose and porous. The epoxy resin is mainly distributed in the gaps between the microspheres, and as the pressure on the powder gradually increases, the gaps between the microspheres gradually shrink, and the density inside the skeleton slowly becomes larger. At this time, the epoxy resin matrix is mainly distributed inside the hollow microspheres, the interfacial gaps between the microspheres are also fully filled, and the composite material is very dense as a whole. Figure 4.14f–h shows the microscopic morphology of EP/BNMB composite with different BN sizes under 1 MPa pressure for the same mass of hollow powder. Both EP/BNMB-10 μm and EP/BNMB-15 μm are very dense inside, and the closely packed spherical structure is visible, which proves that the epoxy resin is fully infiltrated into the porous skeleton. In addition, the EP/BNMB composites showed a clear horizontal orientation trend with increasing boron nitride size. The most obvious horizontal orientation trend of EP/BNMB-35 μm composites is mainly due to the largest deformation of the hollow BNMB-35 μm skeleton under pressure. The excessive deformation leads to the overdense buildup of boron nitride sheets on the surface layer of microspheres, which blocks the deep penetration of liquid epoxy resin and increases the internal defects of the composite. This matrix-filler bicontinuous structure can maintain the good mechanical strength of the composite on the one hand and provide a pathway for the heat transfer on the other hand, which can significantly improve the phonon transfer efficiency compared to the random distribution.

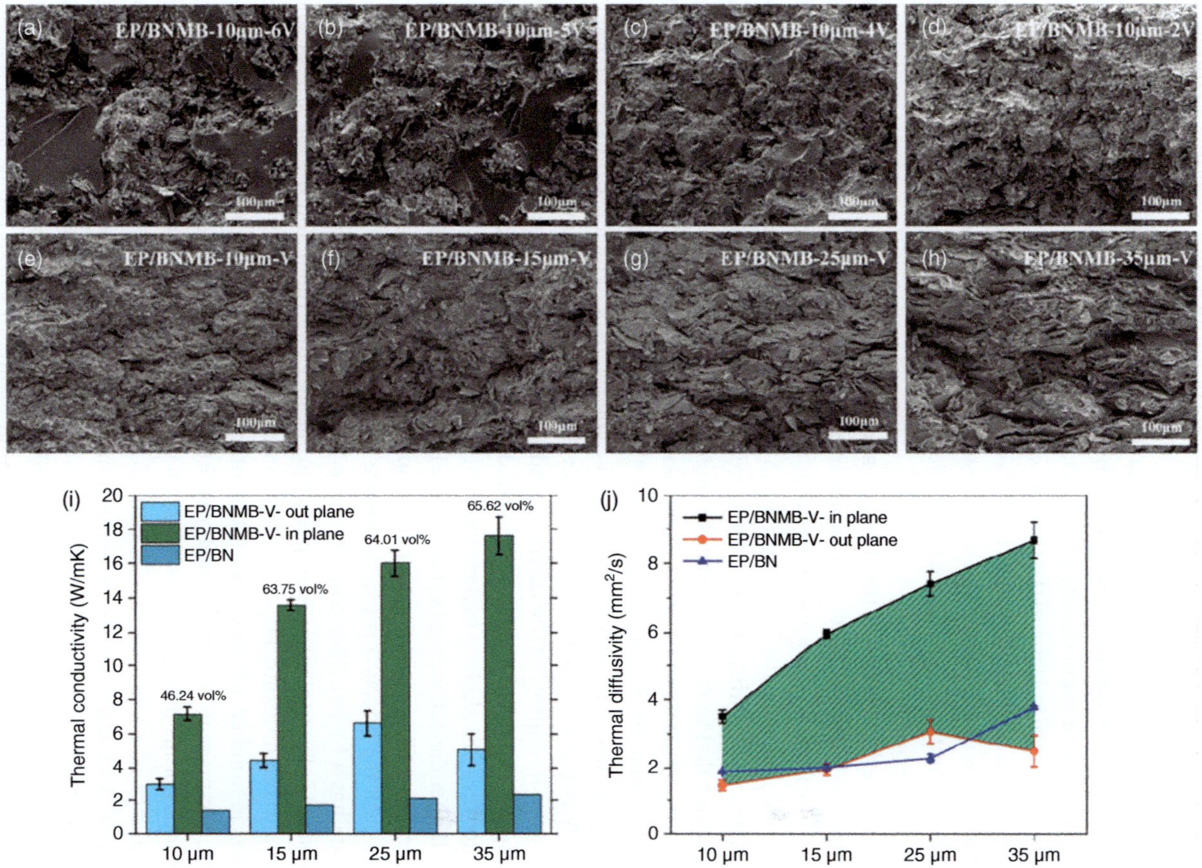

Figure 4.14 Cross-sectional SEM of EP/BNMB composites. (a–e) Are the microscopic morphologies of EP/BNMB-10 μm composites corresponding to different compression volumes. (f–h) are the EP/BNMB composites with the same compression volume and different BN sizes; (i) Comparison of thermal conductivity and (j) comparison of thermal diffusion coefficient. Source: Reproduced with permission from Ref. [14]; © 2019 Elsevier.

Figure 4.14i shows the thermal conductivity plots of EP/BNMB composites corresponding to different BN sizes. It can be found that both in-plane and out-of-plane thermal conductivity of the composites increase significantly with the increase in BNMB skeleton densities, and the difference between the two increases gradually. A similar variation pattern is shown in the composites corresponding to different sizes of BN.

There are two reasons for this trend: first, with the dramatic reduction of BN skeleton volume, the thermal conductivity network is more perfect, and the BN sheets are also stacked more closely, the interfacial thermal resistance between the fillers decreases significantly, and the thermal conductivity of the composites improves significantly; second, the smaller the BN skeleton volume is, the more significant the deformation of the hollow microspheres is, which leads to more and more obvious BN horizontal orientation, and the in-plane and out-of-plane thermal conductivities of the composites.

The difference between the in-plane and out-plane thermal conductivities of the composite is getting larger and larger. In addition, comparing the thermal conductivity of EP/BNMB composites corresponding to different boron nitride sizes at the same volume, it can be found that the in-plane thermal conductivity of the composites increases significantly with the increase in boron nitride size. This is mainly due to the fact that as the size of the BN increases, the number of filler interfaces in the thermal conductivity network decreases significantly, and in addition, the contact area between the BN lamellae is larger, the average phonon free range is longer, and the energy loss is less, so that the thermal conductivity in the horizontal direction is greatly improved. Such a large increase shows the great advantage of the isolation structure in improving the thermal conductivity of the composite.

Figure 4.15 shows that the values of dielectric constant and dielectric loss of the composites are slightly higher than those of the pure epoxy resin matrix, due to the higher dielectric constant of BN itself, which is about 6.85.[15] Normally, as the filler content increases, the dielectric constant of the composites also becomes larger. However, the opposite is true in Figure 4.15a, where the EP/BNMB-35 μm-V composite has a higher filler content compared to the other samples, but its dielectric constant is lower. This is mainly due to the fact that the boron nitride size in this sample is significantly larger than the other samples, and the filler–matrix interface is less in number, which weakens the interfacial polarization phenomenon. In addition, the presence of air also reduces the dielectric constant of the composite with obvious porous defects visible in the SEM image of this sample. At an external field frequency of 1 MHz, the EP/BNMB-35 μm-V composite has a low dielectric constant of 3.92 and a dielectric loss of only 0.0209, which is very close to the value of pure epoxy resin and shows excellent application prospects. As a comparison, we also characterized the dielectric properties of the EP/BN composites prepared by the homogeneous mixing method, as shown in Figure 4.15c,d. It can be found that the dielectric constants of all samples are lower than those of the pure epoxy matrix, which is due to the presence of a large number of porous defects inside the samples. When the filling volume fraction of BN reaches more than 50%, the common mixing method is no longer able to obtain uniformly dispersed and dense samples.

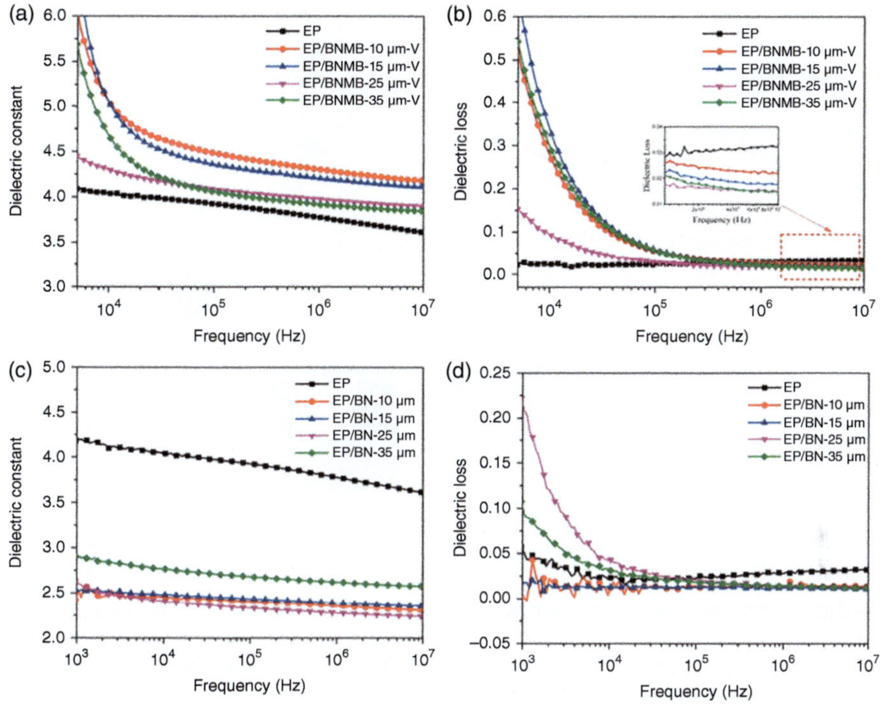

Figure 4.15 (a, b) Plots of dielectric constant and dielectric loss of EP/BNMB-V composites with frequency; (c, d) Plots of dielectric constant and dielectric loss of EP/BN composites with frequency. Source: Reproduced with permission from Ref. [14]; © 2019 Elsevier.

When the direct filling method is used to improve the thermal conductivity of composites, the enhancement effect is not obvious when the filler volume fraction is low, the thermally conductive particles are isolated by the polymer matrix wrapping, and the thermal conductivity network is discontinuous. Therefore, in order to further improve the thermal conductivity, it is often necessary to significantly increase the filling amount of filler to make the inorganic particles lap each other to form a continuous thermal conductivity network. The dramatic increase in and processing viscosity and deterioration of mechanical properties brought by high filler has been troubling researchers.

4.3.2 Three-Dimensional Skeleton and Its Epoxy Resin Composite

A large number of BNNS nanosheets were successfully prepared by heat treatment-assisted ultrasonic exfoliation, and the number of oxygen-containing groups on the surface of the nanosheets was increased to provide reaction sites for coupling agent grafting, as shown in Figure 4.16. Then a simple and efficient vacuum filtration-assisted self-assembly process was proposed to successfully prepare BNNS/SiCw thermally conductive skeletons with highly vertically oriented porous structures. Vacuum filtration enables rapid horizontal orientation of the filler, and

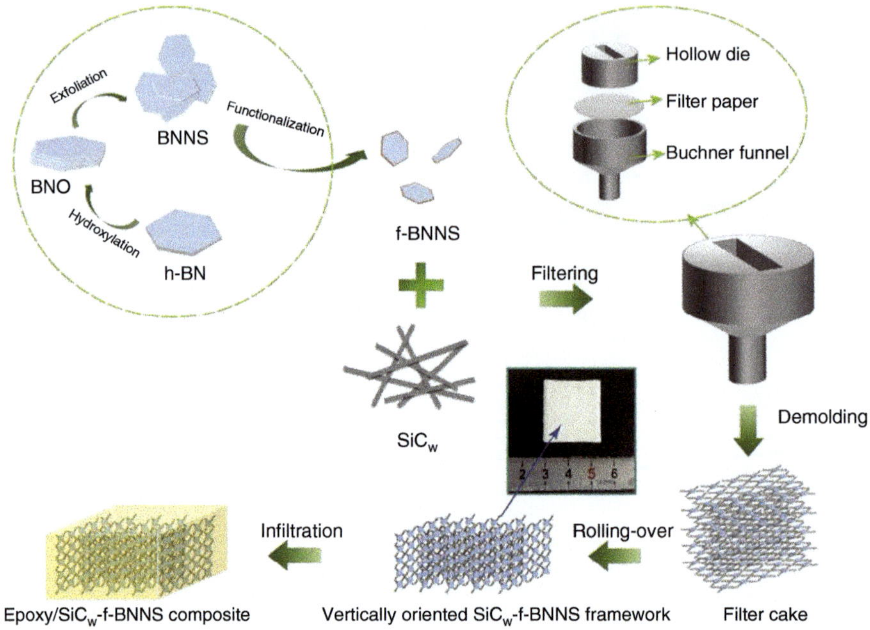

Figure 4.16 Preparation of f-BNNS, porous SiCw/f-BNNS skeleton, and its epoxy resin composite. Source: Reproduced with permission from Ref. [16]; © 2020 Elsevier.

the horizontally oriented skeleton can be easily converted to vertical orientation by a hollow mold. The method constructs a heat sink-like shaped thermally conductive network with epoxy composites showing excellent out-of-plane thermal conductivity. In addition, the thermal management capability of the composite is analyzed in detail by theoretical simulations and chip heat sink applications.

Figure 4.17a shows the microscopic morphology of pure BN, and it can be found that the BN filler shows a typical lamellar structure with smooth surface and variable thickness. Figure 4.17b shows the microscopic morphology of BN after heat treatment in a muffle furnace, and its appearance is significantly different compared with that in Figure 4.17a. The BN lamellae swell out from the axial direction and unfold into multilayer lamellae with a significant reduction in thickness. The book-like lamellar structure can also be found in Figure 4.17c, which further confirms that BN undergoes expansion under heat treatment and the layer spacing increases significantly, which greatly facilitates the subsequent liquid-phase ultrasonic peeling process.

Figure 4.17d shows the microscopic SEM image of BNNS after ultrasonic peeling, the laminar BNO is peeled into a large number of thin flakes at this time, and the size has also decreased. The size statistics of the fragments in the SEM images reveal that the average size of BNNS is about 1.35 μm. In addition, the red laser in the inset of Figure 4.17d undergoes a medium typical Tyndall effect when penetrating the BNNS/IPA solution, which further proves that BNNS has been dispersed to the nanoscale in the isopropanol solution and exhibits colloidal properties. Figure 4.17e

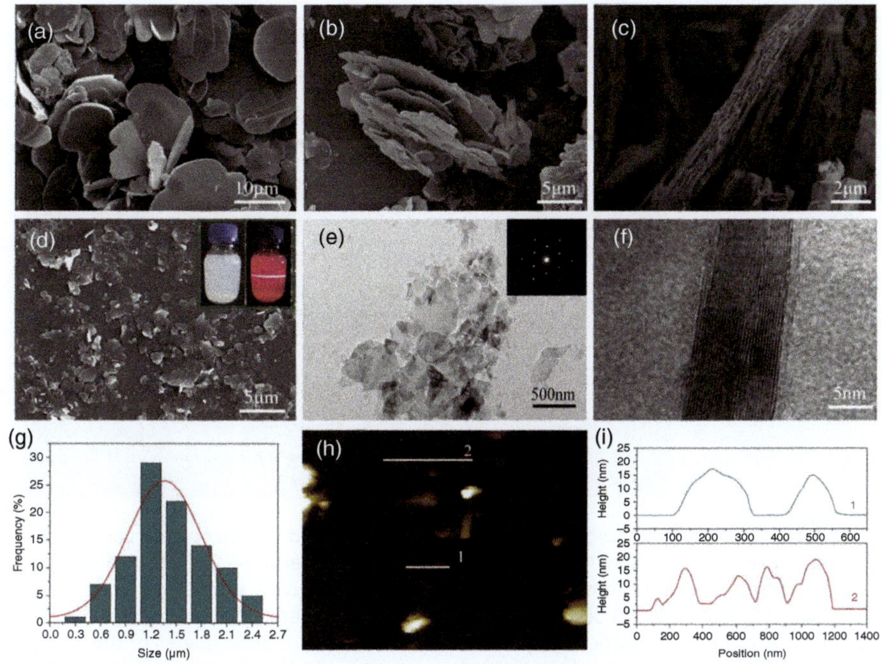

Figure 4.17 Microscopic morphological analysis of BN and BNNS. (a) SEM image of BN before heat treatment; (b, c) are the morphologies of BN after heat treatment; (d–f) are SEM, TEM, and HRTEM images of BNNS, respectively; (g) BNNS size distribution; (h) AFM image of BNNS; (i) the thickness distribution of BNNS nanosheets on two straight lines in (h). Source: Reproduced with permission from Ref. [16]; © 2020 Elsevier.

shows the morphology of BNNS under transmission electron microscopy, which shows more visually that BNNS has been exfoliated into very thin fragments, which further proves the effectiveness of the thermally assisted liquid-phase ultrasonic exfoliation method. The inset shows a typical hexagonal electron diffraction pattern of BN, which indicates that the exfoliation process has no effect on the crystal structure of BNNS. Figure 2.17f shows the high-resolution TEM image of BNNS, which reveals that the thickness of BNNS is about 10 nm, and this result is also consistent with the AFM characterization results, as shown in Figure 4.17h,i. In summary, we successfully obtained BNNS nanosheets by the heat treatment-assisted liquid-phase ultrasonic exfoliation method. This method has substantially improved the stripping efficiency compared with the normal liquid-phase ultrasound method.

Figure 4.18 shows the microscopic morphology of the porous skeleton and its epoxy resin composite in different directions. During the pumping process, the SiC_w fibers were stacked in layers in the horizontal direction due to the negative pressure driven, and the porous skeleton was turned over after its demolding, showing a very flat fiber entanglement structure when observed from the y-axis direction, and the previous layer stacking morphology turned into a regular vertically oriented morphology when observed from the x-axis direction. Figure 4.18d shows the skeleton obtained by pure f-BNNS pumping. Due to the two-dimensional

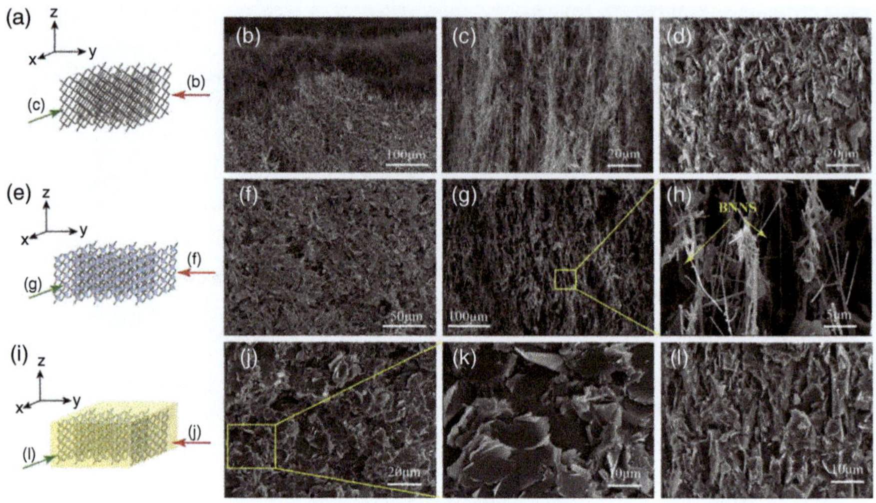

Figure 4.18 Microscopic morphology of porous skeleton and its epoxy resin composites in different orientations. (a) Schematic diagram of vertically oriented SiC_w skeleton; (b, c) SEM images of different orientations of the SiC_w skeleton, respectively, with the directions shown in the arrow in (a); (d) SEM images of vertically oriented f-BNNS skeleton. (e) Schematic diagram of vertically oriented SiC_w/f-BNNS; (f, g) SEM images of the $3SiC_w$/7f-BNNS skeleton in different orientations, respectively, with the directions shown by the arrows in (e); (h) partial enlargement of (g). (i) Schematic diagram of EP/SiC_w-f-BNNS composite; (j–l) SEM images of EP/SiC_w-f-BNNS composites in different orientations, respectively, with directions as indicated by the arrows in (i). Source: Reproduced with permission from Ref. [16]; © 2020 Elsevier.

characteristics of BNNS nanosheets, the lamination effect is produced under the negative pressure of pumping, just like the fibers. BNNS thermally conductive skeletons with vertically oriented structures can be obtained after flipping. Using this property of one-dimensional thermally conductive fibers and two-dimensional thermally conductive nanosheets, this experimental method can easily construct thermally conductive networks in the vertical direction.

Then, we composite the two fillers and use the synergistic effect of both to construct a vertically layered composite filler thermal conductivity network as shown in Figure 4.18e. Figure 4.18f shows the morphology of the SiC_w/f-BNNS composite filler thermal conductive skeleton observed from the y-axis direction after turning over, and it can be clearly seen that the porous skeleton presents a flat section. The flatter the composite filler in this direction, the higher the regularity in the x-axis direction. As shown in Figure 4.18g, SiC_w and f-BNNS form an ordered, vertically oriented porous structure. The local magnification (h) reveals that the SiC_w fibers are tightly entangled in the vertical direction, and the BNNS nanosheets are constrained in it. Due to the self-supporting effect of the rigid SiC_w fibers, the composite filler layers then maintain a certain pore structure between them, providing space for the infiltration of epoxy resin. This fully demonstrates that we successfully prepared porous SiC_w/f-BNNS thermally conductive skeletons with vertically oriented structures by a facile vacuum-assisted self-assembly method.

In addition, SiC_w and f-BNNS show good synergy in the composite thermal conductive network: on the one hand, the tight entanglement of fibers and nanosheets substantially improves the strength of the skeleton; on the other hand, by regulating the mixing ratio of SiC_w and f-BNNS, the porosity of the porous skeleton can be changed, which directly affects the content of the thermal conductive filler inside the final composite. Figure 4.18i shows the schematic structure of the composite, and its real SEM image is in good agreement with the schematic. As shown in Figure 4.18j, the scale-like horizontal distribution of f-BNNS nanosheets can be clearly observed in the cross-section of the composite in the y-axis direction, and the local magnification also reveals that the edges of f-BNNS have different degrees of curling, and this flexibility further proves that BN is successfully exfoliated. As shown in Figure 4.18l, the composite material shows a shell-like "brick mud" laminate structure in the x-axis direction, with the thermally conductive skeleton arranged vertically in a regular and orderly manner and the epoxy resin filled closely in between, forming a dense composite material.

Figure 4.19a shows the variation trend of the thermal conductivity of EP/SiCw-f-BNNS composites in the horizontal and vertical directions with the composite proportion. It can be found that the prefabricated vertical thermal conductive skeleton greatly improves the out-of-plane and in-plane thermal conductivity of the composite. The out-of-plane thermal conductivity of EP/SiCw composites with vertically oriented SiC_w skeleton reached 0.88 W/m K, and the in-plane thermal conductivity reached 0.51 W/m K, which were significantly higher than that of the common mixed samples with the same content (3.17 vol%). This is mainly due to the continuous heat conduction network formed by the pre-built SiC_w skeleton in the epoxy resin, which provides a shortcut for phonon transmission. At the same time, the layered stacking structure reduces the interface thermal resistance in the vertical direction of SiC_w fiber and greatly improves the out-of-plane heat transfer capability of the composite. With the increase in the proportion of f-BNNS, the out-of-plane thermal conductivity of the composite was significantly improved. This is due to the substantial increase in f-BNNS content and orientation degree of high thermal conductivity filler.

As shown in Figure 4.19c, the comparison chart of SiC_w/f-BNNS skeleton porosity and corresponding composite material density is shown. It can be seen from the illustration that the skeleton volume of SiC_w/f-BNNS decreases significantly with the increase in f-BNNS ratio under the same quality of composite fillers, so the skeleton porosity also decreases significantly. This directly leads to a significant increase in the density of the composite and the content of internal filler, which also means that the heat conduction network is more perfect. In addition, comparing the in-plane and out-of-plane thermal conductivity of the composite, it can also be found that the difference between the two increases gradually with the increase in filler content, which further proves that the prefabricated vertical skeleton can provide a more efficient channel in the vertical direction to avoid unnecessary scattering of phonons in the transmission process.

Simple polymer materials are often very sensitive to heat accumulation and the heating size changes greatly, while the thermal expansion coefficient of

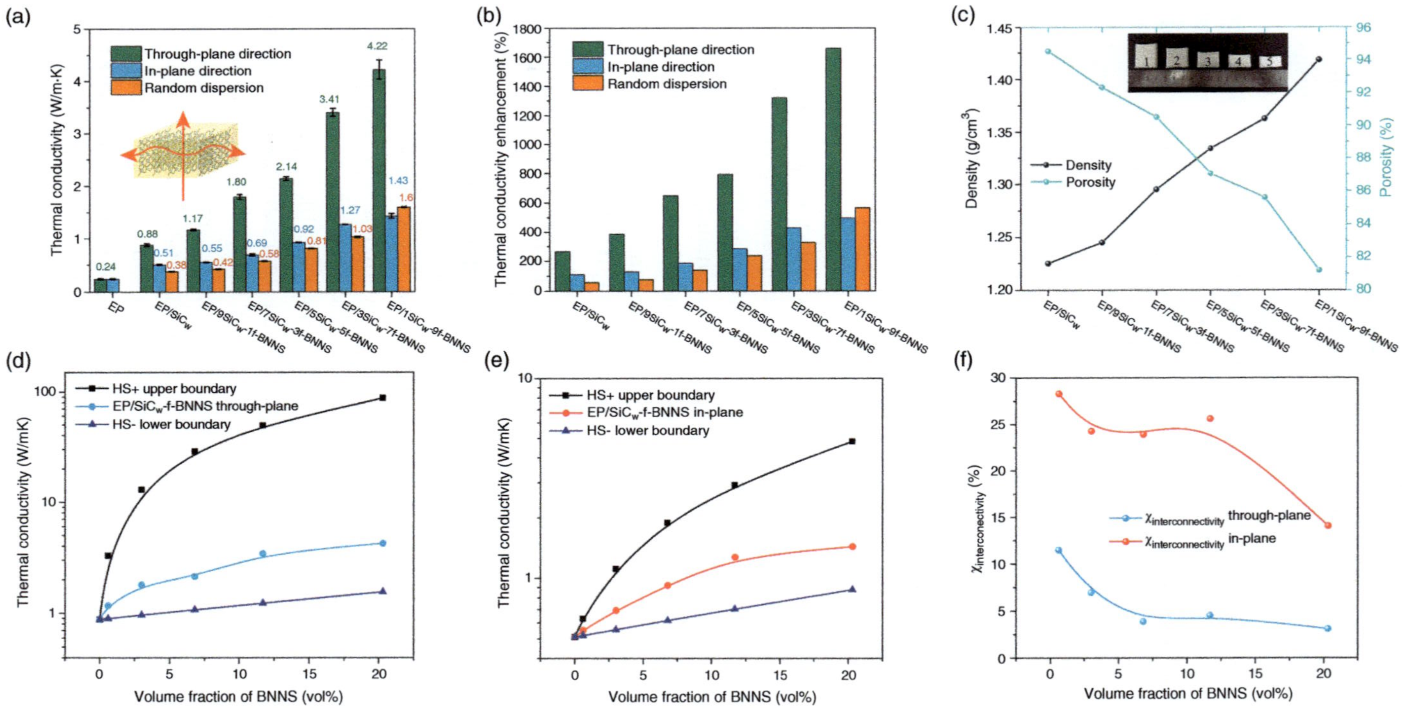

Figure 4.19 Analysis of anisotropic thermal conductivity of EP/SiC_w-f-BNNS composite. (a) Comparison diagram of thermal conductivity of EP/SiC_w-f-BNNS composite and EP/SiC_w-BNNS composite; (b) Comparison chart of thermal conductivity of composite materials compared with that of epoxy resin; (c) Comparison diagram of porosity of SiC_w/f-BNNS skeleton and density of composite materials, the illustrations show SiC_w/f-BNNS skeleton objects of the same quality but different proportions, from left to right, 9SiC_w/1f-BNNS to 1SiC_w/9f-BNNS; (d) Comparison curve between out-of-plane thermal conductivity of composite and HS model results; (e) Comparison curve between composite in-plane thermal conductivity and HS model results; (f) Variation curve of f-BNNS entanglement degree in different directions inside the composite. Source: Reproduced with permission from Ref. [16]; © 2020 Elsevier.

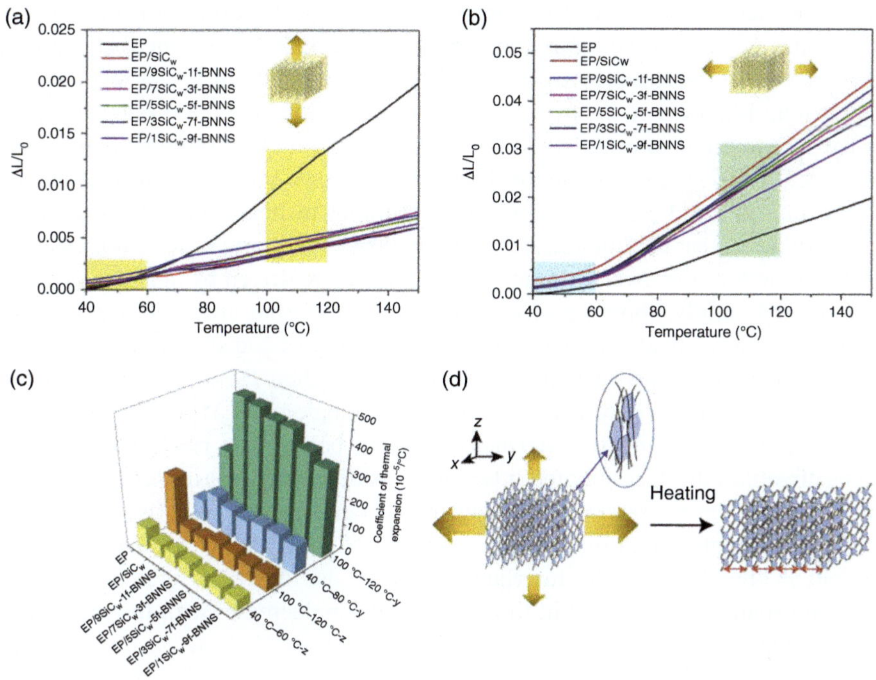

Figure 4.20 Thermal expansion performance characterization of EP/SiC$_w$-f-BNNS composite. (a, b) Are the curves of thermal expansion behavior versus temperature along the z- and y-axes of the composites; (c) The coefficient of thermal expansion of the composite at 40–60 °C and 100–120 °C along the z- and y-axes; (d) Schematic diagram of SiC$_w$ and BNNS on the thermal expansion behavior. Source: Reproduced with permission from Ref. [16]; © 2020 Elsevier.

semiconductor electronic devices is very low. The mismatch of thermal expansion behavior between devices and substrate materials will lead to thermal stress accumulation and lead to a series of problems such as desoldering, cracking, and warping, which will seriously affect the reliability of equipment. Therefore, low thermal expansion coefficient is another important indicator of high-performance electronic packaging materials [17]. Considering the anisotropy of the heat conduction network in this experiment, we measured the thermal mechanical properties of the sample in both the *z*- and *y*-directions, as shown in Figure 4.20a,b. The thermal expansion coefficient of the composite between 40–60 °C (glassy state) and 100–120 °C (high elastic state) is shown in Figure 4.20c. It can be found that the embedded three-dimensional vertical thermal conductive framework effectively inhibits the thermal expansion behavior of epoxy resin in the *z*-axis direction, and the higher the filler content, the more obvious the inhibition effect. The results show that the vertically oriented SiC$_W$/f-BNNS thermal conductive skeleton effectively inhibits the movement of epoxy molecular segments in the vertical direction.

On the one hand, the tight entanglement between fibers restricts the degree of chain segment freedom; on the other hand, the coupling agent modification of BNNS surface improves the interaction between filler and matrix, further improving the dimensional stability of the composite. The thermal expansion behavior of composites in y-axis direction is contrary to the above. Although the CTE value of the epoxy resin decreased with the increase in filler content, it was significantly higher than that of pure epoxy resin. This abnormal phenomenon may be caused by the interlaminar internal stress during the preparation of the heat conduction network. In the process of suction filtration self-assembly, SiC_w and f-BNNS are tightly intertwined to form a continuous composite heat conduction network, and multilayer heat conduction networks are stacked to form a three-dimensional heat conduction framework. Although SiC_w fibers are driven by negative pressure and stacked tightly, their rigidity generates interlaminar internal stress in the three-dimensional heat conduction framework. Figure 4.20d shows the interaction between SiC_w and f-BNNS inside the composite. When the sample temperature rises, the epoxy resin is restricted by tightly entangled SiC_w/f-BNNS network in the z-axis direction, and the thermal expansion behavior is not obvious. In the y-axis direction, the accumulated internal stress between the layers of the network instead promotes the lateral expansion of the sample, so the thermal expansion coefficient is higher than that of pure epoxy resin. With the decrease of SiC_w fiber content in the skeleton, the interlaminar stress also decreases gradually, so the thermal expansion coefficient in the y-axis direction decreases with the increase in filler content.

4.4 Improved Thermal Conductivity of Fluids and Composites Using Boron Nitride Nanoparticles Through Hydrogen Bonding

4.4.1 Preparation and Characterization of Improved Thermal Conductivity of Fluids and Composites Using Boron Nitride Nanoparticles

A number of different sizes of boron nitride (BN) particles, ranging from the nano- to microscale, in a variety of different shapes (nanotubes, nanosheets, etc.) were obtained from Lower Friction Inc. http://www.lowerfriction.com/ and BNNT Inc. http://www.bnnt.com/.

Because it has been previously reported that surface treatments can change the adhesion characteristics of BN particles with other materials and in this way, enhance the thermal performance, a Harrick plasma cleaner was used to treat BN particles in the current investigation. This created a more hydrophilic surface and thereby enhanced the overall contact. An alternative method was also used to treat the BN surface with an aqueous NaOH solution, which due to the reaction between BN and the OH ion, formed OH-BN nanoparticles.

The carbon fibers utilized were purchased from Pyrograf Products Inc. http://pyrografproducts.com, and polyalphaolefin (PAO), Glycerol, etc., were acquired from Sigma Aldrich Inc. Self-curing commercially available polymeric solutions (paints) such as polyurethane, polyacrylic, and epoxy were all available and obtained locally.

Scanning electron microscopy (SEM) and transmission electron microscopy (TEM) were used to characterize the BN particles described above, in their "as received" condition and also after treatment, to compare the particle size, shape, purity, and functional group. The surface-enhanced BN particles (hydrophilic) were used in the experiments described herein.

The thermal conductivity data were obtained using a Hot Disk™ thermal constants analyzer, with the following parameters: measurement depth -6 mm, room temperature, power -0.012 W, measurement time -15 seconds, sensor radius -3.189 mm, TCR -0.0471/K, disk-type kapton, and temperature drift rec. A three-roll mill (Ross Engineering Inc. New York) was used to prepare the homogeneous fluids and polymer composite. Each sample we measured three times and the resulting values averaged. This resulted in a measurement uncertainty of ±2–3%. The measured temperature is room temperature (25 °C).

We use the standard deviation to discuss the uncertainty.

$$\sigma = \sqrt{\sum_{i=1}^{N} (q_i - \overline{q})^2 / N}$$

where σ is the standard deviation, q_i is the measured thermal conductivity value, and N refers to the trials that thermal conductivity was measured.

The standard deviation is found by taking the square root of the average of the squared deviations of the thermal conductivity values subtracted from their average thermal conductivity (TC) value.

Some samples were produced and used to verify the hydrogen bonding assumption. Figure 4.21 demonstrates that the following sample preparation procedure was used for all of the tests:

1. The nanoparticles were slowly added to the liquid media, that is, PAO or Glycerol, while stirring.
2. Once all of the BN particles had been added, stirring continued for another 2 minutes.

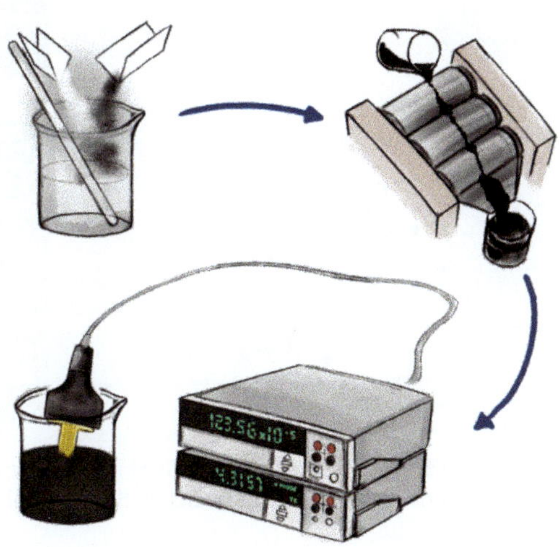

Figure 4.21 Illustration of experimental procedure. Source: Haiping Hong, Bin Yao, Ding Lou, Mingyang Mao, Zhong Hu.

3. The sample was then allowed to cool while stirring, until it was safe to handle.
4. The sample was then "manufactured" through a standard "three-roll mill" at least eight times to obtain a smooth consistency.

4.4.2 Discussion and Analysis of BN Composites as Thermal Interface Materials

Hydrogen bonding is defined as the interaction between a hydrogen atom attached to an electronegative atom and an electronegative atom that possesses a lone pair of electrons. The strongest hydrogen bonds involve the first-period elements B, F, O, or N (see Figure 4.22).

Therefore, due to the nature of the boron (B) and nitrogen (N) atoms, BN, the hydrogen bonding necessary for this investigation is easily formed. Figure 4.23 illustrates the SEM images of the BN particles and the carbon fibers. As clearly indicated, the BN particles are well exfoliated.

Table 4.1 lists the thermal conductivity of the fluids containing the BN particles of varying BN wt%. As shown, with the same amount of BN concentration (samples 1 and 2, 14 wt% BN loading), the glycerol fluid showed an enhanced thermal

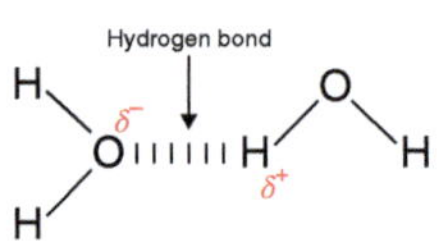

Figure 4.22 Diagram of hydrogen bonding. Source: Haiping Hong, Bin Yao, Ding Lou, Mingyang Mao, Zhong Hu.

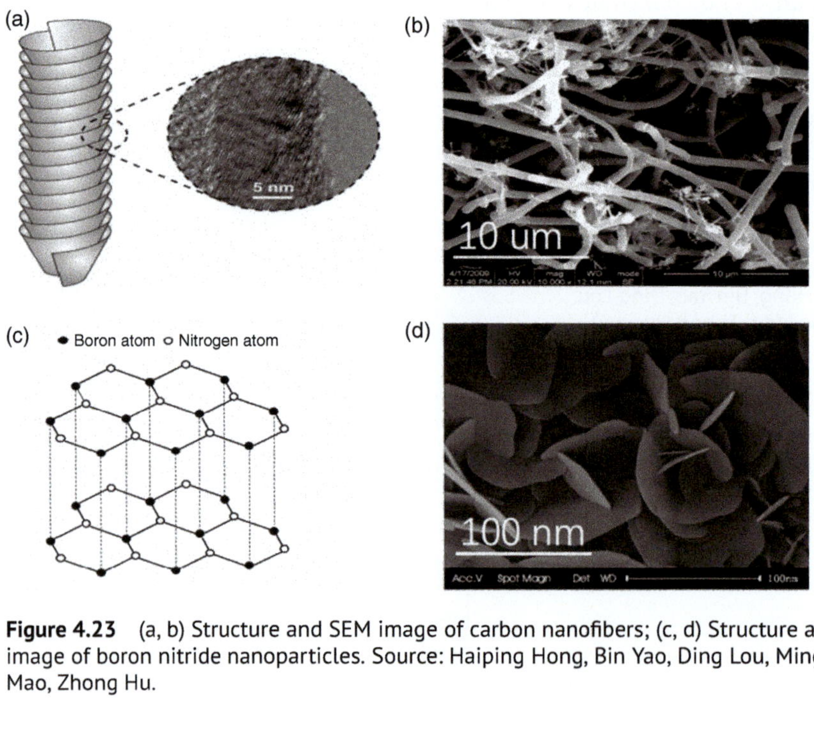

Figure 4.23 (a, b) Structure and SEM image of carbon nanofibers; (c, d) Structure and SEM image of boron nitride nanoparticles. Source: Haiping Hong, Bin Yao, Ding Lou, Mingyang Mao, Zhong Hu.

Table 4.1 Thermal conductivity of BN particle-contained fluids.

Sample no.	Base fluid	Second base fluid	Particles' weight percentage	TC of base fluids (W/m K)	TC (W/m K)	TC increase	TC increase per weight percent
1	Glycerol	N/A	BN (14 wt%)	0.31	0.46	47%	3.3
2	PAO	N/A	BN (14 wt%)	0.17	0.23	37%	2.6
3	Glycerol (50%)	Water (50%)	BN (14 wt%)	0.45	0.65	45%	3.2
4	Krytox XHT 750	N/A	BN (9.1%)	0.11	0.15	32%	3.6
5	Silicon oil	N/A	Silica (5.1%)	0.18	0.19	8%	1.5

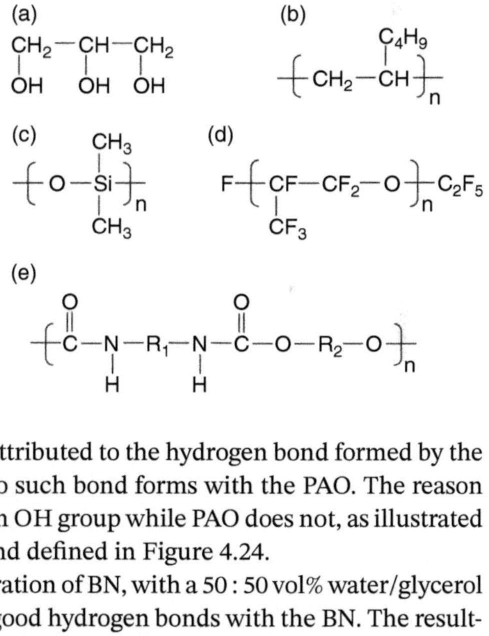

Figure 4.24 Chemical structure of (a) glycerol, (b) Polyalphaolefin (PAO), (c) Silicon oil, (d) Krytox XHT 750 oil, and (e) Polyurethane. Source: Haiping Hong, Bin Yao, Ding Lou, Mingyang Mao, Zhong Hu.

conductivity. This enhancement was attributed to the hydrogen bond formed by the glycerol and BN particles. Note that no such bond forms with the PAO. The reason for this difference is that glycerol has an OH group while PAO does not, as illustrated in the chemical structure illustrated and defined in Figure 4.24.

Sample 3 contains the same concentration of BN, with a 50 : 50 vol% water/glycerol mixture, in which both solvents form good hydrogen bonds with the BN. The resulting thermal conductivity enhancement was similar to that observed with sample 1. Figure 4.25 clearly shows the thermal conductivity value of various base fluids with 14 wt% boron nitride loading. The blank bar represents the thermal conductivity of the base fluid, and the striped bar is the thermal conductivity of the fluids with the addition of the BN nanoparticles.

It is important to note that sample 5 contained silicon oil and silica nanoparticles, and both the solvent and the particles were unable to form hydrogen bonds. Therefore, the measured thermal conductivity enhancement was very low, only 8% at 5 wt% particle loading. Sample 4 contained krytox oil, which has a fluorine atom

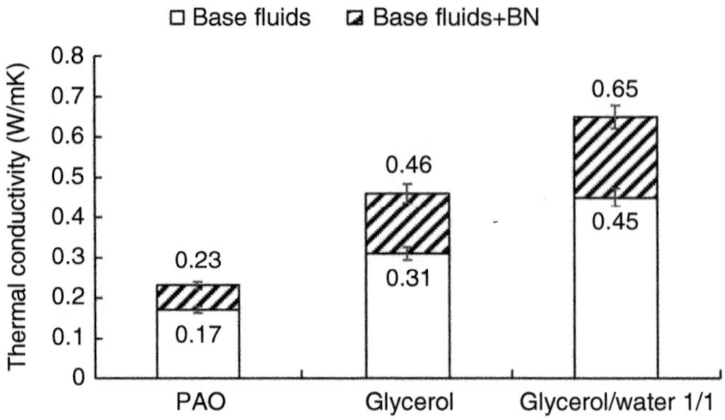

Figure 4.25 Thermal conductivity value of various base fluids with 14 wt% boron nitride loading. Source: Haiping Hong, Bin Yao, Ding Lou, Mingyang Mao, Zhong Hu.

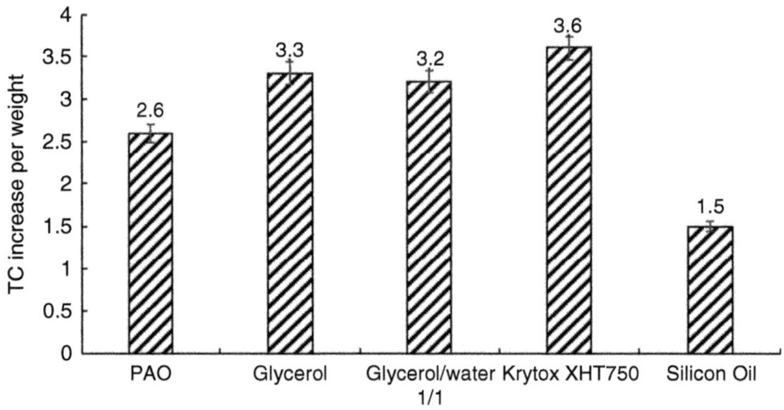

Figure 4.26 Thermal conductivity increases per weight percentage loading of BN with various base fluids. Source: Haiping Hong, Bin Yao, Ding Lou, Mingyang Mao, Zhong Hu.

and as a result can form a strong hydrogen bond with BN. With 9.1 wt% BN loading, the thermal conductivity was enhanced by as much as 32%. The resulting thermal conductivity increase per weight was the highest value recorded.

Figure 4.26 shows the thermal conductivity increase per weight percentage loading of boron nitride with various base fluids. It is clearly seen that with hydrogen boning, the thermal conductivity increase per weight percentage is higher (3.2–3.6) than that of non-hydrogen bonding (2.6 and 1.5, respectively). These results confirm that hydrogen bonding plays an important role in the thermal conductivity enhancement of the resulting nanofluids. These results raised the question of whether this concept could extend from fluids to polymer composites. To understand this, self-curing polymer solutions (paints) containing hydrogen bonding formed solvents, such as ethanol, water, and isopropanol, were used. These types of polymer solutions are commercially available, for example, polyurethane, polyacrylic, epoxy, etc. Table 4.2 lists the thermal conductivity of the BN particle containing polyurethane composites evaluated.

Table 4.2 Thermal conductivity of BN particle-contained polyurethane composite.

Sample no.	Particles' weight percentage	TC of composite (W/m K)	TC percent increase	TC increase per weight percent
1	BN (0 wt%)	0.36	N/A	N/A
2	BN (20 wt%)	1.02	183%	9.1
3	BN (35 wt%)	1.78	394%	11.2

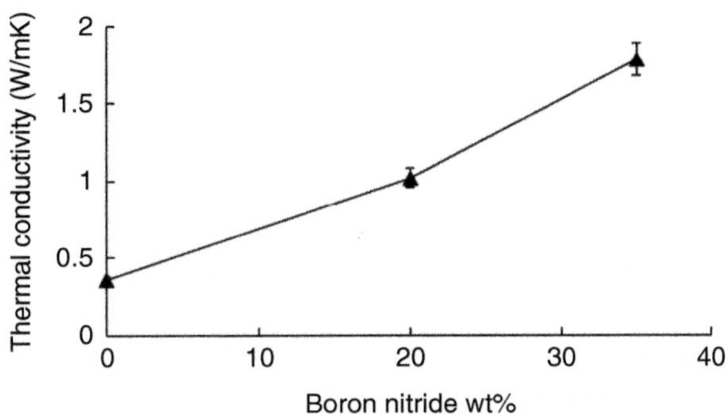

Figure 4.27 Thermal conductivity value of polyurethane composites containing boron nitride. Source: Haiping Hong, Bin Yao, Ding Lou, Mingyang Mao, Zhong Hu.

As clearly shown in Table 4.2, the hydrogen bond formed polymer composites demonstrated significantly enhanced thermal conductivity values, that is, a 20 wt% loading leads to a thermal enhancement of 183% and a 35 wt% loading leads to a 394% enhancement, also see Figure 4.27. Of course, all composite samples are electrically insulating. These results verify the overall concept proposed in this investigation of these composites. In addition, it is worth noting here that the thermal conductivity value is believed to be due to the use of a three-roll milling method that squeezes the particles tightly together, enhancing the physical connection. As indicated, the higher the BN loading percentage, the greater the thermal conductivity enhancement. With increased BN concentration (>20 wt%), the polymer film is easily cracked.

The results illustrate the thermal conductivity enhancement resulting from the hydrogen bonding which is superior to that caused by alignment. Wong et al. [18] reported that magnetic alignment of hexagonal boron nitride (hBN) platelets leads to 104% higher than that of unaligned counterpart, both of which were observed at a relatively low filler loading of 20 wt%. Here, the experimental results indicate that a greater than 183% increase was obtained. In addition, the method described herein is both simple and cost-effective.

In summary, a series of spherical BN-related fluids and polymer composite samples were developed. The experimental results indicate that hydrogen bonding can significantly enhance the thermal conductivity of the samples. It remains to be seen whether it is possible to further improve the thermal conductivity of the samples by adding additional materials. It is well known that nanotubes, nanosheets, and nanofibers have long-chain and broad surface contacts, which

Table 4.3 Thermal conductivity of BN particle-contained fluids with the addition of carbon nanofiber.

Sample no.	Base fluid	Second base fluid	First particles	Second particles	TC of base fluids (W/m K)	TC (W/m K)	TC percent increase	TC increase per weight percent
1	Glycerol	N/A	BN (7%)	CNF (7%)	0.31	1.44	366%	52.3
2	Glycerol	N/A	BN (10.5%)	CNF (3.5%)	0.31	0.90	188%	53.7
3	Glycerol (50%)	Water (50%)	BN (7%)	CNF (7%)	0.45	1.79	298%	42.7
4	Silicon oil	N/A	CNF (4.9%)	N/A	0.18	0.51	181%	36.9

may provide additional opportunities to enhance the thermal conductivity even further [19]. In addition, some interesting work and results using polymers such as polyamide-imides and boron nitride [20, 21].

To this end, additional carbon nanofibers were added to the fluids and polymer composite system. Table 4.3 lists the thermal conductivity of the BN particle fluid mixtures with the addition of carbon nanofibers. As indicated, it is clear that adding carbon fibers significantly enhances the thermal conductivity of the nanofluids. Although the BN contribution to the thermal conductivity is not obvious, the thermal conductivity data of samples 1 and 2 help provide an explanation. Sample 1 contains 7 wt% BN and 7 wt% carbon fibers, which yields a thermal conductivity of 1.44 W/m K, or an increase of 366%. Sample 2 contains 10.5 wt% BN and 3.5 wt% carbon fibers, which yields a thermal conductivity of 0.90 W/m K, or an increase of 188%. It is obvious that the carbon fiber contributes to the thermal conductivity enhancement. Because silicon oil does not have hydrogen bonding, sample 4 shows the least increase in the thermal conductivity.

Figure 4.28 shows the thermal conductivity increase per weight percentage with the addition of the carbon nanofibers. The data concur with the conclusion that hydrogen bonding leads to higher thermal conductivity enhancement per weight percentage. From Table 4.4, it is clear that with the addition of carbon nanofibers, the thermal conductivity increases significantly (BN 20 wt%, CNF 5 wt% yields an enhancement of 480% enhancement, sample 2). However, the polymer composite becomes conductive, due to the nature of the carbon nanofibers (>1000 ohm). With greater BN loading (e.g., 35 wt%), cracking was observed; hence, the measured thermal conductivity values were not stable. The data from Table 4.4 are presented in Figure 4.29.

As observed in Tables 4.3 and 4.4, it is apparent that the addition of long-chain carbon nanofibers can significantly increase the thermal conductivity of the fluids and composite. The addition of nanofibers will also increase the electrical conductivity value. Through the appropriate balance of these factors, one can maintain the insulating properties which are critical for many applications.

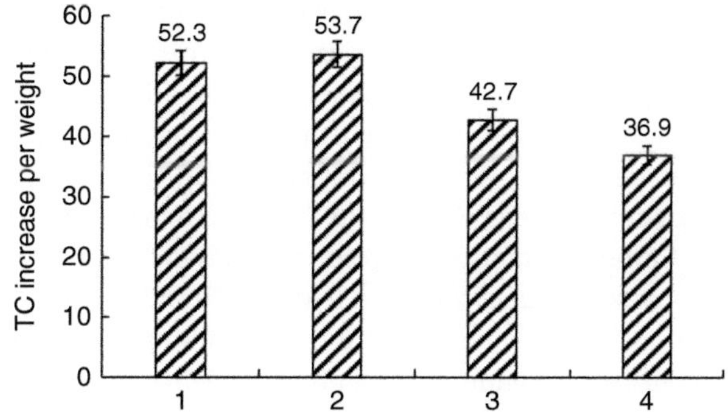

Figure 4.28 Thermal conductivity increases per weight percentage with the addition of carbon nanofiber. Source: Haiping Hong, Bin Yao, Ding Lou, Mingyang Mao, Zhong Hu.

Table 4.4 Thermal conductivity of BN particle containing polyurethane composite with the addition of carbon nanofiber.

Sample no.	Composite	Particles weight percentage	TC of composite (W/m K)	TC percent increase
1	Polyurethane	BN (0 wt%)	0.36	N/A
2	Polyurethane with 5 wt% carbon fiber	BN (20 wt%)	2.1	483%
3	Polyurethane with 5 wt% carbon fiber	BN (35 wt%)	3.5–4	N/A

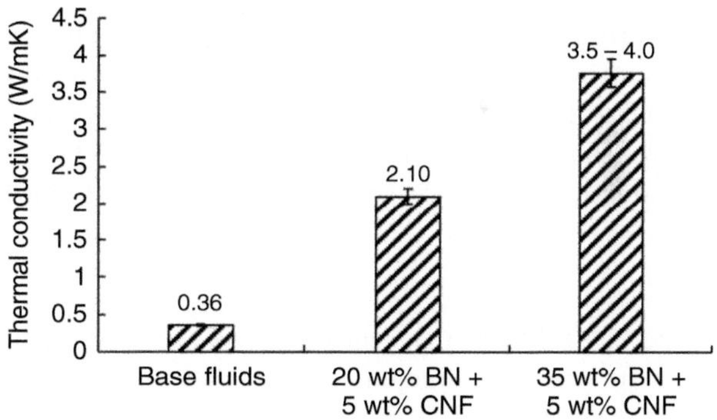

Figure 4.29 Thermal conductivity of BN particle containing polyurethane composite with the addition of carbon nanofiber. Source: Haiping Hong, Bin Yao, Ding Lou, Mingyang Mao, Zhong Hu.

4.5 Improved Thermal Conductivity of PEG-Based Fluids Using Hydrogen Bonding and Long Chain of Nanoparticle

4.5.1 Preparation and Characterization of Thermal Conductivity of PEG-Based Fluids Using Hydrogen Bonding and Long Chain of Nanoparticle

A number of different sizes of BN particles, ranging from the nano- to microscale, in a variety of different shapes (nanotubes, nanosheets, etc.) were obtained from Lower Friction Inc. http://www.lowerfriction.com/ and BNNT Inc. http://www.bnnt.com/.

It has been previously reported that surface treatments can change the adhesion characteristics of BN particles to other materials and, in this way, enhance the thermal performance. In the current investigation, a Harrick plasma cleaner was used to treat BN particles to create a more hydrophilic surface. An alternative method used was to treat the BN surface with an aqueous NaOH solution, which resulted in a reaction between BN and the OH ion, to form OH-BN nanoparticles.

Carbon nanofibers were purchased from Pyrograf Products Inc. http://pyrografproducts.com, and hydroxyl functionalized multiwall nanotubes were purchased from Nanostructure & Amorphous Materials Inc. https://www.nanoamor.com/carbon-nanotubes-nanofibers. Polyethylene glycol (PEG 400) was acquired from Sigma Aldrich.

Scanning electron microscopy (SEM) images were obtained by a ZEISS SUPRA™ 40VP. The characterized samples are boron nitride nanoparticles and nanotubes, carbon nanofibers, and multiwall nanotube with OH group. The samples are described above in their "as received" condition and also after treatment, to compare the particle size, shape, purity, and functional group. The surface-enhanced BN particles (hydrophilic) were used in the experiments described herein.

The thermal conductivity data were obtained using a Hot Disk™ thermal constants analyzer, with the following parameters: measurement depth –6 mm, room temperature, power –0.012 W, measurement time –15 seconds, sensor radius –3.189 mm, TCR –0.0471/K, disk-type Kapton, and temperature drift recorder. A three-roll mill (Ross Engineering Inc. New York) was used to prepare the homogeneous fluids and polymer composite.

During the measurement, the sensor was embedded in the sample. An electrical current was then input and passed through the sensor at a level high enough to increase the sensor temperature between a fraction of a degree up to several degrees. At the same time, the resistance as a function of heating time was recorded. Thermal conductivity value was calculated through the following equation:

$$k = \frac{Qdx}{AdT}$$

where Q refers to the heat flux, A is the cross-sectional area, dx/dT is the thickness/temperature gradient, and k is defined as the thermal conductivity.

Several samples were fabricated to verify the hydrogen bonding assumption. Figure 4.30 shows that the following sample preparation procedure was used for all of the tests:

1. BN and carbon nanomaterial were slowly added to the liquid media (PEG400) while stirring.
2. Once all particles had been added, stirring continued for another 2 minutes.
3. The mixture was then milled at least eight times using the "three-roll mill" equipment to obtain a smooth and homogeneous fluid or grease-like mixture.
4. Thermal conductivity was measured accordingly.

The role of the milling machine is to squeeze the sample in the micro–nanosize and enhance the intermolecular force and dispersion.

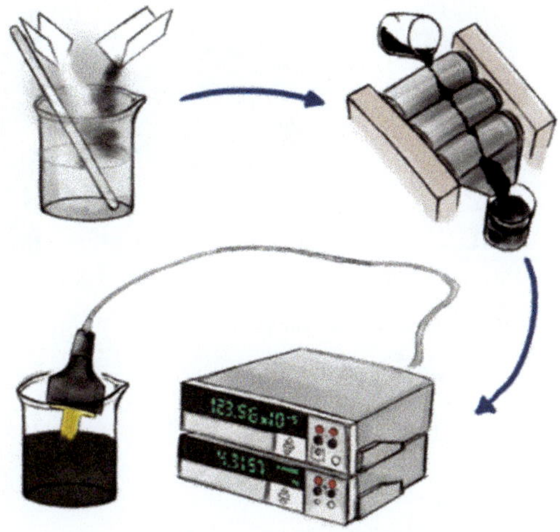

Figure 4.30 Illustration of experimental procedure. Source: Haiping Hong, Bin Yao, Ding Lou, Mingyang Mao, Zhong Hu.

4.5.2 Discussion and Analysis of PEG-Based Fluids Using Hydrogen Bonding and Long Chain of Nanoparticle

Hydrogen bonding is defined as the interaction between a hydrogen atom and an electronegative atom that possesses a lone pair of electrons. The strongest hydrogen bonds involve the first-period elements B, F, O, or N (see Figure 4.31).

Therefore, due to the nature of the boron (B) and nitrogen (N) atoms, and the resulting BN, the hydrogen bonding necessary for this investigation is easily formed. Figure 4.32 illustrates the SEM image of hex-BN nanoparticles, BN nanotube, MWNT-OH, and carbon nanofibers. As clearly indicated, the BN particles are well exfoliated, and size is around 50–80 nm.

Table 4.5 lists the thermal conductivity of PEG 400 fluids loaded with hexagonal BN (hBN) nanoparticles. It is clearly seen that with the increase in BN weight percentage (wt%), the thermal conductivity increases nonlinearly. With a 7.5 wt% of BN, the thermal conductivity of PEG400 reached 0.237 W/m K, which was an 18.5% increase compared to the pure PEG fluid. With a BN loading of 14 wt%, the thermal conductivity reached 0.285 W/m K and a 47.7% increase compared to the pure PEG fluid. Below 1 wt% loading, the thermal conductivity values of the mixtures did not show obvious enhancement.

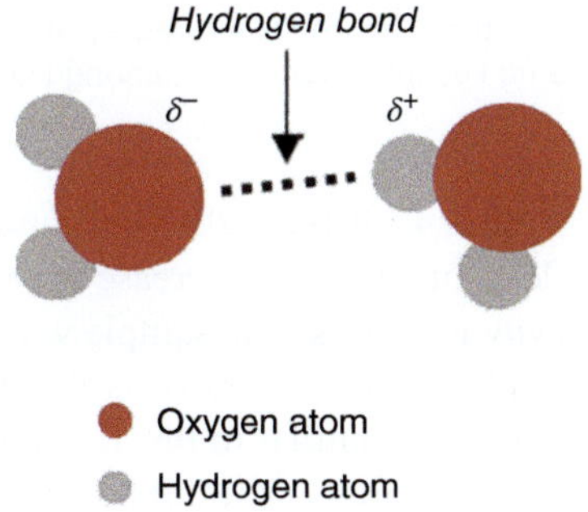

Figure 4.31 Diagram of hydrogen bonding between two water molecules. Source: Haiping Hong, Bin Yao, Ding Lou, Mingyang Mao, Zhong Hu.

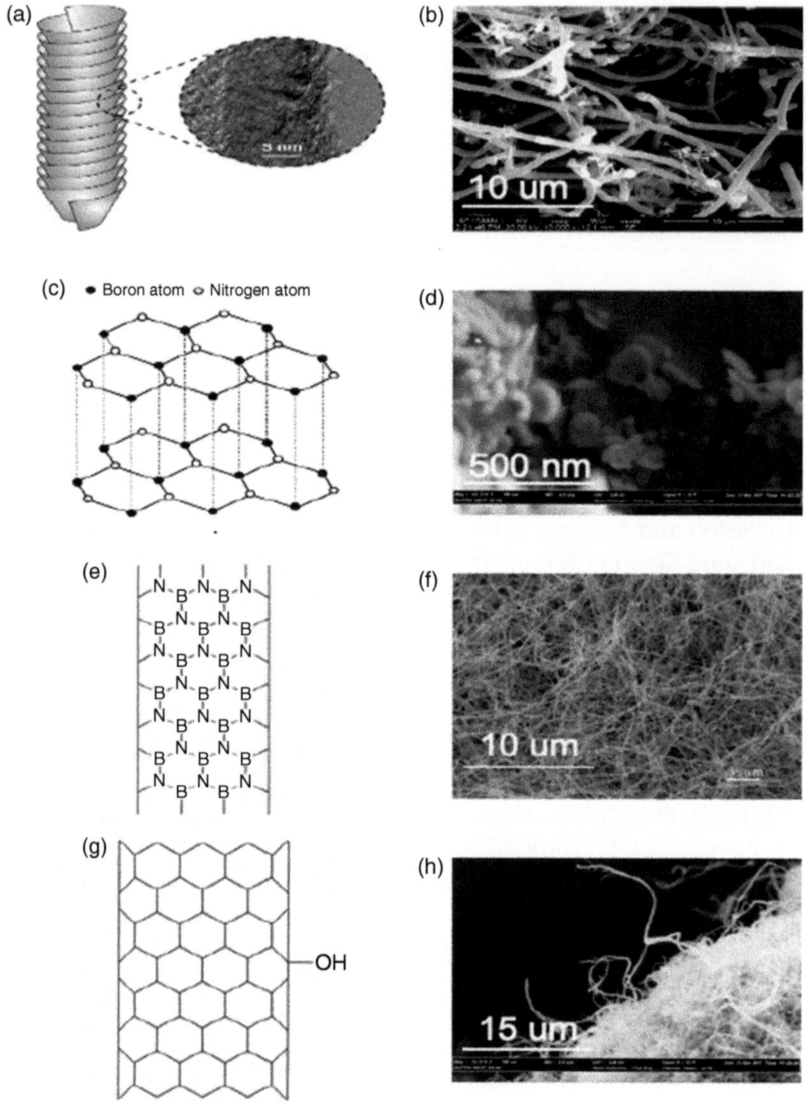

Figure 4.32 (a, b) Structure and SEM image of carbon nanofibers; (c, d) Structure and SEM image of boron nitride nanoparticles; (e, f) Structure and SEM image of boron nitride nanotubes; (g, h) Structure and SEM image of MWNT-OH. Source: Haiping Hong, Bin Yao, Ding Lou, Mingyang Mao, Zhong Hu.

Table 4.6 lists the thermal conductivity of MWNT-OH-loaded PEG 400 fluids. It is clear that with the increase in MWNT-OH weight percentage, the thermal conductivity increases. The sample with MWNT-OH loading of 7.5 wt% yielded a thermal conductivity of 0.387 W/m K, which was an 89.7% increase compared to the pure PEG fluid. Similar to BN, the thermal conductivity of the mixture with MWNT-OH loading below 1 wt% did not show obvious enhancement.

Table 4.5 Thermal conductivity of hex-BN nanoparticle-contained PEG 400 fluids.

hBN weight percentage	TC (W/m K)	TC increase	TC increase per weight percent
14 wt%	0.285	47.7%	3.41
7.5 wt%	0.237	18.5%	2.47
4.5 wt%	0.216	8%	1.78
2 wt%	0.207	3.5%	1.75
1 wt%	0.199	−0.5%	
0.5 wt%	0.203	1.5%	
0.1 wt%	0.198	−1%	
Pure PEG400	0.200		

Table 4.6 Thermal conductivity of MWNT-OH contained PEG 400 fluids.

MWNT-OH	TC (W/m K)	TC increase	TC increase per weight percent
7.5 wt%	0.387	89.7%	12.0
4.5 wt%	0.315	54.4%	12.1
2 wt%	0.239	17.1%	8.5
1 wt%	0.216	5.9%	5.9
0.5 wt%	0.212	3.9%	
0.1 wt%	0.214	4.9%	
Pure PEG400	0.204		

Figure 4.33 illustrates the relationship between the thermal conductivity and the weight percentage of nanomaterials (up to 14 wt%) in the PEG 400 fluid. As observed, the slope of the plot of the MWNT-OH data is significantly steeper than the slope of the BN data. In terms of the increase in the thermal conductivity, PEG400 with 7.5 wt% BN showed 2.47 of thermal conductivity increase per weight, while for the PEG400 loaded with 7.5 wt% MWNT-OH, thermal conductivity increase per weight was 12.

One possible explanation for these thermal conductivity enhancements may be due to the length, stereo, and space configuration. Although both MWNT-OH and BN contain hydrogen bonds, MWNT-OH has a relatively long length (around 10–30 μm) and a large space configuration, while BN is spherical in shape with a diameter on the scale of 70 nm.

From the data presented in Tables 4.5 and 4.6, it is apparent that the MWNT-OH nanoparticles are more effective in enhancing the thermal conductivity than the

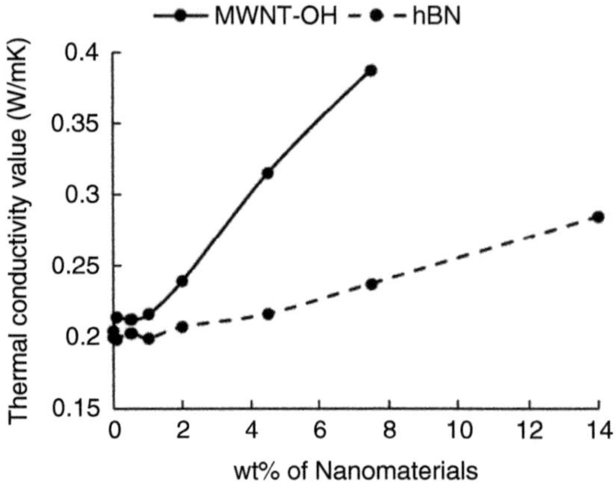

Figure 4.33 Thermal conductivity as a function of weight percentage of nanomaterials in PEG 400 fluids. Source: Haiping Hong, Bin Yao, Ding Lou, Mingyang Mao, Zhong Hu.

BN nanoparticles. Previous investigations have indicated that carbon nanofibers (CNFs) can significantly enhance the thermal conductivity in the fluids due to the long chain (structure) and relatively large dimension and space configuration [19]. Therefore, it may be interesting to adopt CNF into the current system (PEG 400 fluids) and see how this material influences (increase) the thermal conductivity values.

Table 4.7 illustrates the thermal conductivity of PEG400 fluids with introduction of the combination of BN and carbon nanomaterials. The two combinations are BN+MWNT-OH and BN+CNF, with weight percentages of 0.5 wt% and 2 wt%, respectively. It is seen from the data that CNF enhances the thermal conductivity of the fluid mixture more efficiently than MWNT-OH, due to the relatively longer chain and larger diameter of CNF. The thermal conductivity increase per weight indicates that CNF has twice efficiency MWNT-OH. Of course, adding

Table 4.7 Thermal conductivity of PEG400 with the addition of the combination of hBN and carbon nanomaterials.

	TC (W/m K)	TC increase (%)	TC increase per weight percent
hBN			
0.5 wt%	0.203	1.5	3
2 wt%	0.207	3.5	1.75
hBN + MWNT-OH			
0.5 wt% + 0.5 wt%	0.207	7.2	14.4
2 wt% + 2 wt%	0.236	22.3	11.1
hBN + CNF			
0.5 wt% + 0.5 wt%	0.220	14.0	28
2 wt% + 2 wt%	0.318	64.8	32.4

Table 4.8 Thermal conductivity of PEG400 fluids with the addition of the combination of BNNT and MWNT-OH.

MWNT-OH + BN nanotube	TC (W/m K)	TC increase (%)	TC increase per weight percent
1.5 wt% + 1.5 wt%	0.252	30.6	20.4
2 wt% + 2 wt%	0.284	47.1	23.5
2 wt% + 2 wt% (after mill)	0.230	19.1	9.5
MWNT-OH + hBN	TC (W/m K)	TC increase	
2 wt% + 2 wt%	0.236	22.3	11.1

either MWNT-OH or CNF led to significant increase in the thermal conductivity, compared to BN.

Recently, boron nitride nanotubes (BNNTs) have attracted considerable interest from investigators working in this area, largely because BNNTs typically form long chains (around 50 μm) and are therefore expected to result in significantly enhanced thermal conductivity when compared to regular hexagonal boron nitride (hBN) nanoparticles. In addition, the BNNT is a non-electrically conductive (insulating) material so that it will have unique applications. Therefore, the experimental program was expanded to include the addition of BN nanotubes to fluids containing other materials such as BN nanoparticles, MWNT-OH, and CNF to determine the influence/effect of these nanoparticles on the thermal conductivity.

Table 4.8 shows the thermal conductivity of PEG400 with the addition of the combination of MWNT-OH and BNNT: 1.5 wt% MWNT-OH and 1.5 wt% BNNT led to a thermal conductivity of 0.252 W/m K, which was a 30.6% increase over the base fluid, and the addition of 2 wt% MWNT-OH and 2 wt% BNNT led to a thermal conductivity of 0.284 W/m K, which was a 47.1% increase over the base fluid. When the three-roll mill was used to mix the 2 wt% MWNT-OH and 2 wt% BNNT sample, scratches were observed on the mill rollers, indicating that the BN nanotubes might have been heavily damaged due to the ultrahigh interlayer friction. The thermal conductivity of the milled sample was remeasured, and the thermal conductivity was reduced to 0.23 W/m K. Similar results were observed for the boron nitride sample (0.236 W/m K). The results again indicated that the BNNT was damaged by the mechanical rolling process. The ultimate results, however, indicated that the introduction of BNNT can enhance the thermal conductivity of the samples.

Table 4.9 lists the thermal conductivity of PEG400 with the addition of the combination of BNNT and hBN/CNF. As shown, the addition of 1.5 wt% hBN and 1.5 wt% BNNT resulted in a thermal conductivity of 0.224 W/m K, or a 16.1% increase over the base fluid, while the addition of 2 wt% hBN and 2 wt% BNNT led to a thermal conductivity of 0.244 W/m K, or a 26.4% increase over the base fluid. Similarly, PEG400 with a combination of 1.5 wt% CNF and 1.5 wt% BNNT led to a thermal conductivity of 0.298 W/m K, or a 54.4% increase over the base fluid, and the addition of 2 wt% CNF and 2 wt% BNNT led to a thermal conductivity of 0.334 W/m K, or a 73.1%

Table 4.9 Thermal conductivity of PEG400 with the addition of the combination of BNNT and hBN/CNF.

hBN + BN nanotube	TC (W/m K)	TC increase	TC increase per weight percent
1.5 wt% + 1.5 wt%	0.224	16.1%	10.7
2 wt% + 2 wt%	0.244	26.4%	13.2
CNF + BN nanotube	TC (W/m K)	TC increase	TC increase per weight
1.5 wt% + 1.5 wt%	0.298	54.4%	36.3
2 wt% + 2 wt%	0.334	73.1%	36.5

increase over the base fluid. The results clearly indicate that BNNTs can be used to significantly enhance the thermal conductivity of the samples, and the CNFs can increase thermal conductivity more efficiently than BNs.

4.6 Conclusion

To sum up, the topic focuses on the synthesis of high thermal conductivity fillers, the construction of composite interface structure, the design of thermal conductivity network, and the regulation of segment structure of epoxy resin matrix to achieve the goals of improve the reliability of electronic devices and extended their service life. The surface modification of the nanothermally conductive fillers is used to control their distribution in the polymer matrix, and on the premise of ensuring the excellent properties of the polymer, the good thermal conductivity of the polymer is achieved. Based on the integrated design idea, a variety of new multi-functional packaging thermal management materials have been developed, such as high-frequency communication substrate materials with high thermal conductivity and low loss, and fast heat dissipation packaging materials with high thermal conductivity and heat storage.

References

1 Wang, H.X., Wang, N., Cheng, T.S. et al. (2019). Halides-assisted low-temperature synthesis of hexagonal boron nitride nanosheets. *Particle* 36 (10): 1900278.
2 Liu, K., Chen, L., Chen, Y. et al. (2011). Preparation of polyester/reduced graphene oxide composites via in situ melt polycondensation and simultaneous thermo-reduction of graphene oxide. *Journal of Materials Chemistry* 21 (24): 8612–8617.
3 Shen, X., Lin, X., Yousefi, N. et al. (2014). Wrinkling in graphene sheets and graphene oxide papers. *Carbon* 66: 84–92.

4 Su, Z., Wang, H., Tian, K.H. et al. (2018). The combination of π-π interaction and covalent bonding can synergistically strengthen the flexible electrical insulating nanocomposites with well adhesive properties and thermal conductivity. *Composites Science and Technology* 155: 1–10.

5 Novoselov, K.S., Mishchenko, A., Carvalho, A. et al. (2016). 2D materials and van der Waals heterostructures. *Science* 353 (6298): aac9439.

6 Hunter, C.A. and Sanders, J.K.M. (1990). The nature of π-π interactions. *Journal of the American Chemical Society* 112 (14): 5525–5534.

7 Sinnokrot, M.O., Valeev, E.F., and Sherrill, C.D. (2002). Estimates of the ab initio limit for π–π interactions: the benzene dimer. *Journal of the American Chemical Society* 124 (36): 10887–10893.

8 Su, Z., Wang, H., Ye, X. et al. (2018). Anisotropic thermally conductive flexible polymer composites filled with hexagonal born nitride (h-BN) platelets and ammine carbon nanotubes (CNT-NH2): effects of the filler distribution and orientation. *Composites Part A: Applied Science and Manufacturing* 109: 402–412.

9 Su, Z., Wang, H., He, J. et al. (2018). Fabrication of thermal conductivity enhanced polymer composites by constructing an oriented three-dimensional staggered interconnected network of boron nitride platelets and carbon nanotubes. *ACS Applied Materials & Interfaces* 10 (42): 36342–36351.

10 Su, Z., Wang, H., Tian, K. et al. (2018). The combination of pi-pi interaction and covalent bonding can synergistically strengthen the flexible electrical insulating nanocomposites with well adhesive properties and thermal conductivity. *Composites Science and Technology* 155: 1–10.

11 He, J., Guo, Y.L., Tian, K.H. et al. (2017). Surface growth of MG (multilayer graphene)–SiC nanofillers/poly(vinylidene fluoride) composites for improving thermal conductivity and maintaining electrical insulation. *Materials Research Express* 4: 085016.

12 Qin, T.F., Wang, H., He, J. et al. (2019). Preparation and application of high thermal conductivity TMBPDGE-DDM@h-BN composites. *Materials Research Express* 6: 095320.

13 Fang, L., Wu, C., Qian, R. et al. (2014). Nano–micro structure of functionalized boron nitride and aluminum oxide for epoxy composites with enhanced thermal conductivity and breakdown strength. *RSC Advances* 4 (40): 21010–21017.

14 Xiao, C., Chen, L., Tang, Y. et al. (2019). Enhanced thermal conductivity of silicon carbide nanowires (SiCw)/epoxy resin composite with segregated structure. *Composites Part A: Applied Science and Manufacturing* 116: 98–105.

15 He, J., Wang, H., Qu, Q.Q. et al. (2020). Three-dimensional network constructed by vertically oriented multilayer graphene and SiC nanowires for improving thermal conductivity and operating safety of epoxy composites with ultralow loading. *Composites Part A* 139: 106062.

16 Xiao, C., Guo, Y., Tang, Y. et al. (2020). Epoxy composite with significantly improved thermal conductivity by constructing a vertically aligned three-dimensional network of silicon carbide nanowires/ boron nitride nanosheets. *Composites Part B: Engineering* 187: 107855.

17 Hong, H., Jung, Y.H., Lee, J.S. et al. (2019). Anisotropic thermal conductive composite by the guided assembly of boron nitride nanosheets for flexible and stretchable electronics. *Advanced Functional Materials* 29 (37): 1902575.

18 Lin, Z., Liu, Y., Raghavan, S. et al. (2013). Magnetic alignment of hexagonal boron nitride platelets in polymer matrix: toward high performance anisotropic polymer composites for electronic encapsulation. *ACS Applied Materials Interfaces* 5: 7633.

19 Feng, L., Xie, N., and Zhong, J. (2014). Carbon Nanofibers and Their composites: a review of synthesizing, properties and applications. *Materials* 7: 3919–3945.

20 Yu, C., Zhang, J., and Tian, W. (2018). Polymer composites based on hexagonal boron nitride and their application in thermally conductive composites. *RSC Advances* 8: 21948–21967.

21 Jiang, F., Cui, S., and Song, N. (2018). Hydrogen bond-regulated boron nitride network structures for improved thermal conductive property of polyamide-imide composites. *ACS Applied Materials Interfaces* 10: 16812–16821.

5

Optimal Design of High Thermal Conductive Metal Substrate System for High-Power Devices

Hong Guo[1,2], *Zhongnan Xie*[1,2], *and DingBang Xiong*[3]

[1] *GRINM Group Co., Ltd., State Key Laboratory of Nonferrous Metals and Processes, No. 2 Xinjiekou Outer Street, Xicheng District, Beijing 100088, China*
[2] *GRIMAT Engineering Institute Co., Ltd., No. 11 Xingke East Street, Huairou District, Beijing 100088, China*
[3] *Shanghai Jiao Tong University, State Key Laboratory of Metal Matrix Composites, No. 800 Dongchuan Road, Minhang District, Shanghai 200240, China*

5.1 Power Devices and Thermal Conduction

The research and breakthrough of new semiconductor materials and devices have promoted the development of new industries. The wide bandgap semiconductor materials represented by SiC and GaN are new semiconductor materials developed rapidly after Si and GaAs materials. SiC has many outstanding advantages, such as wide bandgap, high saturation drift velocity, high thermal conductivity, and high critical breakdown electric field. It is especially suitable for making high-power, high-voltage, high-temperature, and radiation-resistant electronic devices. GaN material has the characteristics of high two-dimensional electron gas concentration, high breakdown field strength, and high electron saturation speed. It has obvious performance advantages in the application field of microwave high-power devices. The third-generation semiconductor materials show great application potential in many fields such as power electronics, information communication, and so on.

However, with the development of miniaturization and integration of devices, the device's reliability is facing severe challenges. The performance advantages of GaN power devices are far from being fully utilized. The main reason is that GaN microwave power chip has self-heating effect during operation. While the chip outputs high power, the heat in the active area gradually accumulates. The output power density and efficiency of GaN devices deteriorate rapidly, which makes its performance advantages far from being fully exerted. The heat dissipation problem has become one of the main technical bottlenecks limiting the further development and application of GaN microwave power device technology.

Figure 5.1 shows a typical package diagram of flip chip ball grid array (FC-BGA). There are multiple layers of thermal resistance from the heat source to the heat sink, including the thermal resistance caused by the heat sink layer and the thermal interface materials layer. As the key position of the chip-cooling channel, the heat sink

Thermal Management Materials for Electronic Packaging: Preparation, Characterization, and Devices, First Edition. Edited by Xingyou Tian.

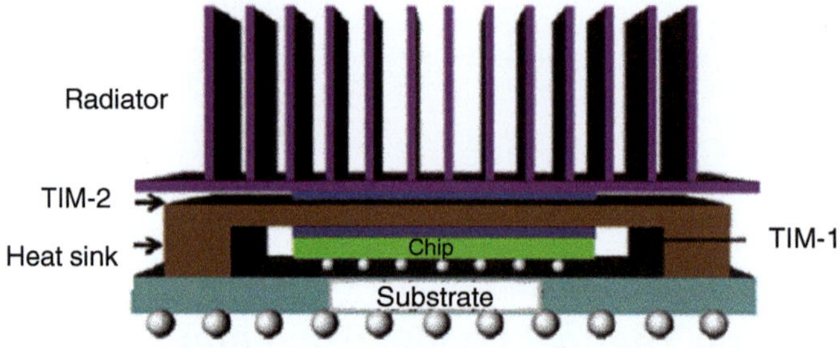

Figure 5.1 Typical package diagram of flip chip ball grid array. Source: Hong Guo, Zhongnan Xie, DingBang Xiong.

brings thermal resistance which accounts for more than 20% of the total thermal resistance, which is an important bottleneck of high-power chip cooling. At present, the commonly used heat sink materials include metal and ceramic.

Metal thermal management materials can be used in key parts such as substrate materials, heat sink materials, and airtight packaging bases. Its function is not only to conduct heat but also to support and protect chips and electronic components. Therefore, in addition to high thermal conductivity, metal thermal management materials also need to have a thermal expansion coefficient matching with semiconductor materials and a certain strength. New thermal management materials should have the following properties.

(1) High thermal conductivity. Thermal conductivity is the main performance index of thermal management materials. The thermal conductivity is generally required to be over 500 W/mK. For every 10°C increase in the working temperature of the electronic device, the device reliability reduces by twice. Thermal management materials with high thermal conductivity can accelerate the electronic device's heat conduction.
(2) The thermal expansion coefficient. Generally, metal thermal management materials are directly connected to the chip. In order to avoid thermal stress caused by excessive difference in thermal expansion coefficient between thermal management materials and semiconductor materials, the thermal expansion coefficient of thermal management materials is required to match with GaAs, GaN (4–6 ppm/K), and other semiconductor materials.
(3) Mechanical properties. Thermal management materials should have a certain strength to overcome the damage caused by pressure, vibration, impact, and other external forces. Therefore, thermal management materials need to have excellent mechanical properties to protect and support electronic devices.
(4) Weldability. Metal thermal management materials are mainly connected with chips or other heat dissipation components by welding, which requires good weldability.

The development of thermal management materials is a process to continuously meet the application requirements of electronic devices. In the choice of thermal management materials, pure metal with high thermal physical properties is the first choice for people. However, the thermal expansion coefficient of pure metal is relatively large, and the compatibility with semiconductor materials is poor.

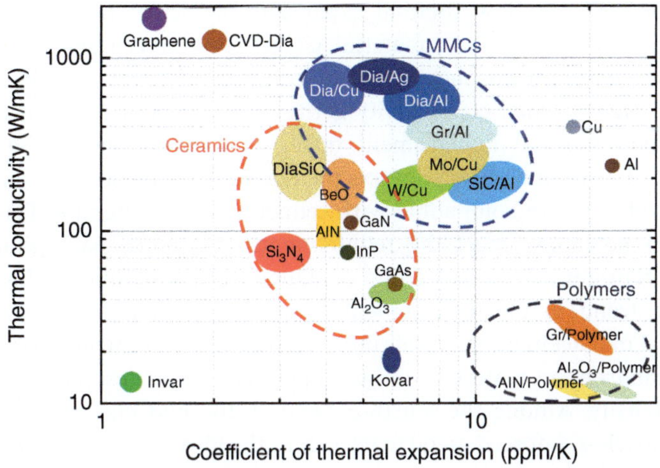

Figure 5.2 Thermal conductivity–thermal expansion coefficient distribution of common thermal management materials. Source: Hong Guo, Zhongnan Xie, DingBang Xiong.

Therefore, the thermal expansion coefficient requirements of thermal management materials are proposed. With the application of thermal management materials in rail transit, aerospace, and other fields, the requirements of lightweight thermal management materials are put forward. At present, thermal management materials have been developed for four generations. As shown in Figure 5.2, it is the thermal conductivity–thermal expansion coefficient performance distribution of common thermal management materials.

The first generation of thermal management materials are mainly metal materials, alloy materials, etc. Aluminum has small mass, high thermal conductivity, easy processing, and low cost, and its thermal expansion coefficient is as high as 23 ppm/K, which cannot match the substrate. Kovar alloy has a thermal expansion coefficient matching the substrate, but its thermal conductivity does not exceed 20 W/mK. It is difficult to meet the heat dissipation requirements of high-power electronic devices. Tungsten–copper alloy has both low expansion of tungsten and high thermal conductivity of copper. Its thermal conductivity can reach about 10 times of Kovar alloy. And its thermal expansion coefficient and thermal conductivity can be adjusted according to the material composition, so it is widely used in the field of high-temperature resistance. However, the density of W-Cu alloy is nearly twice that of Kovar alloy, so its application in lightweight field is limited. SiC/Al is the most widely used of the third-generation thermal management materials. SiC/Al composites have a thermal expansion coefficient matching the substrate. It can also be adjusted by adjusting the percentage content of reinforcement particles. But its thermal conductivity is generally between 160 and 250 W/mK. Due to the limitation of SiC volume fraction, the thermal conductivity of SiC/Al composites is difficult to be further improved. Therefore, researchers focus on the enhancement phase with higher thermal conductivity.

The fourth-generation thermal management materials are metal matrix composites with diamond as reinforcing phase. Single-crystal diamond has high thermal conductivity (>1500 W/mK) and low thermal expansion coefficient (2 ppm/K), which makes it an ideal reinforcing phase material. Compared with the first three representatives, diamond-reinforced metal matrix composites show excellent thermal conductivity. High thermal conductivity is the most typical performance of the fourth-generation thermal management materials different from other materials. In addition, its thermal expansion coefficient can be adjusted by the diamond content, which is more compatible with semiconductor materials. Traditional materials can no longer meet the requirements of modern packaging technology for high comprehensive properties of materials. Compounding has become an inevitable trend in the development of new thermal management materials. New thermal management materials must overcome the shortcomings of traditional thermal management materials in thermophysical properties, mechanical properties, and density. Among the fourth-generation thermal management materials, diamond particle-reinforced metal materials, as the ideal heat dissipation materials for the electronic industry can meet the needs of future high-power density electronic devices.

5.2 Optimization and Adaptability Design, Preparation and Modification of High Thermal Conductive Matrix and Components

5.2.1 Preparation and Thermal Conductivity of Gr/Cu Composites

Since the high thermal conductivity of suspended monolayer Gr was reported in 2008, Gr has been considered as a novel reinforcement to improve the thermal conductivity of composites. In fact, there are more factors affecting the intrinsic thermal conductivity of Gr. While the defects are introduced in the preparation process of the Gr composites, the coupling effect of Gr-matrix contact reduces its thermal conductivity. In addition, the distribution of Gr in the composite is a key factor affecting the thermal conductivity of the composite. Thus, the research on the thermal conductivity of Gr-reinforced metal matrix composites is more complicated. Gr is currently reported to have an enhanced effect on the thermal conductivity of polymer matrix composites. While an increase in thermal conductivity has been reported in Gr/Cu heterostructure materials or Gr/Cu thin-film materials, the effect of Gr on the thermal conductivity of metal matrix composites is unclear.

Some reports on Gr/metal matrix composites use the addition of Gr oxide to prepare the composites, and the functional groups of Gr oxide reduce the intrinsic thermal conductivity. Besides, the preparation of composites by mechanical mixing of Gr is reported, which can damage the structure of Gr. This problem can be avoided by using an in situ catalytic growth of Gr on Cu substrates. It has been shown that monolayer Gr can be grown on Cu foil by using polymethylmethacrylate (PMMA) as a solid carbon source [1]. Such in situ growth of Gr on Cu powder has also been shown to be feasible [2]. In this research [3], Gr/Cu-based composites were prepared by in situ growth of Gr on Cu flake powders and hot-press sintering.

The advantage of in situ catalytic growth of Gr is to avoid the use of Gr oxide with low structural integrity and Gr structural damage during the dispersion process and improve the interface bond strength between Gr and metal matrix. Moreover, the highly aligned Gr in such laminated composites would maximize the contribution of the high in-plane thermal conductivity of Gr to the thermal conductivity of the composite [4]. The thermal conductivity and thermal expansion coefficient of the laminated composites were investigated in the in-plane and cross-sectional directions. The interfacial thermal resistance and thermal expansion coefficient were analyzed according to some theoretical models. We also evaluated the thermal mismatch strain between Gr/Cu composites and semiconductor silicon.

5.2.1.1 Gr/Cu In Situ Composite Method

The spherical Cu powder was placed in a stainless-steel ball mill jar, and the agate ball was used as the grinding ball with a mass ratio of about 1 : 8. Anhydrous ethanol was used as the ball mill medium, and the spherical Cu powder underwent plastic deformation and was transformed into flake Cu powder. The reduced Cu flake powder annealed at 573 K under 5% H_2/Ar mixed atmosphere was added to the PMMA–anisole solution, mixed well with thorough stirring, centrifuged, and then the lower PMMA-coated Cu flake powder was removed and dried. The in situ growth process was carried out in a tube furnace. The dried PMMA/Cu flake powder was placed in an alumina crucible under H_2 and Ar gas flow and rapidly heated to 1173 K. After holding for 1 hour and cooling to room temperature, the in situ grown Gr/Cu composite powder was obtained. The Gr/Cu composite powder was assembled in a graphite mold and hot pressed to obtain the densified Gr/Cu composite. The whole preparation process is shown in Figure 5.3. For comparison, pure Cu samples without Gr were prepared under the same conditions.

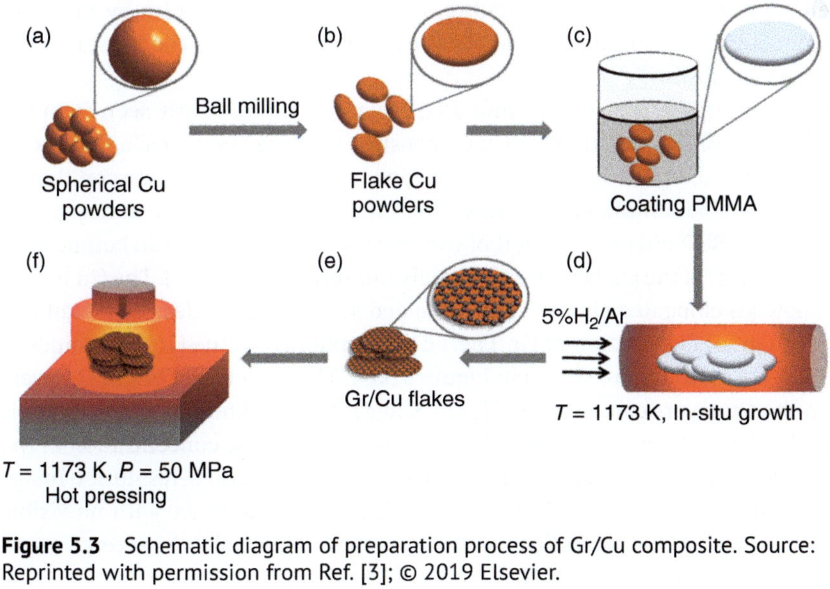

Figure 5.3 Schematic diagram of preparation process of Gr/Cu composite. Source: Reprinted with permission from Ref. [3]; © 2019 Elsevier.

After in situ growth at high temperatures, relatively transparent wrinkled Gr uniformly covers the Cu surface (Figure 5.4a–c). Gr was grown on both surfaces of the flaky Cu powder. It can be observed in the SEM images that the grain boundaries of the flaked Cu are visible under the grown Gr. The number of layers of the grown Gr can be controlled by adjusting the concentration of the carbon source used. The thickness of Gr depends on the concentration of the carbon source used to start the cladding, and the thickness of Gr layers on each surface is about 3–10 nm. Based on the thickness of a single layer of Gr ~0.34 nm, the number of layers of Gr prepared by this method can be estimated to be about 10–30 layers.

The structural quality of Gr in composite powders and Gr obtained from composites by chemical etching was characterized by Raman spectroscopy (Figure 5.4d). The relative intensity between the D- and G-peaks in the Raman spectrum (I_D/I_G) reflects the quality of the Gr, with higher ratios indicating a potentially higher defect density. As shown in Figure 5.4d, the ratios I_D/I_G for the three powders are 0.9, 0.85, and 0.84, respectively, which are lower than most Gr oxide or reduced Gr oxide [5]. Based on the collected Raman spectra, the defect density N_D in Gr can also be quantified according to equation (5.1) [6]:

$$N_D(\mathrm{cm}^{-2}) = \frac{(1.8 \pm 0.5) \times 10^{22}}{\lambda^4} \left(\frac{I_D}{I_G} \right) \tag{5.1}$$

where λ is the laser wavelength. The defect density increases linearly with the ratio of I_D/I_G, and the calculated N_D values of the prepared Gr are $20.3 \pm 5.6 \times 10^{10}\ \mathrm{cm}^{-2}$, $19.2 \pm 5.4 \times 10^{10}\ \mathrm{cm}^{-2}$, and $18.9 \pm 5.3 \times 10^{10}\ \mathrm{cm}^{-2}$, respectively. The higher concentration of carbon sources leads to more defects in Gr. The X-ray photoelectron spectra (XPS) of the prepared Gr/Cu composite powders are shown in Figure 5.4e,f. The high-resolution energy spectral splitting curves of C1s peak and Cu2p confirm the formation of Gr by sp^2 hybridization of carbon atoms formed after the high-temperature catalytic decomposition of the coated PMMA film by the flaky Cu powder substrate, and there is no chemical bonding between the in situ grown Gr and the Cu substrate.

As shown in Figure 5.5a–c, the laminated structure can be clearly seen from the fracture surface of the prepared Gr/Cu composites. The prepared Gr/Cu flaky powders are stacked along the hot-pressing direction, and Gr is located between the Cu and Cu layers. The laminated structure along the hot-pressing direction was also confirmed by EBSD characterization of the cross-section of the Gr/Cu composites, and the thickness of the Cu layer was relatively uniform (Figure 5.5c). The Cu matrix grains show an elongated shape. This laminated structure provides the conditions for the sequential distribution of Gr. The TEM morphology of the Gr/Cu composite and the Gr–Cu interface are shown in Figure 5.5d–f. The presence of bright Gr layers between the Cu layers can be seen (Figure 5.5d,e). The thickness of the Gr layer is measured to be ~10 nm but can be controlled by adjusting the concentration of the carbon source used for different composites (Figure 5.5e). The high-resolution image of the Gr–Cu interface (Figure 5.5f) shows a clear Gr–Cu interface with no visible impurities, voids, and gaps, indicating the formation of adhesion between Cu and Gr by mechanical intercalation [7].

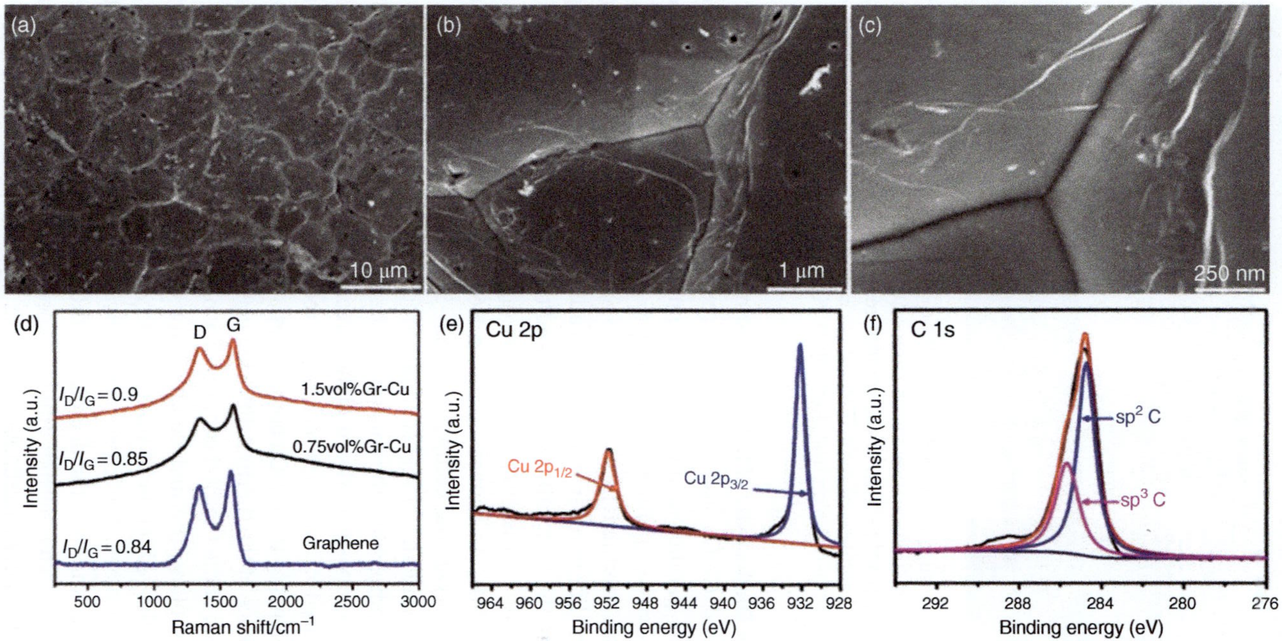

Figure 5.4 (a–c) SEM images of 0.75 vol% Gr/Cu composited powders. (d) Raman spectra and (e, f) X-ray photoelectron spectroscopy (XPS) of Gr/Cu composites. Source: Reprinted with permission from Ref. [3]; © 2019 Elsevier.

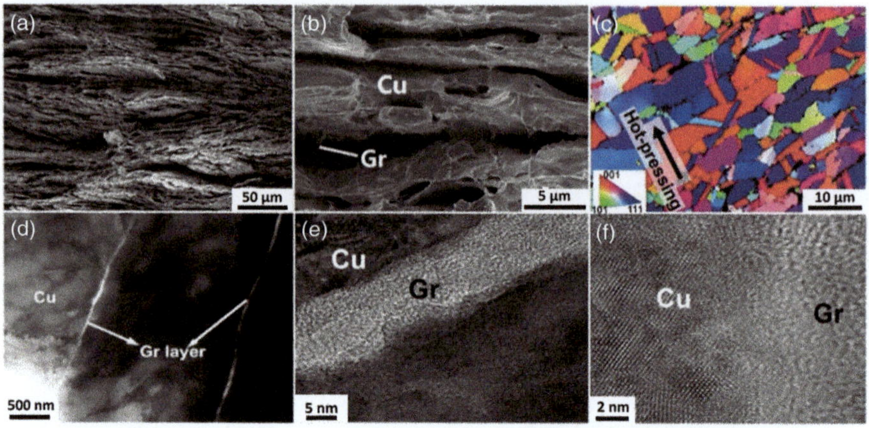

Figure 5.5 Microstructures of the bulk 0.75 vol%-Gr/Cu-laminated composites. (a, b) SEM images of the fractured cross-section of the composite with different magnifications. (c) EBSD image of the cross-section of the composite. (d) TEM image of morphology for the Gr/Cu composite. (e, f) High-resolution TEM images of the Gr–Cu composite interface. Source: Reprinted with permission from Ref. [3]; © 2019 Elsevier.

5.2.1.2 Thermal Conductivity of Gr/Cu Micro–Nano-Laminated Composites

For the laminated Gr/Cu composites, the thermal conductivity was measured in two directions parallel to the thermal compression direction (cross-sectional thermal conductivity $\kappa_{\perp}$) and perpendicular to the thermal compression direction (in-plane thermal conductivity $\kappa_{//}$). Figure 5.6a shows the $\kappa_{//}$ and $\kappa_{\perp}$ for Gr/Cu composites with different Gr contents. Because of the introduction of possible impurities during the ball milling process and the non-complete densification, the thermal conductivity of the pure Cu comparison sample is lower than the standard value of ~390 W/mK. As shown in Figure 5.6a, the addition of Gr reduces the thermal conductivity of the composite by 20–50 W/mK in the in-plane direction and 60–130 W/mK in the cross-sectional direction compared to the Cu matrix. In the prepared composites, the maximum thermal conductivity obtained for the 0.75 vol%-Gr/Cu composite in the in-plane direction was 304 W/mK, while the minimum TC measured in the cross-sectional direction was 154 W/mK. The thermal diffusion coefficients of the Gr/Cu composites are shown in Figure 5.6b, indicating that the dependence of the thermal diffusion coefficient on the Gr content is similar to the dependence of the thermal conductivity on the Gr content. As the Gr content increases, the in-plane thermal conductivity of the Gr/Cu composites initially increases due to the reduction of voids (<0.75 vol% content). After the Gr content is higher than 0.75 vol%, the cross-sectionally thick Gr induces high interfacial thermal resistance, and thus the decrease in thermal conductivity decreases as the Gr content increases. The thermal diffusion coefficient and specific heat determine the thermal conductivity simultaneously. With the introduction of Gr, the specific heat of the composite increases, but the decrease in the thermal diffusion coefficient is significant, causing a decrease in the in-plane thermal conductivity of the composite. The thermal conductivity in the cross-sectional direction of the composite is

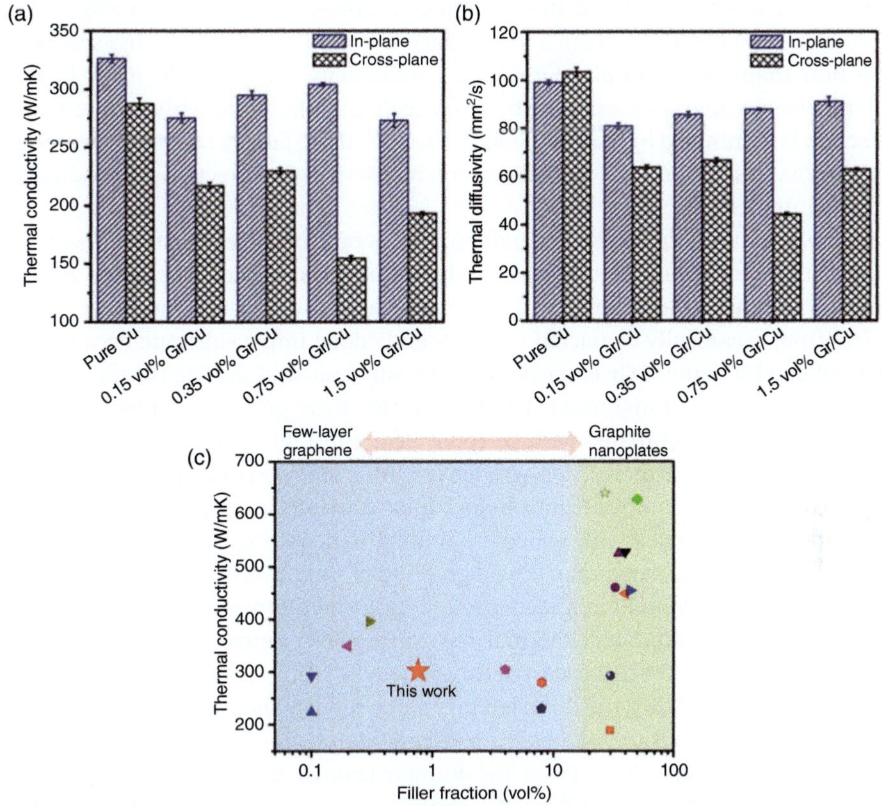

Figure 5.6 (a) Thermal conductivity and (b) thermal diffusivity in the in-plane and cross-plane directions of the Gr/Cu composites with a different content Gr. (c) Comparison of thermal conductivity of Gr/Cu composites in this work with graphite/Cu composites in previously reported work versus filler fraction. Source: Reprinted with permission from Ref. [3]; © 2019 Elsevier.

influenced by a combination of Gr voids, interfacial content, and interfacial thermal resistance. With the introduction of low content of Gr, the thermal conductivity of Cu matrix composites is lower than that of pure Cu material.

The reduction in the in-plane thermal conductivity of Gr/Cu-laminated composites can be attributed to three reasons. First, the grain refinement and more dislocations in the Cu matrix through the doping of Gr lead to stronger electron scattering [8]. Second, although catalytically grown Gr usually has a higher quality than Gr derivatives with terminal functional groups (e.g. Gr oxide), several factors reduce the intrinsic thermal conductivity of in situ grown Gr, such as the growth fold, layer thickness, and Gr–Cu matrix interaction. Third, despite the good interfacial bonding between Gr and Cu, interfacial thermal resistance is always present [9], which reduces the thermal conductivity of the composite. Anisotropic thermal conductivity was observed in the Gr/Cu-laminated composites. The in-plane thermal conductivity was higher than the interfacial thermal conductivity. Due to the two-dimensional structural characteristics of Gr, phonon scattering is stronger in the cross-sectional direction due to weaker interlayer van der Waals forces in the interlayer direction, resulting in a lower thermal conductivity in the cross-sectional direction than in the in-plane thermal conductivity. In addition, more Gr/Cu heterogeneous interfaces exist in the cross-sectional direction, while the in-plane direction is dominated by Cu grain boundaries. All these factors lead to high interfacial thermal resistance and low thermal conductivity in the direction perpendicular to the laminated structure.

Figure 5.6c summarizes the thermal properties of composites reinforced with Gr or graphite nanosheets reported in the literature. Suspended Gr exhibits higher in-plane thermal conductivity than scaled graphite or graphite nanosheets, but it differs from organically contacted Gr or Gr embedded into a substrate because the thermal conductivity of Gr is sensitive to the substrate and defects (lateral dimensions, wrinkles, and roughness) [10]. In contrast, flake graphite or nanoflake has stable morphology compared with Gr (poor flexibility), so its thermal conductivity is more stable. Furthermore, high-volume fractions of Gr generate pores during the densification of the composite, but higher filler fractions are more easily achieved in composites reinforced with scaled graphite or nanosheets. In addition, Gr has a higher specific surface area than scaled graphite or nanosheets, leading to a higher Gr/metal interface content and higher total interfacial thermal resistance.

To evaluate the interfacial thermal resistance (ITR) between Gr and Cu, calculations were performed based on three models [10, 11], namely the acoustic mismatch model (AMM), the diffusion mismatch model, and the H–J model. In the corresponding composites, κ_m is 287.6 W/mK and κ_c is 154 W/mK. κ_g depends on the defect density (N_D). From the Raman results, the calculated N_D ranges from 18.9×10^{10} to $20.3 \times 10^{10}\ cm^{-2}$, resulting in a κ_g of ~400 W/mK. Taking 0.75 vol% Gr as an example, the values calculated by them are extreme boundary conditions and the actual values of ITR are between $2.12 \times 10^{-9}\ m^2K/W$ (DMM) and $38.5 \times 10^{-9}\ m^2K/W$ (AMM) according to the assumptions of AMM and DMM. The H–J model takes into account the effect of actual size and Gr volume content, the calculation is closer to the actual value, and the value of ITR calculated by this model is $3.01 \times 10^{-9}\ m^2K/W$.

Therefore, most composites reinforced with scaled graphite or nanosheets have higher filler fractions and higher thermal conductivity than Gr/metal matrix composites (as shown in Figure 5.6c). However, the high-volume graphite usually leads to poor mechanical properties of the composites. Gr is grown in situ on the surface of Cu sheets, and the percentage of Gr in Gr/Cu composites is less than 1.5 vol%. Due to the limited Gr content, the inherent reduced thermal conductivity of Gr embedded in Cu [12], and the microstructural defects of the bound Gr negatively affect the thermal conductivity of the Gr/Cu composites.

5.2.1.3 Coefficient of Thermal Expansion of Composite Materials

Figure 5.7a shows the CTE of 0.75 vol%-Gr/Cu composites and pure Cu matrix measured in cross-sectional and in-plane directions. The laminated structure of the composites and the two-dimensional structural features of Gr cause anisotropy in the CTE of Gr/Cu composites in both directions. In the in-plane direction, although Gr

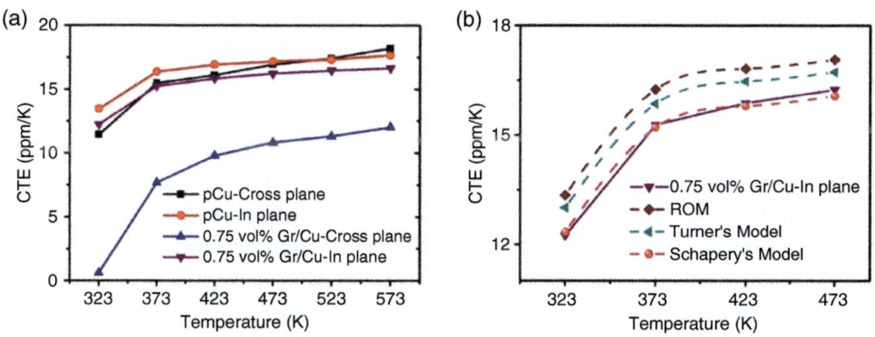

Figure 5.7 (a) Coefficient of thermal expansion (CTE) of 0.75 vol% Gr/Cu composites and Cu matrix as a function of temperature, and (b) comparison of CTE for Gr/Cu composites with theoretical models. Source: Hong Guo, Zhongnan Xie, DingBang Xiong.

or graphite have a negative CTE (Gr: −2 ppm/K, graphite: −1 ppm/K), the CTE of the composites is not significantly different from that of the pure Cu matrix compared to pure Cu, and the in-plane CTE of the 0.75 vol% Gr/Cu composites is slightly lower by ~1 ppm/K. In contrast, the cross-sectional direction CTE decreases sharply to 12 ppm/K with the addition of only 0.75 vol% Gr in the cross-sectional direction.

In the Gr/Cu composite system, three models (Role of Mixture, Turner model, and Schapery model) are commonly used to explain the variation of CTE of the composite with temperature. The Role of Mixture (ROM) neglects the interaction between the filler and the matrix, according to which the CTE(α) of the composite is calculated by the following equation:

$$\alpha_c = \alpha_m V_m + \alpha_g V_g \tag{5.2}$$

where V is the volume fraction, and the subscripts c, m, and g refer to the composite, matrix, and Gr, respectively. The calculated results for ROM are higher than the experimental values.

The Turner model [13] is based on an assumption that only uniform strains are present in the composite. The CTE is derived by using an equilibrium of the internal average stresses and the expression is

$$\alpha_{\mathrm{c}} = \frac{\alpha_{\mathrm{m}} V_{\mathrm{m}} K_{\mathrm{m}} + \alpha_{\mathrm{g}} V_{\mathrm{g}} K_{\mathrm{g}}}{V_{\mathrm{m}} K_{\mathrm{m}} + V_{\mathrm{g}} K_{\mathrm{g}}} \tag{5.3}$$

where V is the volume fraction and $K = E/3(1-2v)$ represents the bulk modulus (E is Young's modulus and v is Poisson's ratio) [13].

The Schapery model [13] considers the stress interactions between the components, and the CTE is derived by employing the thermoplastic principle. As predicted by the Schapery model, the in-plane CTE of the laminated composite is

$$\alpha_{\mathrm{c}} = \frac{\alpha_{\mathrm{m}} V_{\mathrm{m}} E_{\mathrm{m}} + \alpha_{\mathrm{g}} V_{\mathrm{g}} E_{\mathrm{g}}}{V_{\mathrm{m}} E_{\mathrm{m}} + V_{\mathrm{g}} E_{\mathrm{g}}} \tag{5.4}$$

where E is Young's modulus, V represents the volume fraction, and α is the coefficient of thermal expansion.

Comparing the CTE of the 0.75 vol%-Gr/Cu composite with the values predicted by the three models mentioned above (shown in Figure 5.7b), it can be seen that the experimental results are closer to the values calculated by the Schapery model. As shown in Figure 5.7a, the CTE in the cross-sectional direction of the composite is 12 ppm/K, which is significantly lower than the in-plane CTE and is smaller than the CTE of Cu (18 ppm/K) and the CTE in the cross-sectional direction of graphite (28 ppm/K). The measured CTEs in the cross-sectional direction of Gr/Cu composites do not comply with the mixing law. A similar phenomenon was observed in graphite/Cu composites [14]. As reported in graphite/Cu-laminated composites [14], the mechanical and thermal interactions between graphite and Cu matrix were simulated by elastic theory, and the CTE of graphite embedded in the Cu matrix was calculated to be negative (−26 ppm/K). According to elastic theory, the coefficient of thermal expansion of graphite is

$$\alpha_{33} = \frac{d\varepsilon_{33}}{dT} = \alpha_{gr,3} + \upsilon_{2D}\Delta\alpha_x - \frac{d\upsilon_{2D}}{dT}\varepsilon_{11}^0 \tag{5.5}$$

where $\upsilon_{2D} = -2C_{13}/C_{33} = -0.83$ is the Poisson's ratio of the two-dimensional material with temperature derivative $dv_{2D}/dT = -4.3 \times 10^{-2}\ K^{-1}$ [15]. The α_{33} and ε_{33} are the thermal expansion coefficient and strain of graphite embedded in Cu in the cross-sectional direction. The $\alpha_{gr,3} = 28$ ppm/K is the intrinsic thermal expansion coefficient of graphite in the longitudinal direction, where $\Delta\alpha_x = 13$ ppm/K and $\varepsilon_{11}^0 \approx -10^{-3}$. The α_{33} can be deduced to be −26 ppm/K. The decrease of CTE in the cross-sectional direction of the composite is caused by the negative thermal expansion coefficient in the cross-sectional direction of Gr.

In contrast, Stephanie Reich's group pointed out in a recent study that neutron scattering experiments demonstrated that graphite has a positive thermal expansion coefficient in the Z-axis direction [16]. Therefore, a new mechanism for the reduction of the thermal conductivity in the Z-axis direction of the composites by Gr (alkene) is proposed. In the embedded metal matrix, the graphite folds and breaks (generates macroscopic strain) during the cooling process after sintering during preparation. As the temperature increases, graphite crystals are stretched to their original shape, reducing their total thickness and the length of the composite along the Z-axis. As a result, the thermal expansion coefficient of graphite (alkene) in the cross-sectional direction is negative, reducing the thermal expansion coefficient of the composite.

Compared to metals with higher CTE (when they are used in heat sinks), the 0.75 vol%-Gr/Cu composite in-plane orientation should induce less thermal strain in the semiconductor due to its relatively low cross-sectional CTE. To evaluate the difference, devices were prepared by bonding a single-crystal slice of Si (001) to the 0.75 vol%-Gr/Cu composite (in-plane orientation) or pure Cu. As shown in Figure 5.8a (the enlarged view is shown in Figure 5.8a with the Si slice perpendicular to the in-plane orientation of the sample), and the Raman spectra (Figure 5.8b,c) of Si were collected by increasing the device temperature and the change in peak position was quantified as thermal strain. The phonon frequency of Si on Cu decreased from 520.5 cm^{-1} for unstressed Si at room temperature to 514 cm^{-1} at 473 K. Most of the shift originated from the anharmonic nature of the vibration and the thermal

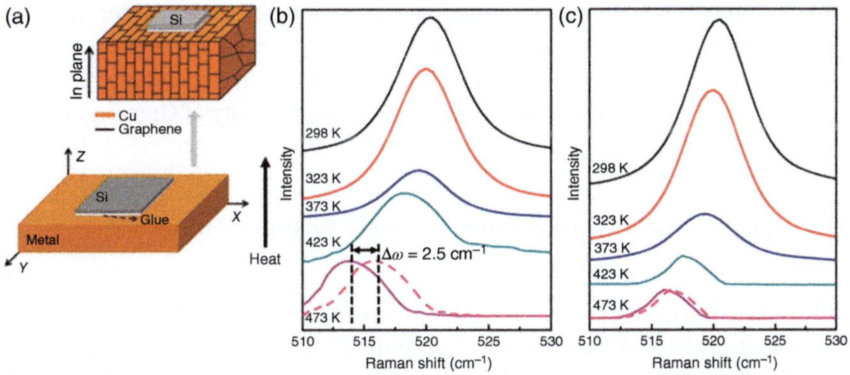

Figure 5.8 Thermal strain at the Si–heat sink interface. (a) Schematic of the Si–heat sink sample and temperature dependence of the Raman mode in Si placed on (b) pure Cu and (c) the 0.75 vol% Gr/Cu composites. The dashed spectrum was recorded on free-standing Si. Source: Reprinted with permission from Ref. [3]; © 2019 Elsevier.

strain caused by the thermal expansion mismatch resulted in $\Delta\omega(473\,\text{K}) = 2.5\,\text{cm}^{-1}$ (Figure 5.8b). In contrast, the frequency shift of $\Delta\omega(473\,\text{K}) = 0.5\,\text{cm}^{-1}$ generated by the device on the 0.75 vol%-Gr/Cu composite is smaller and five times smaller than the Raman peak shift of silicon glued to pure Cu (Figure 5.8c). The phonon spectral frequency shift of the Gr/Cu–Si interface due to thermal strain deduced from the experimental results is $-2.8 \times 10^{-3}\ \text{cm}^{-1}\ \text{K}^{-1}$, which is lower than that of the Cu–Si interface ($-1.1 \times 10^{-2}\ \text{cm}^{-1}\ \text{K}^{-1}$). Based on the phonon frequency shift, we calculated the average thermal stress $\text{d}\sigma/\text{d}T = 0.67$ MPa/K at the Gr/Cu–Si interface, which is significantly lower compared to Cu (2.58 MPa/K). The stresses induced by the Gr/Cu composite (maximum stress of 134 MPa at 473 K) are lower than the critical stress level of silicon in the operating temperature range of 298 to 473 K, while Cu induces a stress of 516 MPa.

In this research, oriented Gr/Cu composites were prepared by in- situ growth of Gr on the surface of micro–nano-sheeted Cu powder using PMMA as a solid carbon source, and further self-assembled and densified by sintering. The performance research showed that the thermal conductivity of the Gr/Cu composites was lower than that of the pure Cu matrix in the studied Gr concentration range (0–1.5 vol%), showing an increasing trend followed by a decreasing trend. The high interfacial thermal resistance between Gr and matrix and the high interfacial content in the composites were the main reasons for the decrease in the thermal conductivity of the composites. Compared with the in-plane direction, the thermal expansion coefficient in the cross-sectional direction is significantly lower, which effectively reduces the thermal stress between the composite and the semiconductor. Although the in situ growth of introduced layer less Gr cannot improve the thermal conductivity of Cu matrix, it can effectively reduce the coefficient of thermal expansion to match the coefficient of thermal expansion of semiconductor and is expected to be a low density, the high thermal conductivity of thermal management material to replace W-Cu and Mo-Cu. This research provides an experimental object and theoretical basis for clarifying the response of Gr quality and content, composite interface, and composite configuration (ordered/disordered dispersion) on the thermal conductivity and thermal expansion coefficient of Gr/Cu composites, and provides a reference for the subsequent design and study of the thermal conductivity of Gr/metal matrix composites.

5.2.2 Preparation and Thermal Conductivity of Graphite/Cu Composites

Graphite flakes (Gf) have excellent thermal conductivity, low cost, and easy processing, and the arrangement distribution of graphite in the composite directly affects the material thermal conductivity. The planar thermal conductivity of Gf has been reported to be higher than 1000 W/mK. Highly oriented pyrolytic graphite (HOPG) has a planar thermal conductivity of 1600–2000 W/mK. In the cross-sectional direction, thermal conductivities of 5–10 W/mK were reported, regardless of the graphite type. In graphite/Cu (Gf/Cu) composites, researchers have proposed several methods such as chemical plating, wet mixing, and high-energy ball milling to achieve the oriented arrangement of graphite in the Cu matrix. However, the preparation of highly oriented graphite/metal composites still requires the development of more efficient methods in terms of the ease of preparation methods and the thermal conductivity of the final composites. Additionally, graphite/Cu composites still have poor interfacial bonding and voids due to the non-wetting between graphite and Cu matrix. Two common methods of alloying graphite and matrix using surface metallization improve interfacial bonding and mechanical properties. However, these methods still face the following challenges: (i) difficulty in selecting the ideal carbide-forming element or interlayer; (ii) difficulty in controlling the thickness and crystallinity of the interlayer; and (iii) difficulty in controlling the interfacial reaction process (carbide formation) in composites. The Gr with intrinsic 2D structure and high planar thermal conductivity can be used as an intermediate layer to improve the thermal conductivity of C/Cu composites. It can bridge phonon transport and avoid direct contact between Cu and graphite. The advantages of Gr as an intermediate layer for interfacial modifications are as follows: (i) the thickness of Gr uniform nanolayer is tunable; (ii) Gr has lower density and higher thermal conductivity compared to carbides or metallic elements mentioned above; and (iii) the interfacial composition is single without interfacial reaction products.

In this research [17], the arrangement distribution of graphite in the composite is designed and controlled based on the morphology control of Cu powder. Considering the geometric morphology of graphite, better (i) highly oriented distribution of flake graphite and (ii) geometric morphology compatibility with flake graphite can be achieved by using flake metal powders as the matrix. In addition, the relationship between the graphite orientation distribution and the thermal conductivity of the composites was quantitatively analyzed. In addition, unlike surface metallization and matrix alloying, in situ grown Gr was used as an intermediate layer to achieve interfacial modification and to improve the thermal conductivity of Gf/Cu composites.

5.2.2.1 Variations in the Intrinsic Thermophysical Properties of Graphite Sheets During the Compounding Process

The SEM morphology of the sieved graphite flakes is shown in Figure 5.9a,b. The graphite flakes have a flake diameter ranging from 450 to 550 μm and a thickness of about 25 μm. The XRD patterns given in Figure 5.9c confirm the physical phase of the original graphite flakes. The diffraction peaks at $2\theta = 26.38°$ and $2\theta = 54.72°$ correspond to the (002) and (004) crystal planes of graphite, respectively. According to the Bragg equation, the d-spacing of the (002) crystal plane of the graphite flake is 3.378 Å. The Raman spectrum in Figure 5.9d shows that there are only typical G and 2D peaks and no D peaks in the graphite sample, indicating that the pristine graphite flake presents high quality. The TEM images in Figure 5.10a,b show the flake structure of graphite, and the high-resolution image clearly shows the lamellar structure of graphite. The SEM images of graphite flake before and after hot pressing and the Raman spectra at different positions are shown in Figure 5.10c–f. By comparing the Raman spectra of graphite flakes before and after hot pressing, the defects introduced into the graphite flakes can be quantified. The results show that the hot-pressing process helps to reduce the I_D/I_G ratio on the edges of the graphite flakes. This indicates that the heat treatment in the Ar gas environment helps to reduce the defects in the graphite flakes at the high temperature, thus improving the intrinsic thermal conductivity of the graphite flakes.

The structural model and surface morphology of the graphite/Cu composite in the X–Y plane and Z plane (perpendicular and parallel to the hot-pressing direction,

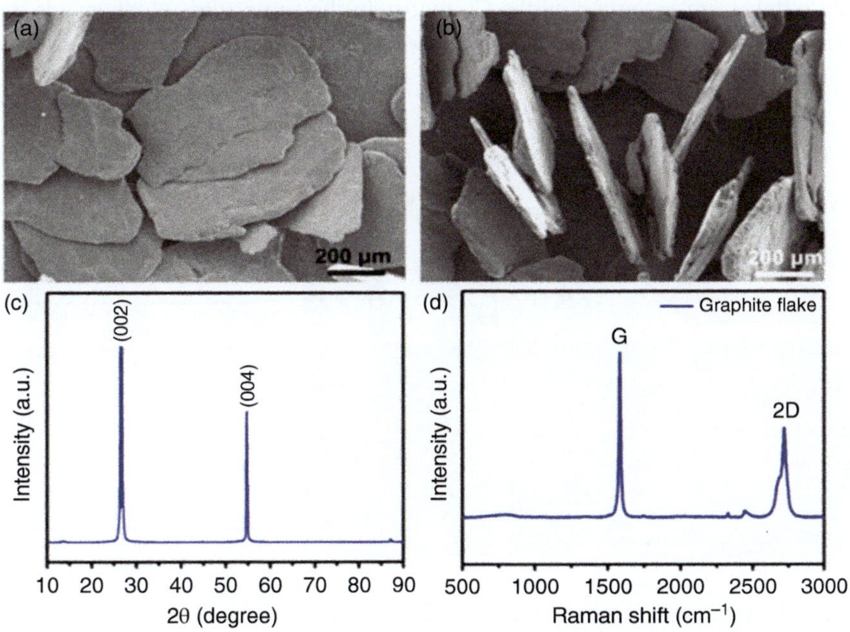

Figure 5.9 (a, b) Morphology of flake graphite. (c) XRD of flake graphite. (d) Raman spectrum of flake graphite. Source: Reprinted with permission from Ref. [17]; © 2020 Elsevier.

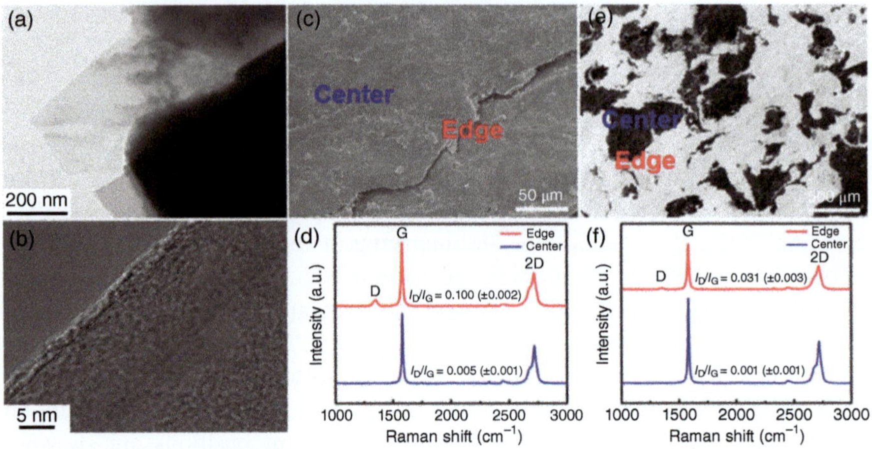

Figure 5.10 (a, b) TEM and HRTEM images of graphite flakes. (c–f) microscopic morphology and Raman spectra of the original graphite flakes (c, d) and graphite flakes (e, f) in the composite after hot pressing in the central and edge regions. Source: Reprinted with permission from Ref. [17]; © 2020 Elsevier.

called X–Y plane and Z plane directions, respectively) are shown in Figure 5.11. In Figure 5.11b,c, the black dark area represents the graphite flake, while the light gray bright area represents the Cu substrate. In the X–Y plane, the large-size graphite flakes are randomly distributed in the Cu matrix. In the Z plane, it is clear that the graphite flakes are separated by the Cu matrix in the composite, which avoids the direct stacking contact between them. As shown in Figure 5.11d, the EDS elemental

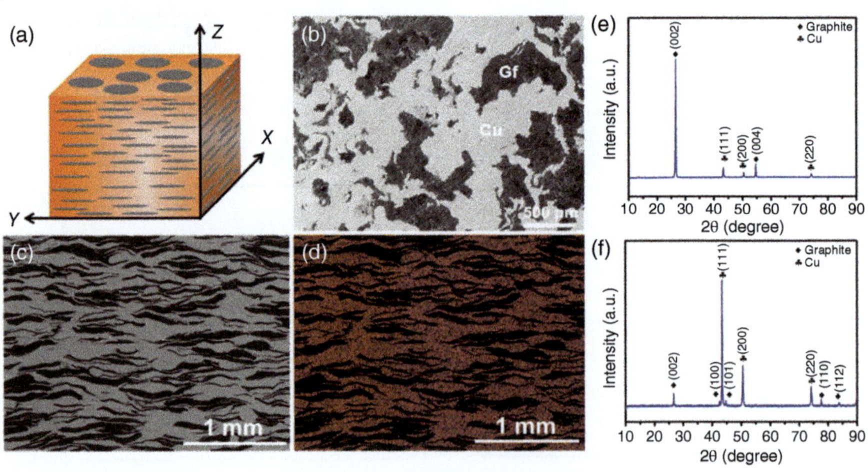

Figure 5.11 (a) Structural model of graphite/Cu composites. SEM images of graphite flake distribution in Cu matrix along (b) X–Y and (c) Z planes. (d) The corresponding EDS elemental distribution of Gf/Cu composites along Z-direction. XRD spectra of Gf/Cu composites in (e) X–Y plane and (f) Z-plane directions. Source: Hong Guo, Zhongnan Xie, DingBang Xiong.

distribution shows Cu and C elemental distributions. The anisotropic characteristics of the Gf/Cu composites were further determined using XRD. In the X–Y plane direction, the XRD of the Gf/Cu composites is shown in Figure 5.11e, where the diffraction peaks detected at $2\theta = 26.5°$ and 54.7° can be attributed to the (002) and (004) crystallographic planes of the hexagonal graphite. In addition, the three significant diffraction peaks corresponding to the (111), (200), and (220) crystal faces of Cu indicate the isotropy of Cu. As shown in Figure 5.11f, four diffraction peaks were detected in the Z plane of the Gf/Cu composite at $2\theta = 26.5°$, 42.5°, 44.6°, and 77.4°, corresponding to the (002), (100), (101), and (110) crystal planes of graphite. Generally, the disappearance of the (004) crystalline plane peak in the Z plane indicates the orientation of graphite in the composite. Moreover, the low intensity (002) diffraction peak in the Z plane (almost 1/35 of the X–Y plane) further confirms the strong orientation structure of the graphite flakes in the composite. The presence of the weak diffraction peak at the Z-plane (002) is attributed to the fact that some of the flake graphite is not perfectly oriented in the matrix and to a small amount of scratch damage of the graphite in the Z-plane.

According to Adams's study, the graphite in-plane thermal conductivity has a more significant correlation with the graphite in-plane spacing d (002). A smaller d (002) leads to a higher graphite in-plane thermal conductivity. Besides, the increased layer plane stacking is associated with the removal of defects (vacancies, gap filling, and impurities). Therefore, the decrease in d (002) indicates an increase in the thermal conductivity of the graphite flakes. For the original graphite flake, the (002) graphite crystal plane was detected at $2\theta = 26.38°$ (Figure 5.9c). After hot pressing, the diffraction peak of the (002) crystalline plane of graphite was detected at $2\theta = 26.5°$ in the X–Y plane (Figure 5.11e). According to the Bragg equation, the (002) crystal plane spacing of the graphite flakes was reduced from 3.378 Å (pristine graphite) to 3.363 Å after hot-pressing under argon atmosphere. Based on the relationship between the in-plane thermal conductivity and d (002) found by Adams, the in-plane thermal conductivity of the graphite flakes in the composites after hot-pressing was estimated to be close to 1020 W/mK. While the in-plane thermal conductivity of the pristine graphite flakes was about 500 W/mK, which is consistent with the Raman measurements, and the lower defects indicate an increase in the thermal conductivity of the composite. A. Morvan et al. proposed a reduction treatment of graphite flakes at 400 °C, and the graphite in-plane thermal conductivity was increased to 1100 W/mK. The (002) crystal plane spacing of the treated graphite flakes was reduced to 3.360 Å. The use of other methods, such as chemical pretreatment to remove adhering contaminants on the graphite surface, can improve the thermal conductivity. In this chapter, the hot-pressing process at a high temperature (1173 K) and high pressure (50 MPa) can improve the graphite thermal conductivity, which may be due to the improved graphite crystallinity and reduced defects.

5.2.2.2 Orientation Modulation of Graphite Sheets in Composites

To regulate the orientation distribution of graphite flakes in the Cu matrix, the effect of the morphology and size of the Cu matrix powder on the graphite flake

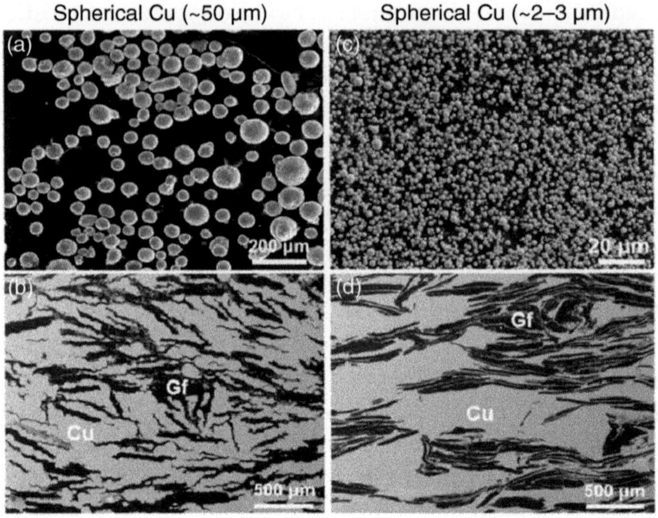

Figure 5.12 Original morphology of spherical Cu powder as matrix and graphite distribution along Z-axis in 50 vol%-Gf/Cu composites. (a, b) Spherical Cu powder with a diameter of about 50 µm. (c, d) Spherical Cu powder with a diameter of about 2–3 µm. Source: Reprinted with permission from Ref. [17]; © 2020 Elsevier.

distribution was investigated. First, spherical Cu powders with diameters of about 50 and 2–3 µm were used as the matrix. The original morphology and graphite distribution of the Cu powder in the composites are shown in Figure 5.12. In the case of spherical Cu (~50 µm) as the matrix, the large size of Cu powder as the matrix resulted in the almost irregular orientation of graphite flakes in the composites (Figure 5.12a,b). As shown in Figure 5.12c,d, when smaller spherical Cu powder (~2–3 µm) is used as the matrix, the orientation of graphite is improved, but the graphite flakes accumulate in some regions and show the poor distribution in the composites. The results indicate that spherical powder as a matrix cannot achieve a uniform highly oriented distribution of graphite flakes in the composites.

Considering the morphological compatibility, the orientation of graphite flakes in the composites was therefore changed using flake Cu powder. Also, the effect of the thickness of the flake Cu powder on the graphite orientation was investigated. For comparison, three types of flake Cu powders with average thicknesses of 20–30 µm, 10–20 µm, and 1–2 µm were used. As shown in Figure 5.13, the flaky Cu powder enables a highly oriented arrangement of graphite. Moreover, the thicker flake Cu powder (20–30 µm, 10–20 µm) resulted in a smaller spacing between two graphite flakes (marked in Figure 5.13b,d) than in the thinner flake Cu powder (1–2 µm) (marked in Figure 5.13e). The larger spacing at the same content of graphite indicates a better alignment of graphite in the composite due to the smaller projected cross-section. Thus, the flake Cu powder (Figure 5.13) effectively resulted in a highly oriented alignment of graphite in the composite compared to the spherical Cu powder as the matrix. The flake Cu powder has a larger specific surface area and better morphological compatibility with the graphite flakes during the mixing process and

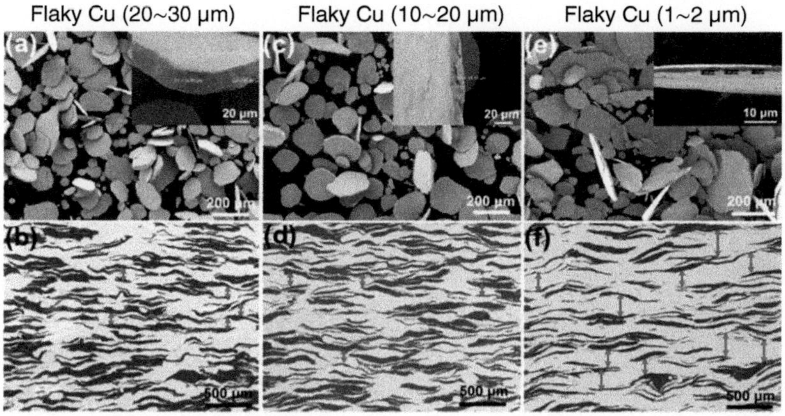

Figure 5.13 Original morphology of ball-milled flake Cu powder and graphite distribution along the Z-axis in 50 vol%-Gf/FCu composites. The thicknesses of the flake Cu powder are (a, b) ~20–30 μm, (c, d) ~10–20 μm, and (e, f) ~1–2 μm. Source: Reprinted with permission from Ref. [17]; © 2020 Elsevier.

hot-pressing, which contributes to the uniform dispersion and highly oriented distribution of graphite in the composites.

Polarization Raman spectroscopy was used to characterize the graphite alignment in the composites. The schematic diagram of the measurement equipment and the test results are shown in Figure 5.14. The Raman experimental setup is shown in Figure 5.14a. The in-plane orientation of the fractured composites is in the X–Y plane, while the polarization varies in the X–Z plane. The variation of G-peak intensity with polarization angle for randomly oriented and perfectly aligned graphite flakes is shown in Figure 5.14b. Theoretically, for the random distribution of graphite (red curve), the intensity of the G-peak is constant when the polarization angle ε is changed. For perfectly aligned graphite (blue curve), the intensity of the G-peak $I_{tot} \propto f^2\cos^4\varepsilon$ is maximum at $\varepsilon = 0$, 180°, and 360°, and zero at $\varepsilon = 90°$ or 270°. Figure 5.14c,d shows the variation of the normalized intensity of the graphite flake G-peak in the composite measured with different Cu powders as a matrix with respect to the polarization angle. The intensity of the G-peak is maximum when the polarization angle $\varepsilon = 0$, 180°, and 360°, while it is minimum when the polarization is perpendicular to the X–Y plane. This qualitatively verifies the orientation direction of graphite in the Cu matrix as observed with SEM. As shown in Figure 5.14c, comparing the G-peak intensity ratio (I_{min}/I_{max}) of the graphite flakes in the composite, the results indicate that the graphite in the flake Cu powder has the best orientation arrangement. Furthermore, the graphite in the Cu flakes with a thickness of about 1 to 2 μm showed the best orientation compared to the other two thicker Cu flakes as the matrix (Figure 5.14d). Furthermore, it is clear that the Gf/Cu composites with different thicknesses of Cu flakes as matrix show small differences in orientation and the difference between the maximum and minimum values is less than 0.06 (~6%). The thickness of flake Cu powder has a small effect on the orientation of graphite flakes in graphite/Cu composites with flake Cu powder as the matrix. Therefore, the highly oriented arrangement of

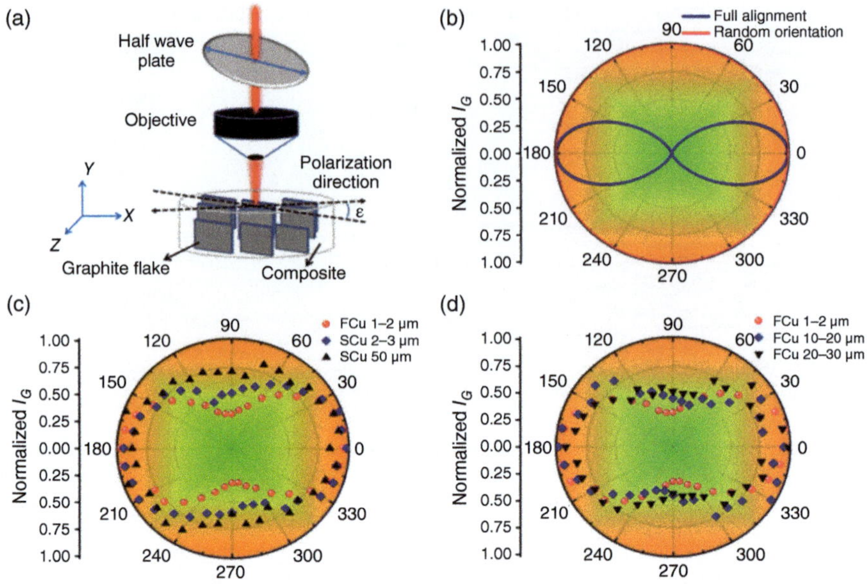

Figure 5.14 (a) Schematic diagram of the measurement of polarization Raman spectra. (b) G-peak intensity patterns of randomly and perfectly oriented graphite flakes. (c, d) Raman intensity of graphite G-peaks versus the angle (ε) between the polarization direction of light and the in-plane direction of 50 vol% Gf/Cu composites. (c) The flake Cu (1–2 μm) and spherical Cu powder with a diameter of about 50 μm and 2–3 μm. (d) The flake Cu powder (1–2 μm, 10–20 μm, and 20–30 μm) as composite matrix. Source: Reprinted with permission from Ref. [17]; © 2020 Elsevier.

graphite flakes in graphite/Cu composites can be achieved by using flake copper powder as the matrix and hot-pressing.

5.2.2.3 Effect of Graphite Sheet Orientation on the Thermal Conductivity of Graphite/Cu Composites

The thermal conductivity of Gf/Cu composites in the X–Y plane ($\kappa_{//}$) and Z plane ($\kappa_{\perp}$) directions are shown in Figure 5.15. The composites with sheet Cu (1–2 μm) as matrix show the highest $\kappa_{//}$(520.1 ± 9.4 W/mK) and the lowest $\kappa_{\perp}$(78.6 ± 1.4 W/mK) in 50 vol%-Gf/Cu composites. 60 vol%-Gf/Cu composites show high $\kappa_{//}$(570.2 ± 13.9 W/mK) and low $\kappa_{\perp}$(66.9 ± 1.7 W/mK). For all Gf/Cu composites, $\kappa_{//}$ is higher than $\kappa_{//}$ due to the anisotropic thermal conductivity and orientation distribution of the graphite flakes. Figure 5.15b shows the anisotropy factor ($\kappa_{//}/\kappa_{\perp}$) of the composites, where the high graphite sheet orientation (<$\cos^2\theta$>) and volume fraction lead to a high anisotropy factor. The maximum anisotropy factor (8.52) is obtained at 60 vol% Gf content.

5.2.3 Preparation and Thermal Conductivity of Graphite/Gr/Cu Composites

In addition to the graphite arrangement in the Gf/Cu composites, the interfacial bonding between graphite and Cu also has an important effect on thermal

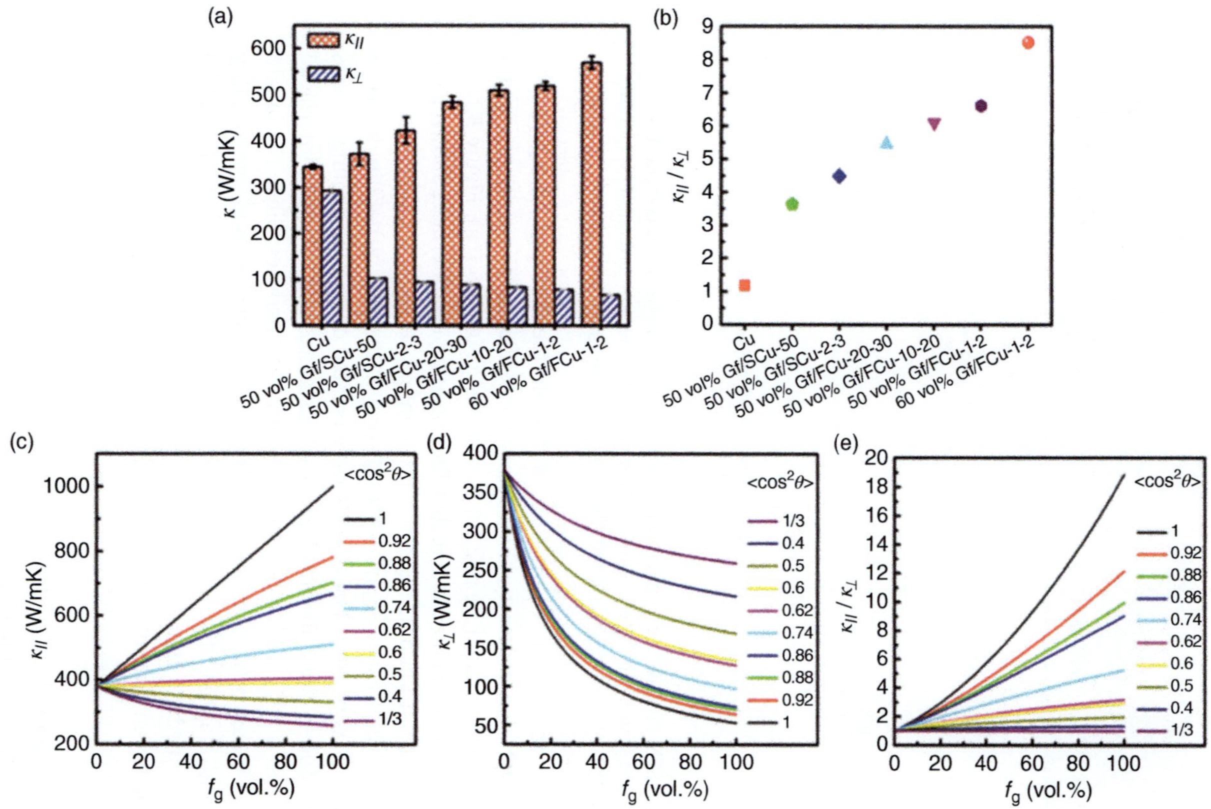

Figure 5.15 Anisotropic thermal conductivity of Gf/Cu composites. (a) Thermal conductivity of different Gf/Cu composites in the X–Y plane direction ($\kappa_{//}$) and Z plane direction ($\kappa_{\perp}$). (b) Anisotropy factor ($\kappa_{//}/\kappa_{\perp}$) of thermal conductivity of different composites. (c, d) Changes of (c) $\kappa_{//}$ and (d) $\kappa_{\perp}$ with fg of the Gf/Cu composites calculated by EMA in different $<\cos^2\theta>$. (e) The change of anisotropy factor ($\kappa_{//}/\kappa_{\perp}$) with fg by EMA calculated at different orientations. Source: Hong Guo, Zhongnan Xie, DingBang Xiong.

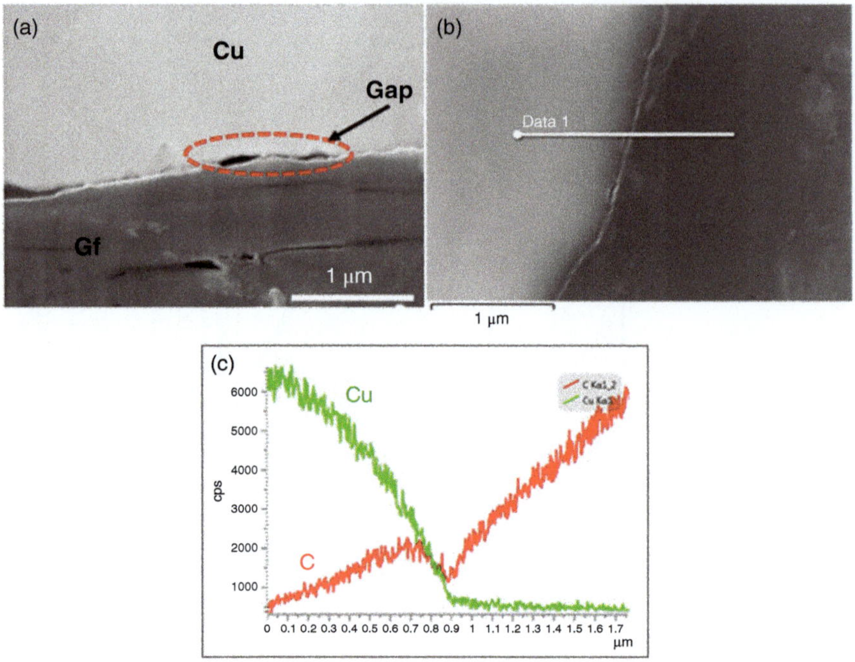

Figure 5.16 (a) SEM images of the microscopic morphology of the Gf/Cu composite interface. (b, c) The results of the cross-interface elemental line scan area and elemental distribution. Source: Hong Guo, Zhongnan Xie, DingBang Xiong.

conductivity. In the Gf/Cu composites, although the hot pressing gives the composites a high dense density, the observed interface is discontinuous and a significant gap between graphite and Cu is observed (Figure 5.16a). Figure 5.16b,c represents the results of elemental line distribution at typical interfaces showing the discontinuous distribution of C elements at the interface in the Gf/Cu composites due to the presence of gaps (Figure 5.16c). It proves that the wettability between graphite and Cu is very poor. These voids were created during the cooling process after sintering due to the different degrees of shrinkage and the poor interfacial bonding between the Cu matrix and graphite in the composites. Since the in situ grown Gr and Cu matrix have good interfacial bonding and also good compatibility between Gr and graphite, the in situ grown Gr in the previous section may be used as an intermediate transition layer to solve the problem of non-wetting at the interface between graphite and Cu matrix. For this reason, we carried out the design preparation and thermophysical properties study of the graphite/Gr/Cu composite system.

The surface morphology of the in situ grown Gr/Cu powder is shown in Figure 5.17a,b. Relatively thin Gr films with some folds are found on the Cu surface. The Raman spectra of the in situ grown Gr/Cu powder are shown in Figure 5.17c. The D- and G-peaks in the Raman spectra of the Gr/Cu powder confirm the structure of the in situ grown Gr, and the presence of the D peak indicates the presence of defects in the Gr. The microstructure of the Gf-Gr-Cu composite interface is shown

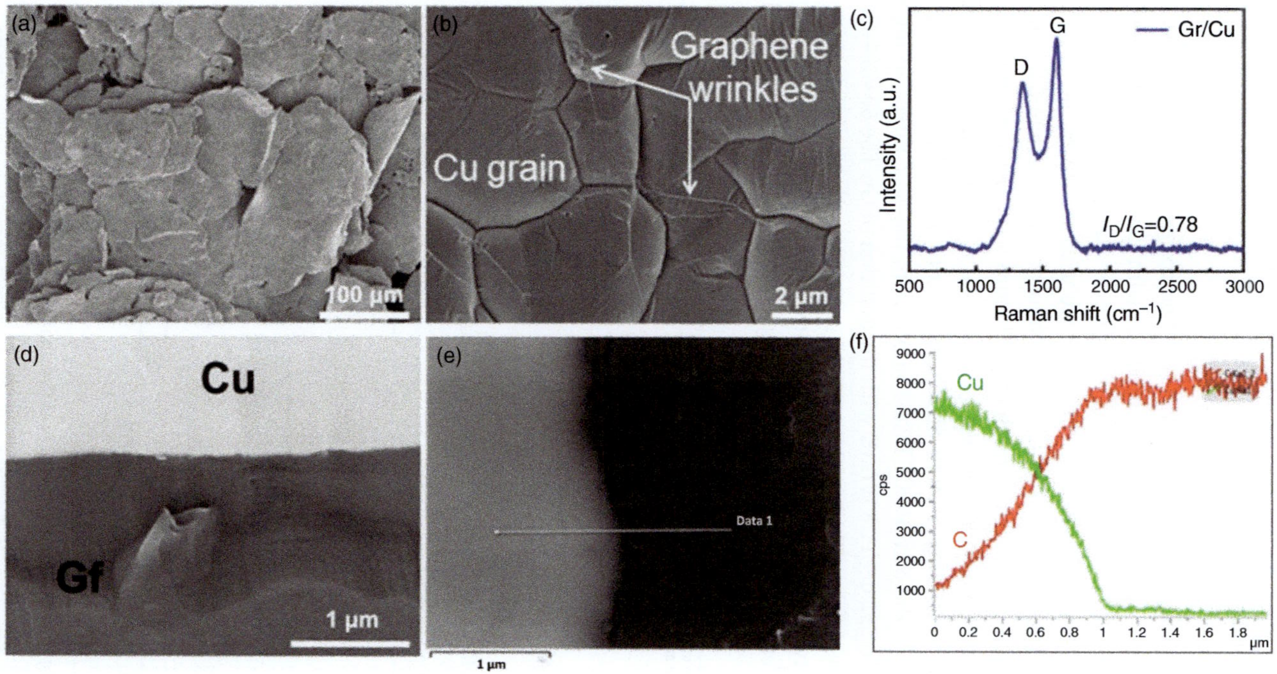

Figure 5.17 (a, b) SEM images of the in situ grown Gr/Cu (flake Cu, 1–2 μm) composite powder. (c) Raman spectra of the in situ grown Gr/FCu powder. (d) SEM images of the microscopic morphology of the Gf-Gr-Cu composite interface. (e, f) Gf-Gr-Cu cross-interface elemental line scan regions and elemental distribution results. Source: Reprinted with permission from Ref. [17]; © 2020 Elsevier.

in Figure 5.17d. By introducing the in situ grown Gr on the flake Cu, apparently tight contact and the continuous interface is observed with no obvious gap between Cu and graphite (Figure 5.17d). The results of elemental line distribution on the interface indicate that the Gf-Gr-Cu composite interface shows a continuous elemental distribution (Figure 5.17e,f).

5.2.3.1 Thermal Conductivity of Graphite/Gr/Cu Composites

The thermal conductivity values of the composites with spherical Cu powder and relatively thick flake Cu powder as the matrix are lower than those of the composites with 1–2 μm flake Cu as the matrix. Therefore, we used the flake Cu with a thickness of 1–2 μm as the matrix. The Gf/Gr/Cu composites were prepared by the same preparation method after in situ growth of Gr. Figure 5.18 shows the thermal conductivity of Gf/Gr/FCu composites, as well as Gr/Cu and Cu composites prepared by sheet powder metallurgy in both directions. Figure 5.18a shows that the thermal conductivity of the 50 vol%-Gf/FCu composite in the X–Y direction increases from 467.5 ± 9.7 to 552.6 ± 11.8 W/mK with the addition of Gr. The 60 vol%-Gf/FCu composite increases from 570.2 ± 13.9 to 597.1 ± 20.9 W/mK. The improvement in thermal conductivity can be attributed to the following factors. Since Cu does not wet the graphite, defects such as tiny voids and air gaps between Gfs and Cu matrix are

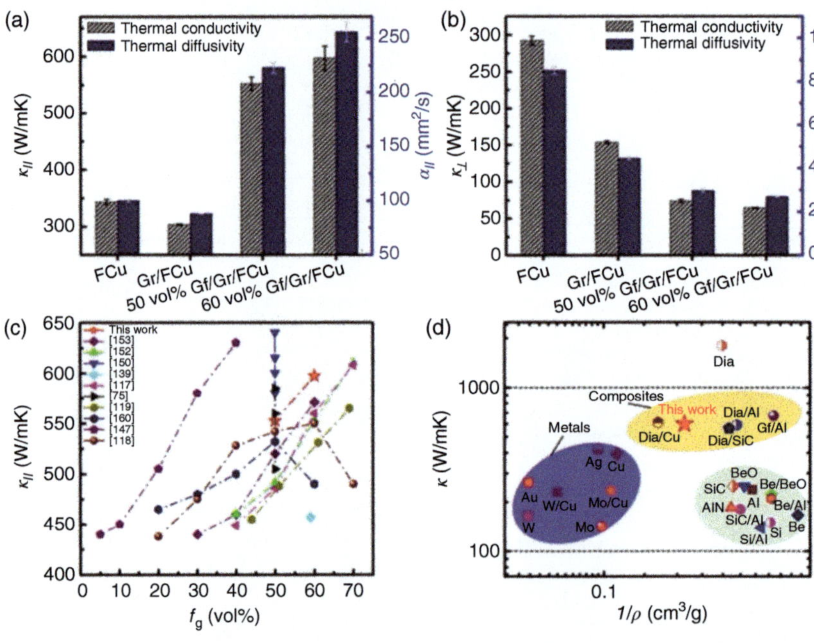

Figure 5.18 (a, b) Thermal conductivity and thermal diffusion coefficient of Cu, Gr/FCu and Gf/Gr/FCu composites (a) in the X–Y plane direction ($\kappa_{//}$, $\alpha_{//}$) and (b) in the Z plane direction ($\kappa_{\perp}$, $\alpha_{\perp}$). (c) Comparison of the thermal conductivity of Gf/Gr/FCu composites with that of Gf/Cu composites reported in the literature. (d) Comparison of the composite-specific thermal conductivity of the composites compared with that of materials commonly used in electronic packaging. Source: Hong Guo, Zhongnan Xie, DingBang Xiong.

unavoidable after sintering, which causes phonon scattering and is extremely detrimental to the heat transfer capability. When a small amount of Gr is added, the thermal conductivity of the in situ grown Gr is higher than that of the void and air layers, replacing the defects during hot-pressing sintering, reducing phonon scattering and improving the interfacial thermal conductivity. The increment of thermal conductivity of the composites after Gr modification is about 30 W/mK, which is much lower than that after carbide modification of diamond/Cu composites due to the higher increment of relative density of diamond/Cu composites (~4%). On the one hand, interfacial modification in Gf/FCu composites has no significant enhancement in the relative density. Although some voids still exist in the unmodified Gf/FCu composites, the relative density is close to 99%, which leads to the higher thermal conductivity of the Gf/FCu material. On the other hand, when the orientation of graphite in the Cu matrix is high ($<\cos^2\theta \gg 0.9$), the low interfacial thermal resistance induced by interfacial modification will only lead to a lower increment of thermal conductivity (30–50 W/mK) because the thermal conductivity of the composite is mainly determined by the orientation of graphite rather than the interfacial thermal resistance. When the graphite orientation is low, larger thermal conductivity increments can be achieved by reducing the interfacial thermal resistance. Interfacial modification with 0.35 vol% Gr still has a positive effect on TC. However, the too-thick Gr layer will lead to higher interfacial thermal resistance and reduce the thermal conductivity of Gf/Gr/FCu composites. The $\kappa_{//}$ for 60 vol%-Gf/Gr/FCu is close to the theoretical value (640 W/mK) due to the improved interfacial bonding. The difference between the experimental and theoretical values can be attributed to the following factors. On the one hand, the poorly crystalline graphite region at the Gr/graphite interface exacerbates the scattering of phonons, leading to a decrease in the heat transfer capacity. On the other hand, the residual Gr layer in the Cu matrix will produce phonon scattering, and more or less reduce the thermal conductivity of the composite as studied in the previous chapter. Figure 5.18b shows the thermal conductivity of the composite in the Z-direction. It is clear that the thermal conductivity in the Z-axis direction is significantly lower than that in the X–Y plane direction for the same graphite content due to the high graphite orientation. Also, the thermal conductivity and thermal diffusivity in the Z-axis direction decrease with increasing graphite volume fraction. In the Z-direction, the degree of thermal conductivity change is not significant even though the interfacial bonding is improved.

5.2.3.2 Thermal Expansion Coefficient of Graphite/Gr/Cu Composites

Figure 5.19 shows the variation of the relative length (a, b) and the corresponding CTE (c, d) for four types of composites with temperatures ranging from 25–300 °C in the X–Y plane and Z-direction. The comparison of the thermal expansion curves in the X–Y plane direction shows that the trends of the relative length and the corresponding CTE with temperature are similar for the four composites with 50 vol% graphite flake content as shown in Figure 5.19a,c. This implies that the degree of bonding at the graphite–Cu interface has almost no effect on the thermal expansion behavior of the composites in this direction. This is attributed to the specific morphology of graphite with a high diameter-to-thickness ratio. In addition to the

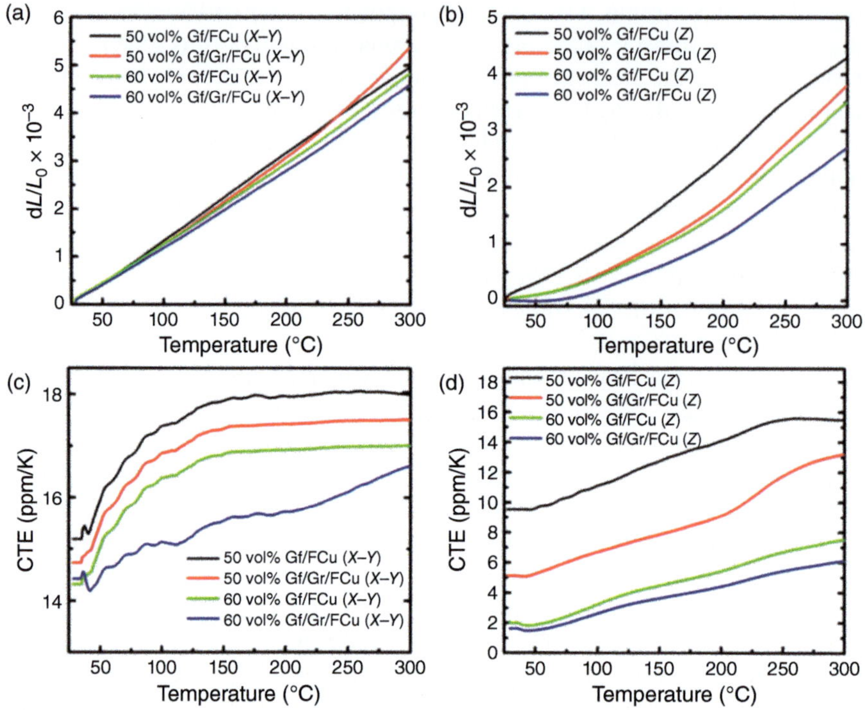

Figure 5.19 Curves of the coefficient of thermal expansion (CTE) corresponding to relative length (a, b) and with temperature (c, d) for four types of 50 vol%-graphite/Cu composites in the temperature range from 25 to 300 °C (a, c: in the X–Y plane direction; b, d: along the Z plane direction). Source: Reprinted with permission from Ref. [17]; © 2020 Elsevier.

volume fraction of graphite, the contact area between Cu and graphite has an influence on the CTE of the composites. The surface area of graphite in the thickness direction is only less than 10% of the total surface area. The effective contact area between Cu and graphite is so small in this direction that the suppression of Cu expansion by graphite is very small. Furthermore, the addition of the Gr layer to the Cu matrix does not change the effective contact area at the interface between Cu and graphite in the X–Y direction, so the four types of composites exhibit similar thermal expansion behavior in this direction.

In the Z-direction, both the relative length and the corresponding CTE of the composites at the same temperature decreased with the addition of in situ grown Gr in the set temperature range, as shown in Figure 5.19b,d. For example, as the added Gr content increases from 0 to 0.75 vol%, the CTE of the 50 vol%-Gf/Cu composite at 200 °C gradually decreases from 14 to 9 ppm/K. The further comparison of the CTE of composites with that of graphite (28 ppm/K in the Z-plane) and Cu (18 ppm/K) shows that the former is not as expected, and is between the latter two. The CTE in the Z-direction is less than each of the latter two, which is contrary to the mixing rule of composite materials ($CTE_{composite} = CTE_{Cu} V_{Cu} + CTE_{graphite} V_{graphite}$, $V = 0.5$ in this experiment). During the CTE measurements, the microcrystals of graphite are stretched to their original shape as the temperature increases, thus reducing the total thickness of graphite and the length of the composite along the Z-plane. As a result, the thermal expansion coefficient of Gr in the Z-plane is negative, thus decreasing the thermal expansion coefficient of the composite.

Interface modification by in situ grown Gr and thermal properties regulation in highly oriented graphite/Cu composites. By controlling the shape and scale of the Cu powder, the effective adjustment of the orientation of the graphite flakes in the Cu matrix is achieved. Spherical Cu powder leads to the disordered and non-uniform distribution of graphite flakes in the matrix, and flaky Cu powder is conducive to the uniform distribution and high orientation of graphite flakes in the matrix. When the thickness of the flaky Cu is 1–2 μm, the orientation of the graphite flakes in the matrix is as high as 0.92, which is close to the theoretical value of 1. The Gr is further in situ grown on the flaky Cu. The interface modification and improvement of wettability make the in-plane TC (parallel to the direction of graphite flake) of the highly oriented 60 vol%-graphite/Cu composite increase to 597 W/mK, and CTE along the longitudinal direction decrease to 6 ppm/K.

5.3 Formation and Evolution Rules of High Thermal Conductive Interface and Its Control Method

5.3.1 Theoretical Calculation of High Thermal Conductive Interface Design

Matrix alloying and surface metallization are the two main ways of interfacial modification of diamond/Cu matrix composites. When matrix alloying is used, the interfacial layer needs to be formed by the diffusion of alloying elements, and the reaction of alloying elements with the diamond in sintering. When the number of alloying elements added is small, carbides preferentially form at the active sites on the diamond surface. As the number of alloying elements added increases, carbides grow and expand along the longitudinal direction, resulting in a continuous and thicker interfacial layer. In the case of interfacial modification by matrix alloying, the regulation of the microstructure of the interfacial layer can only be achieved by optimizing the sintering process and the addition of alloying elements, which affects the densification behavior of the composite and the thermal conductivity of the matrix metal. Therefore, it is difficult to guarantee the effect of the regulation of the microstructure of the interfacial layer. The introduction of a transition layer by metallization of the diamond surface is not only efficient and easy to regulate the interfacial modification, but also can avoid the damage of the modified elements on the thermal conductivity of Cu matrix to a large extent. This has received more attention in the research of the interfacial modification of diamond/Cu matrix composites.

To solve the problems of incompatibility and high interfacial thermal resistance of diamond/Cu matrix composites, the choice of carbide-forming elements such as B, Cr, Mo, Si, Ti, W, and Zr for interfacial modification has become a common choice. At present, more extensive research has been carried out for diamond surface metallization, and the plating methods involve various processes such as chemical vapor deposition (CVD), physical vapor deposition (PVD), molten salt method, and diffusion method. The modified elements involve W, Mo, Cr, Ti, Si, B, Zr, etc. The thickness of the plating layer is mostly in the micron or submicron scale. In contrast, the research on matrix alloying is mainly focused on alloying elements such as B, Cr, Ti, and Zr. In general, on the one hand, the introduction of interfacial modified layers significantly improves the densities of diamond/Cu matrix composites, which are less influenced by the diamond content and can basically reach more than 95%. On the other hand, the interfacial modified layers effectively reduces the interfacial thermal resistance, and the thermal conductivity of diamond/Cu matrix composites can generally reach more than 600 W/mK, and a few can even reach 800 W/mK.

The interfacial modification of diamond/Cu matrix composites by matrix alloying and diamond surface metallization has become the main means to solve the diamond/Cu interfacial incompatibility problem, improve the interfacial bonding, and reduce the interfacial thermal resistance. At present, there are various types of elements used for diamond/Cu interfacial modification, the thickness of the interfacial modified layer varies, and the factors affecting the final composite properties in the experiments are complicated. Even if the same interfacial modified elements are chosen, the thermal conductivity of the resulting diamond/Cu matrix composites may vary greatly. Therefore, it is difficult to judge the effectiveness of the various elements used for interfacial modification in improving the thermal conductivity of diamond/Cu matrix composites based only on the results reported in the literature. Although the purpose of optimizing the composition and thickness of the interfacial modified layer and determining the best interfacial modified layer state can also be achieved by a large number of orthogonal experiments, the experiments are often interfered with by some uncertainties and are time-consuming and impractical.

Due to the synergistic effect of the intrinsic thermal conductivity and phonon velocity, the variation of different interfacial layers with thickness shows specificity. As shown in Figure 5.20a, the thermal conductivity of the composites decreases with the increase in the plating thickness, and the Cr interfacial layer shows a sharp decrease, while the changes of the W, Mo, Ni, and Si interfacial layers are relatively moderate. The Cr, W, Si, Mo, and Ni interfacial layers can all improve the thermal conductivity of the composites to some extent when the thickness is below 750 nm. The enhancement effect of Cr is more significant when the thickness is below 200 nm, while the Ti and B interfacial layers play a slightly positive role only when the plating is very thin. The Ag and Zr interfacial layers are not conducive to the improvement of thermal conductivity.

The phonon velocity of the carbide interfacial layer is higher than that of the Cu matrix, so the introduction of the carbide interfacial layer is beneficial to enhance the interfacial phonon coupling and thus improve the interfacial thermal conductivity. As shown in Figure 5.20b, all carbide interfacial layers are beneficial to improve the composite thermal conductivity at plating thicknesses below 200 nm, and the composite thermal conductivity is insensitive to the plating composition at plating

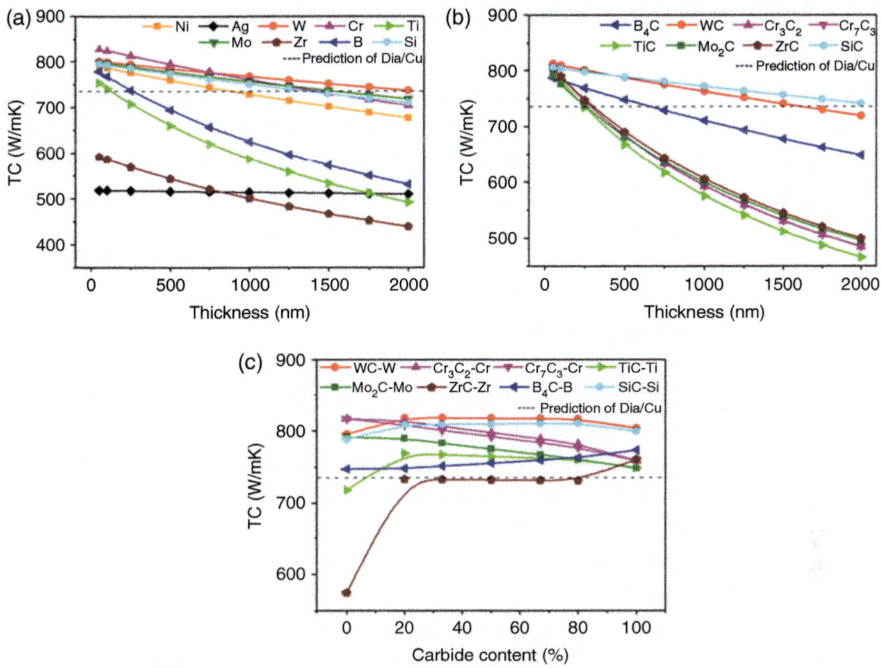

Figure 5.20 Thermal conductivity of diamond–Cu composites in different interfacial layer states. (a) Variations of thermal conductivity with plating thickness when the interfacial layer is non-carbide. (b) Variations of thermal conductivity with plating thickness when the interfacial layer is carbide. (c) Variations of thermal conductivity with plating thickness when the interfacial layer is carbonized. Source: Hong Guo, Zhongnan Xie, DingBang Xiong.

thicknesses as thin as 50 nm. However, with the increase in the plating thickness, the carbide interfacial layers (e.g., ZrC, TiC, Mo_2C, Cr_3C_2, Cr_7C_3) mostly decreased sharply due to the low intrinsic thermal conductivity, while the trend of WC, SiC, and B_4C was relatively moderate. At thicknesses higher than 100 nm, WC and SiC interfacial layers have a clear advantage.

When carbide-forming elements are selected for interfacial modification, carbonization of the interfacial layer is inevitable. In the case that the intrinsic thermal conductivity of carbide is not too low, partial carbonization can combine the high phonon velocity of carbide and the high thermal conductivity of metal to further improve the material thermal conductivity. As shown in Figure 5.20c, the W-WC, Si-SiC, Cr-Cr_3C_2/Cr_7C_3, Mo-Mo_2C, and B-B_4C interfacial layers can play a positive role at the full carbonization level. The Ti-TiC interfacial layer will improve the material thermal conductivity only after the carbonization transformation occurs. For the Zr-ZrC interfacial layer, this interfacial layer is not conducive to thermal conductivity improvement unless it is completely carbonized. With carbonization, the Cr-Cr_3C_2/Cr_7C_3 and Mo-Mo_2C interfacial layers decrease sharply, B-B_4C improves slightly, and W-WC, Si-SiC, and Ti-TiC are almost unaffected by the degree of carbonization.

5.3.2 Study on Interface Regulation of Chromium-Modified Diamond/Cu Composites

When selecting the matrix alloying elements of diamond/copper composites, the following factors are usually considered. First of all, the selected element should be the element easy to form carbide. The alloy elements should have a certain solid solubility in copper. And carbide itself should have high thermal conductivity. The interfacial layer with high thermal conductivity is conducive to reducing the interfacial thermal resistance. The phonon spectrum of carbide should be well matched with copper matrix and diamond. It is beneficial to reduce the acoustic impedance between diamond and copper matrix.

Physical parameters of common carbide-forming elements are shown in Table 5.1. It can be seen from the table that Cr carbide has an intrinsic thermal conductivity similar to ZrC and TiC. But it has higher solid solubility and lower Gibbs free energy in copper. Element Cr is easier to react with C atom to form carbide. At the same time, high solid solubility can provide sufficient Cr atoms for the interface reaction, which is conducive to controlling the interface reaction.

Diamond/Cu-Cr composites with different interface structures were prepared by pressure infiltration. Generally, the diamond/copper composites interface area includes diamond interface area, carbide area, and copper matrix diffusion area. As shown in Figure 5.21, EDS analysis was conducted for the interface area of diamond/copper composites with 0.5 wt% Cr element addition. The carbide layer of chromium is in strip shape. The interface on both sides of carbide is parallel to diamond and matrix, respectively. In order to determine the composition of the interface carbide layer, the exfoliated diamond particles were characterized by XRD. The results are shown in Figure 5.21c. Chromium carbide mainly exists in the form of Cr_3C_2, and there is no diffraction peak corresponding to other chromium carbides.

The interface structure is closely related to the amount of chromium added to the matrix. Therefore, matrix alloys with different Cr concentrations are used to regulate

Table 5.1 Physical parameters of common carbide-forming elements.

Materials	Density (g/cm^3)	Thermal conductivity (W/mK)	Solid solubility (wt%)	Atomic radius (pm)
Cu	8.90	398	—	145
Diamond	3.52	1300–1500	—	67
B4C	2.52	67	0.05	87
TiC	4.93	17	6	176
Cr3C2	6.68	19	0.73	166
Cr7C3	6.92	19.1	0.73	166
ZrC	6.73	21	0.17	206
WC	14.90	120	0	193

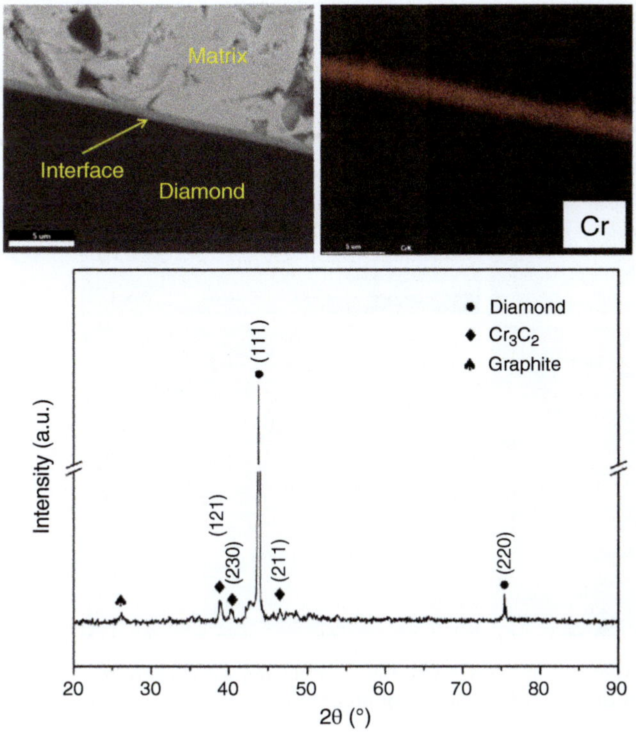

Figure 5.21 Composition analysis of diamond/Cu-Cr composites interface area. Source: Hong Guo, Zhongnan Xie, DingBang Xiong.

the interface structure. As shown in Figure 5.22, the average interface thickness increases with the increase in Cr content. According to the line scanning analysis of SEM, the interface layer thickness is 80, 440, 760, and 1100 nm, respectively. When the amount of Cr added is small, as shown in Figure 5.22a, a thinner Cr_3C_2 can be observed at interface. At some positions, the interface products are discontinuous. This seriously reduces the interface bonding strength. When Cr is added more, as shown in Figure 5.22d, the thickness of Cr_3C_2 interface layer even exceeds 1 μm. Due to the difference in expansion coefficient between diamond and carbide, cracks appear locally between them in the high-temperature cooling process. These defects will lead to a sharp increase in interfacial thermal resistance, resulting in a decline in mechanical properties and thermal conductivity of diamond/copper.

The fracture structures of diamond/copper composites with different interface structures are shown in Figure 5.23. With the increase in Cr content, the proportion of diamond transgranular fracture in fracture surface is increasing. When the Cr content is 0.3 wt%, the transgranular fracture of diamond cannot be observed. The partially extracted diamond surface is smooth, which is not enough to provide strong interface bonding. When Cr content increases to 0.5 wt%, more carbide and copper matrix are adhered to the extracted diamond surface. At the same time,

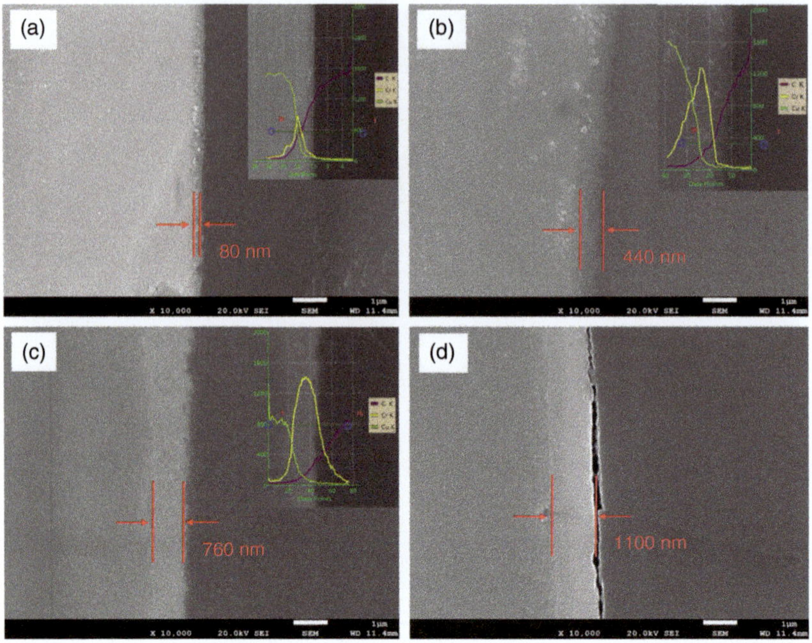

Figure 5.22 Interfacial structures of diamond/copper composites with different thicknesses. Source: Hong Guo, Zhongnan Xie, DingBang Xiong.

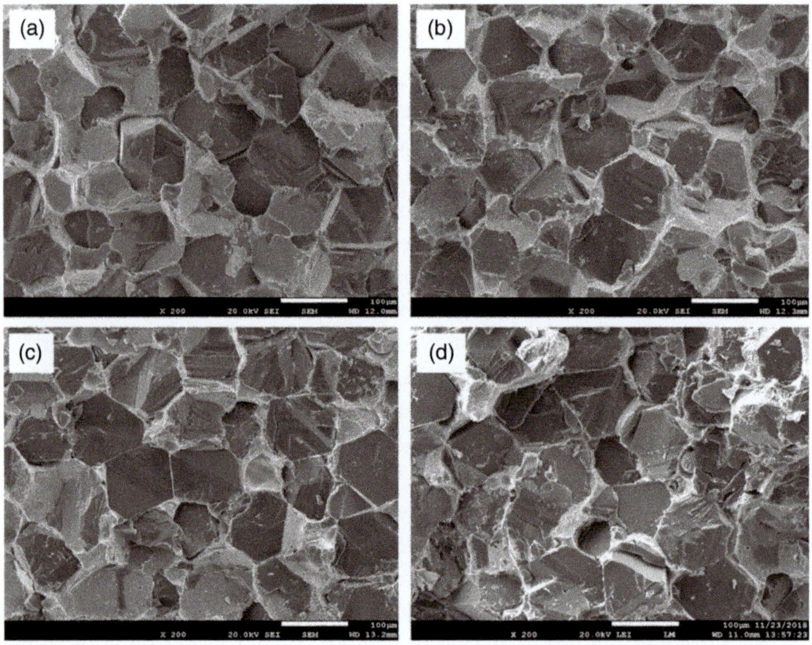

Figure 5.23 Fracture microstructure of diamond/copper composites with different interface structures. Source: Hong Guo, Zhongnan Xie, DingBang Xiong.

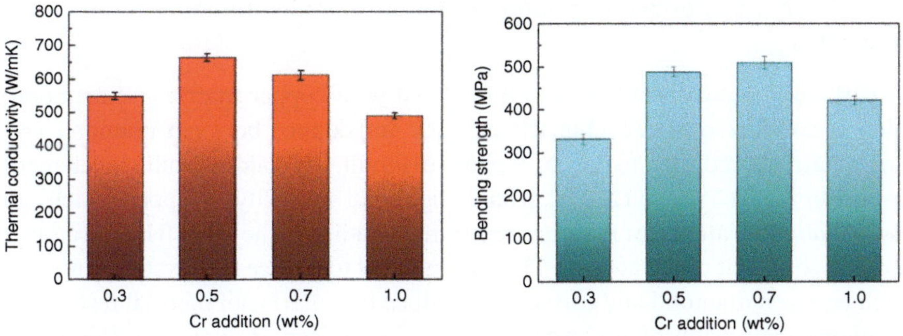

Figure 5.24 Thermal conductivity and bending strength of diamond/copper composites with different interfacial layer thickness. Source: Hong Guo, Zhongnan Xie, DingBang Xiong.

transgranular fracture was observed in some diamonds. When the interface bonding strength is high, the stress concentration will occur at the diamond during the crack propagation process, resulting in the diamond transgranular fracture. Continue to increase Cr content, the fracture structure of the sample is smooth, and the interface bonding between diamond and matrix is strong.

As shown in Figure 5.24, the thermal conductivity of diamond/copper composite varies with the interface layer thickness. With the increase in Cr_3C_2 interface layer thickness, the thermal conductivity of diamond/copper composites first increases and then decreases. The relationship between diamond/copper composites' thermal conductivity and the interfacial layer thickness can be explained by the following reasons. There are phase separation and pores between diamond and copper matrix when the interface layer thickness is small. The defects or voids at the interface will increase the interface thermal resistance and significantly reduce the thermal conductivity. When the interface layer thickness increases to about 440 nm, there is no exposed surface on the diamond surface, and the Cr_3C_2 interface layer closely connects diamond and copper matrix. However, the composites' thermal conductivity began to decrease with the continuous increase of interface layer. This is because the intrinsic thermal conductivity of Cr_3C_2 is low. The increasing interface layer increases the interface thermal resistance. Therefore, in the design of low thermal resistance interface, the thickness of interface layer should be considered as the minimum thickness when there are no exposed defects on the diamond surface.

It is very important to adjust the composition and structure of interface products to improve the interface thermal resistance and interface bonding strength. Appropriate thickness and structure of carbide layer will help to reduce the interfacial thermal resistance and improve the interfacial bonding strength of diamond/copper composites. The interface structure of diamond/copper composites is controlled by Cr, and the optimal thermal conductivity is 665 W/mK. At this time, the thickness of Cr_3C_2 interface layer is 440 nm.

5.3.3 Study on Interface Regulation of Boron-Modified Diamond/Cu Composites

The carbide phonon spectrum is well matched with copper matrix and diamond, which is conducive to reducing the acoustic impedance between diamond and copper matrix. According to the Cu-B phase diagram, the solid solubility of element B in copper is 5 wt% above 1200 °C. B has a high solid solubility in copper, which can provide enough B atoms for interface reaction. In addition, the atom B has an atomic radius close to that of the atom C. This is beneficial to reduce the crystal plane mismatch between diamond and carbide and reduce the interfacial thermal resistance. The phonon spectrum of B_4C was calculated by molecular dynamics. Compared with TiC, ZrC, and other carbides, the phonon spectrum of B_4C matches diamond better in high-frequency band, which is beneficial to lower interfacial thermal resistance.

Diamond/Cu-B composites with different interface structures were prepared by pressure infiltration method. The XRD analysis of diamond/Cu-B composites prepared by Cu-B alloy with 0.5 wt%B element concentration is shown in Figure 5.25. Diamond/Cu-B composites are mainly composed of diamond phase, copper phase, and B_4C phase. XRD shows that the interface carbide of diamond/Cu-B composites is B_4C.

The nucleation and growth of carbide are affected by temperature, boron concentration, and other conditions. At a certain preparation temperature, the boron concentration controls the growth form of carbide, thereby diamond–copper shows different interface structures. The diamond particles surface morphology extracted from diamond/Cu-B composites is shown in Figure 5.26. With the increase of the boron content, first, a nanosized B_4C layer is formed on the diamond surface. With the increasing concentration of B, the structure of B_4C gradually grows from 1 μm dentate structure to 1.5 μm dentate structure. There is not only diffusion bonding between the B4C and the substrate but also physical mechanical occlusion. The above interface structure greatly improves the bending strength of diamond/Cu-B composites and reduces the interface thermal resistance.

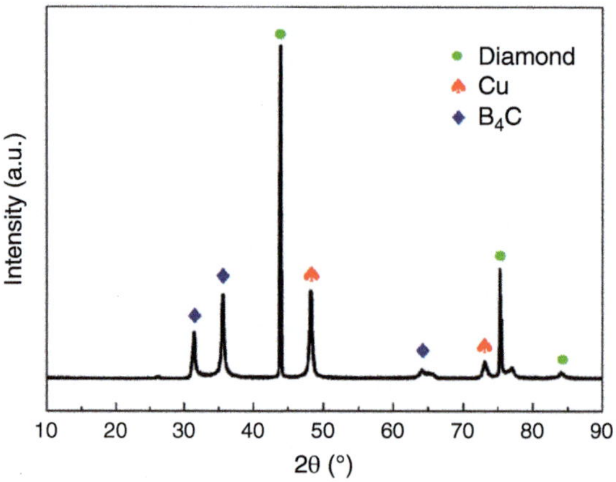

Figure 5.25 XRD analysis of diamond/Cu-B composites. Source: Hong Guo, Zhongnan Xie, DingBang Xiong.

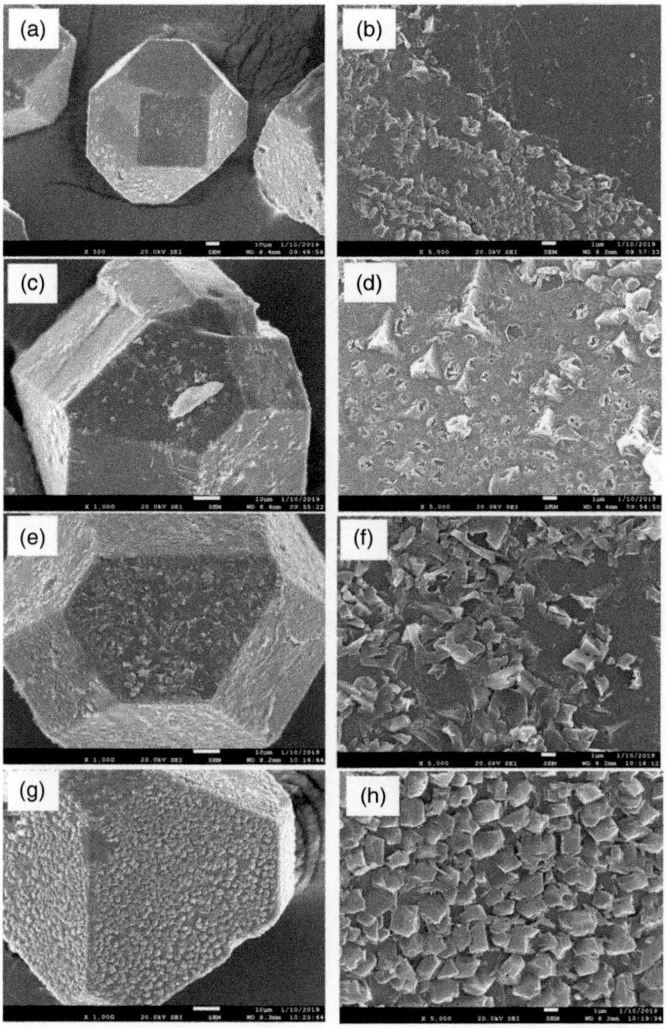

Figure 5.26 Surface morphology of diamond particles extracted from diamond/Cu-B composites. Source: Hong Guo, Zhongnan Xie, DingBang Xiong.

The fracture microstructure of diamond/copper composites with different boron additions is shown in Figure 5.27. With the increase of boron content, the percentage of diamond transgranular fracture in fracture surface is increasing. When the content of boron is lower than 0.5 wt%, the fracture form is mainly diamond extraction. Some carbides are stuck on the extracted diamond surface. Carbide does not form a uniform layer, which is not enough to provide strong interface bonding. When the content of boron is increased to 1 wt%, the diamond basically undergoes transgranular fracture, indicating that the interface adhesion is strong.

The diamond/Cu-B composites' thermal conductivity is shown in Figure 5.28. With the increase of boron content, the thermal conductivity of diamond/Cu-B composites first increased and then decreased. The maximum thermal conductivity

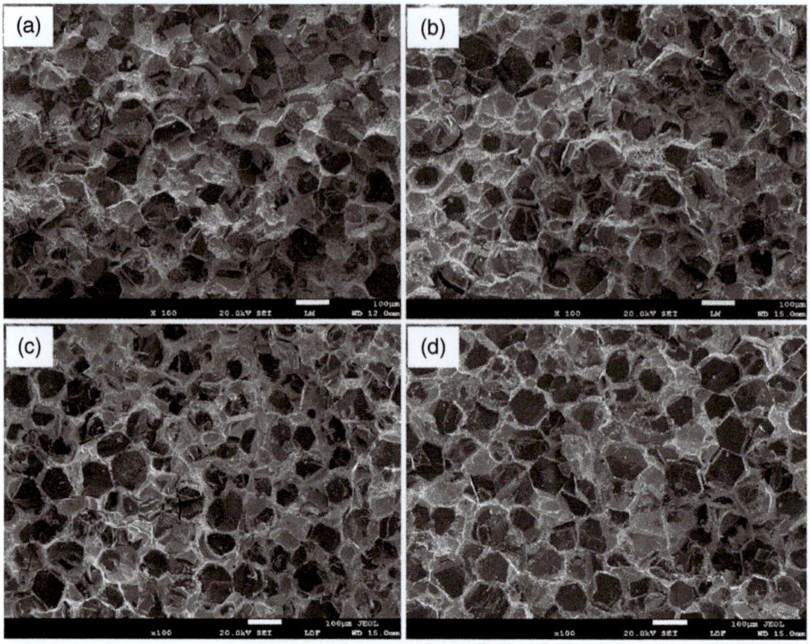

Figure 5.27 Fracture microstructure of diamond/Cu composites with different interface structures. Source: Hong Guo, Zhongnan Xie, DingBang Xiong.

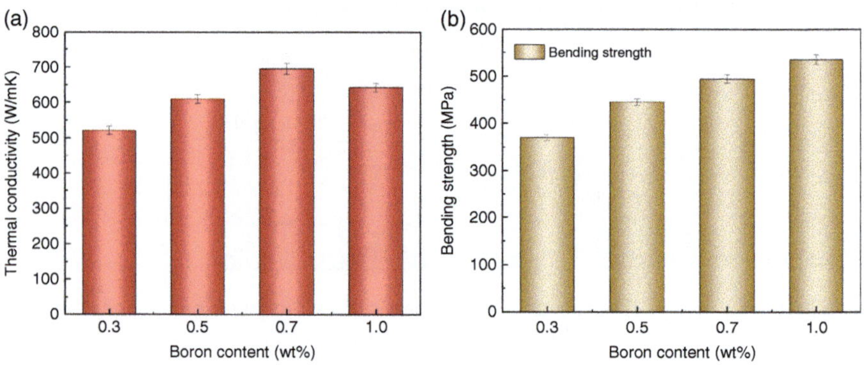

Figure 5.28 Thermal conductivity and bending strength of diamond/Cu-B composites with different interface structures. Source: Hong Guo, Zhongnan Xie, DingBang Xiong.

is 694 W/mK when boron content is 0.7 wt%. The properties of diamond/Cu-B composites are closely related to the interface carbide structure. The larger B_4C grain size increases the interfacial thermal resistance. Therefore, the thermal conductivity of diamond/Cu-B composites will decrease with the further increase of boron content. The bending strength of diamond/Cu-B composites increases with the increase of boron content. With the increase of B_4C thickness, the composite bending strength increases from 368 to 522 MPa. The dentate B_4C extends to the substrate and forms mechanical engagement with the substrate, thus improving the bonding strength between the substrate and B_4C.

5.3.4 Study on Interface Regulation of Gr-Modified Diamond/Cu Composites

As belonging to the above, the matrix alloying and diamond surface metallization are commonly used methods for diamond–Cu interface modification. However, there are challenges in controlling the carbide interlayer. The thermal conductivity of the commonly used metal modification layers and the formed carbides is much lower than that of Cu. In addition, it is difficult to obtain nanosized thickness and uniform distribution by surface metallization. Preferential bonding of carbide-forming elements on some crystalline faces of diamond particles leads to thick and inhomogeneous carbide interlayers in the final composite to achieve an overall coverage of the diamond surface. However, the excessively thick carbide layers with low thermal conductivity can reduce the thermal conductivity of the composite. Moreover, it is difficult to adjust the composition of the carbide interlayer. The coexistence of carbide-forming elements and the formed carbides usually reduce the interfacial thermal conductivity.

The in situ catalytic growth of Gr on the surface of pristine Cu particles avoids the problem of poor wettability between carbon and Cu. This provides strong bonding between Cu and Gr, and Gr is well compatible with its isotope diamond because they have close specific surface energy. Second, the acoustic impedance (Z) of pristine Gr is 42.04×10^6 kg/m^{2}s, similar to that of metal carbides. Defects and stresses can reduce the phonon velocity of Gr, so the Z of in situ grown Gr is less than 42.04×10^6 kg/m^{2}s. Furthermore, the carbide interfacial layer thickness modulation method can easily obtain uniform Gr nanolayers by decomposing the solid carbon source-encapsulated Cu. The thickness can be adjusted by adjusting the concentration of the carbon source compared to the in situ grown Gr. Thus, the in situ grown Gr can improve the interfacial wetting and bridge the acoustic mismatch between diamond and Cu for improving the thermal conductivity of the composite [18].

The method relates to in situ growth of Gr on spherical Cu powder and hot-pressing compound with diamond particles to prepare bulk composites. After annealing and reducing the spherical Cu powder, the solid carbon source PMMA was used to cover the surface of the Cu powder uniformly. The catalytic growth technique was used to convert PMMA in situ to Gr covered on the surface of the Cu powder and used for the preparation of the bulk material in the subsequent study. The specific process is shown in Figure 5.29, and it is similar to the method of growing Gr on flake Cu powder in basic research work. Our experiments confirmed that the in situ growth of Gr on the surface of spherical Cu powder can be controlled by changing the concentration of carbon source, thus controlling the number of layers of the grown Gr and the thickness of the interfacial Gr layer in the final composite, which is conducive to the regulation of the interfacial Gr content in the preparation.

We sintered Cu powder with the diamond in a vacuum hot press furnace after the in situ growth of Gr to obtain the composites. Compared with the diamond/Cu composites without Gr, the diamond/Gr/Cu composites have fewer pores and narrower voids, and the interfacial wetting is improved after the in situ growth of Gr

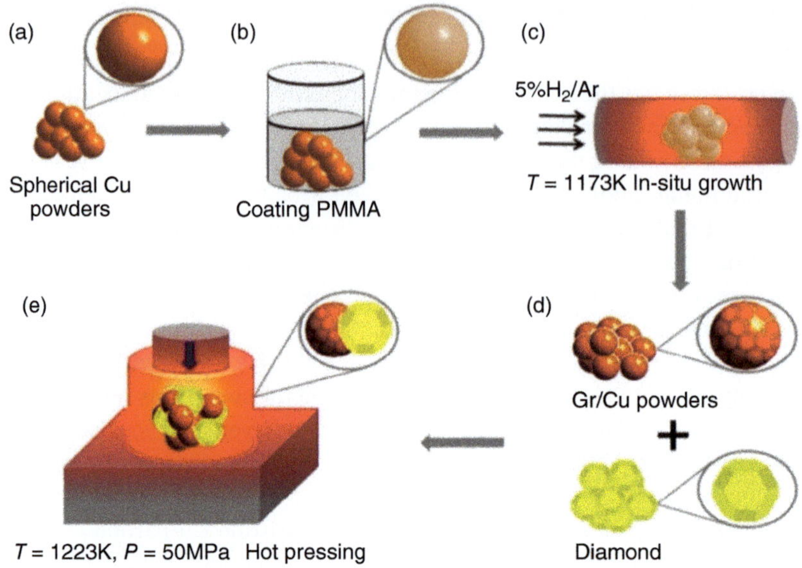

Figure 5.29 Flowchart for the preparation of in situ grown Gr-modified diamond/Cu interface composites. Source: Reprinted with permission from Ref. [18]; © 2019 Elsevier.

on the spherical Cu powder. Meanwhile, the effect of sintering temperature on the densities of diamond/Gr/Cu composites showed that the densities of the materials could reach 98.6% when sintered at temperatures above 950 °C.

The surface morphology analysis of the Dia/Gr/Cu composites after ion polishing shows a tight bond between diamond and Cu and a uniform dispersion of diamond in the Cu matrix. The interfacial structure of the Dia/Gr/Cu composite sample with 0.1 vol% Gr was characterized by TEM. The composites consist of three different regions: Cu, interlayer, and diamond. A smooth and clear interface between Cu and diamond was observed (Figure 5.30). Based on the high-resolution TEM image (Figure 5.30b), the intermediate layer between diamond and Cu was identified as a Gr layer and multilayer Gr streaks were observed in the interfacial layer (Figure 5.30c). The thickness of the Gr layer is approximately ~10 nm. The in situ grown Gr layer prevents direct contact between the diamond particles

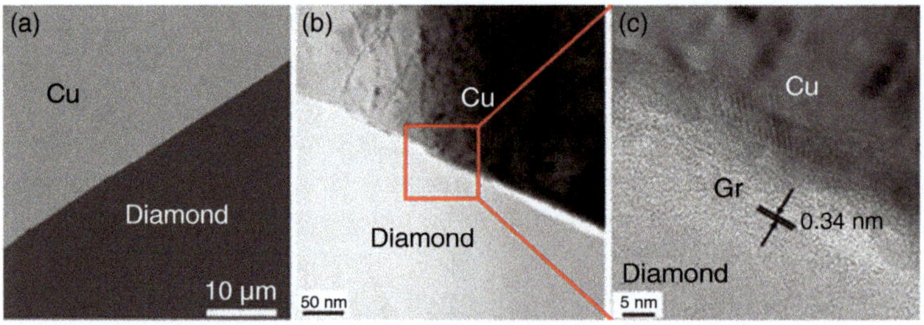

Figure 5.30 (a) SEM image of the ion beam polished surface of Dia/Gr/Cu and (b, c) TEM image of the Dia/Gr/Cu interface. Source: Reprinted with permission from Ref. [18]; © 2019 Elsevier.

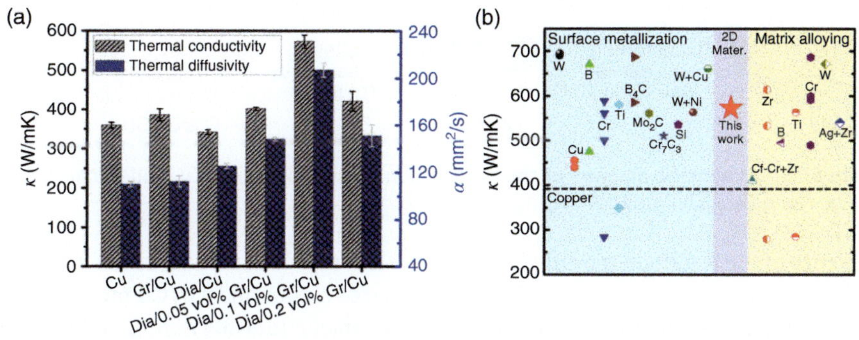

Figure 5.31 (a) Thermal conductivity and thermal diffusion coefficient of Dia/Cu, Gr/Cu and Dia/Gr/Cu composites. (b) Comparison of the thermal conductivity of Gr-modified Dia/Cu composites with the reported thermal conductivity of 50 vol% Dia/Cu composites prepared by diamond or matrix alloying surface metallization methods. Source: Reprinted with permission from Ref. [18]; © 2019 Elsevier.

and the Cu matrix, thus improving the interfacial compatibility between Cu and diamond.

The thermal conductivity (κ) and thermal diffusion coefficient (α) of the composites are shown in Figure 5.31a. It can be seen from the figure that the thermal conductivity of all Dia/Gr/Cu composites is higher than that of Dia/Cu composites without the Gr layer (342 ± 6 W/mK). Therefore, Gr can be used as an effective intermediate layer to improve the thermal conductivity of Dia/Cu composites. Different contents of Gr lead to different thermal conductivities of Dia/Gr/Cu composites. With the adjustment of Gr content, the thermal conductivity TC of the bulk Dia/Gr/Cu composites increases and then decreases, with the highest κ of 573 W/mK at 0.1 vol% Gr content.

The above results indicate that the introduction of Gr at the interface can effectively improve the thermal conductivity of diamond/Cu composites, which is comparable to that of diamond/Cu composites prepared by diamond surface metallization and metal matrix alloying at the same volume fraction of diamond content. To this end, we carried out the laws and mechanisms of the effect of the Gr intermediate modification layer on the thermal conductivity of the diamond–Cu interface.

In recent years, the rapidly developing time-domain thermal reflection (TDTR) method with the spatial nanoscale resolution has been applied in the study of thermal conductivity problems at the diamond/metal and diamond/carbide interfaces. Using a Cu-plated diamond substrate as the model material, different transition layers (composition and thickness) are coated between the two, and the actual preparation process is simulated and treated under the same conditions. The TDTR technique is used to directly determine the interfacial thermal conductivity of the model material under the conditions containing different interfacial modified layers, which can effectively reflect the real change law of interfacial thermal conductivity with the difference of thickness and composition of the interfacial modified layers. Combining the free path of carriers of each component of the material, the thickness of the transition layer is purposefully designed, which is conducive to revealing the

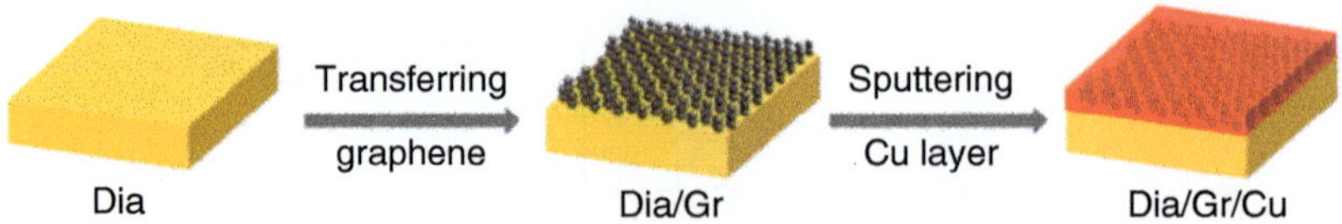

Figure 5.32 Schematic diagram of the preparation process of Dia/Gr/Cu model material. Source: Hong Guo, Zhongnan Xie, DingBang Xiong.

influence of the interface-modified layer on the thermal conductivity of the interface and its internal dominant mechanism. We prepared a diamond/Gr/Cu model material and tested the Gr-modified diamond–Cu interfacial thermal conductivity using the time-domain heat reflection method. The process of preparing the model material experimentally is shown in Figure 5.32.

The TDTR experimental results show that the interfacial thermal conductivity ITC increases from 38.5 to 43.2 MW/m^{2}K due to the introduction of Gr at the interface between diamond and Cu. It should be noted that the studied model material has a higher interfacial bond strength compared to the Dia/Gr/Cu composite prepared by the hot-pressing process. Electrons and phonons are the main heat carriers in metals and diamonds, respectively. The electron–phonon coupling interaction causes strong scattering and reduces the interfacial thermal conductivity. A suitable interfacial modification layer can balance the effect of acoustic impedance between diamond and Cu. The acoustic impedance (Z) of pristine Gr is 42×10^6 Kg/m^{2}s, similar to that of metal carbides. The Gr obtained by in situ catalytic growth cannot avoid defects. During the hot-press densification process, the applied pressure introduces stresses in the Gr. Defects and stresses can reduce the phonon velocity of Gr, so the acoustic impedance of in situ grown Gr is less than 42×10^6 Kg/m^{2}s, which is between Cu and diamond. Therefore, Gr can be used as an intermediate modification layer to act as a transition layer to increase the interfacial phonon transfer coefficient and improve the interfacial thermal conductivity between diamond and Cu (Figure 5.33).

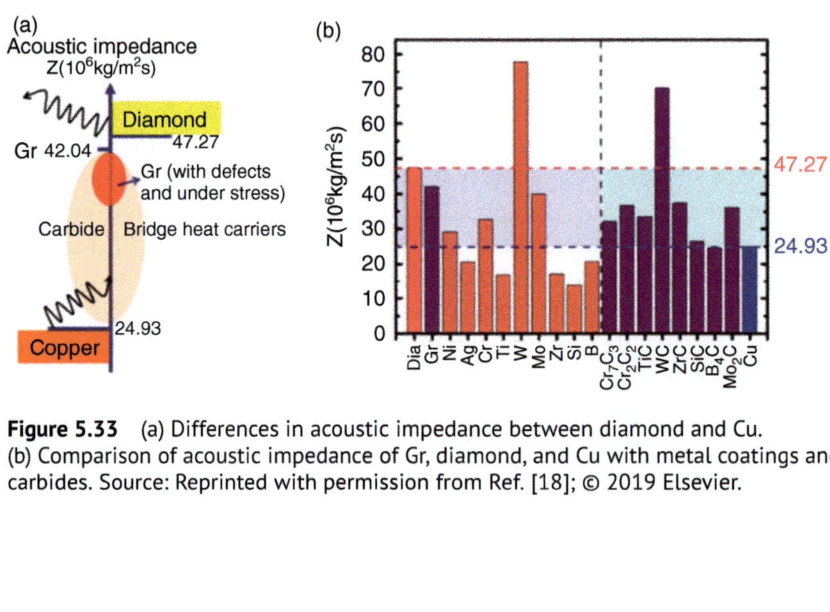

Figure 5.33 (a) Differences in acoustic impedance between diamond and Cu. (b) Comparison of acoustic impedance of Gr, diamond, and Cu with metal coatings and carbides. Source: Reprinted with permission from Ref. [18]; © 2019 Elsevier.

From the existing research results, the interfacial modification makes the wettability between the diamond and Cu matrix significantly improved in most cases. The interfacial bonding is changed from non-bonding or physical bonding to chemical bonding or diffusion bonding, which is beneficial to interfacial heat transfer. After the interfacial modification, the thermal conductivity of diamond/Cu matrix composites is significantly improved, but there is still a certain gap (100–250 W/mK) compared with the theoretical value, mainly because of the low intrinsic thermal conductivity and large thickness of the interfacial layer/phase (generally above 200 nm), which makes the additional interfacial thermal resistance introduced by the interfacial modified layer not negligible. Therefore, the optimized design of the interface modification layer is the key to further improving the thermal conductivity of diamond/Cu-based composites. Theoretical analysis shows that when the thickness of the interfacial modification layer is thin to a certain extent (e.g. below 200 nm), the thermal conductivity of the diamond/Cu matrix composites is enhanced with the reduction of the thickness of the nanosized interfacial modification layer, but the enhancement effect is less obvious and some research results show that this is not the case. For example, when the thickness of the W-W_2C-WC interfacial modification layer is reduced from 240 to 150 nm, the thermal conductivity of diamond/Cu matrix composites increases from 597 to 741 W/mK. And when the thickness is further reduced to 110 nm, the thermal conductivity can reach 907 W/mK. Thus, it can be seen that the thermal conductivity of composites under the nanoscale interfacial layer still has a large space for exploration and development potential, and has attracted attention and research interest in recent years, but there are still some key scientific and technical problems that need to be solved.

5.4 Formation and Evolution Rules of High Thermal Conductive Composite Microstructure and Its Control Method

5.4.1 Configurated Diamond/Metal Composites with High Thermal Conductivity

According to the particle accumulation theory, the total volume of various particles after mixing is related to the following parameters: the initial specific volume v of each particle, the particle size d of different particles, and the particle type S. V is the total volume of particles after mixing.

$$V = f(v, d, S_1 \cdots S_n) \tag{5.6}$$

$$V = \sum_{i=1}^{n} S_i \left(\frac{\partial V}{\partial S_i}\right)_{v,d,S_j} \quad j \neq i \tag{5.7}$$

$$V = \sum_{i=1}^{n} X_i V_i \tag{5.8}$$

For the mixture of two spherical particles with different particle sizes, it can be sorted into the following form:

$$\left(\frac{v - v_l X_l}{v_s}\right)^2 + 2G\left(\frac{v - v_l X_l}{v_s}\right)\left(\frac{v - X_l - v_s X_s}{v_l - 1}\right) + \left(\frac{v - X_l - v_s X_s}{v_l - 1}\right)^2 = 1 \tag{5.9}$$

where v_i is the specific volume of each particle, X_i is the volume ratio of each particle, s and l represent small particle size and large particle size, respectively, G is a parameter related to the particle size ratio of two kinds of particles, R is the ratio of large particle size to small particle size, and G is calculated using the following empirical formula:

$$\frac{1}{G} = \begin{cases} 1.355R^{-1.566}(R \geq 1.214) \\ 1(R \leq 1.214) \end{cases} \tag{5.10}$$

After the grading design of two kinds of diamond particles with different diameters, the total integral number of diamond after introducing secondary diamond can be calculated by formulas 5.9 and 5.10. The total integral number of diamond is closely related to the particle size ratio (PSR) and the secondary diamond volume (SDV) of the two kinds of diamond particles. With the decrease in secondary diamond particle size, the total integral number increases. With the increase of secondary diamond content, the total integral number first increases and then decreases. It reaches the maximum when the secondary diamond is added at about 30 vol%. Due to the large specific surface area of the secondary diamond, more secondary diamonds will introduce more interfaces into the composite, thus increasing the interface thermal resistance. Therefore, when designing the amount of secondary diamond, it is considered to design on the side where less secondary diamond is introduced and larger diamond volume fraction can be obtained. The secondary diamond addition design is shown in Table 5.2.

The diamond particles' close packing form is closer to the orthorhombic packing of hexagonal system. The single-diamond particles' packing volume fraction is usually about 60 vol%. The accumulation rate is consistent with the orthorhombic accumulation of hexagonal system. There are mainly two types of gaps in the stacked form of diamond hexagonal system, namely tetrahedral gap $r_1 = 0.225d$ and octahedral gap $r_2 = 0.414d$. To ensure that the tetrahedral gap is filled, the particle size ratio PSR of

Table 5.2 Secondary diamond addition design table.

Performs code	PSR	SDV (vol%)	Mixture type (μm)
A	8	0	50 + 400
B	8	5	50 + 400
C	8	15	50 + 400
E	8	25	50 + 400
F	8	35	50 + 400

Table 5.3 Diamond particle size ratio design table.

Performs code	PSR	SDV (vol%)	Mixture type (μm)
a	0	25	400
b	2	25	200 + 400
c	4	25	100 + 400
d	6	25	68 + 400
e	8	25	50 + 400
f	10	25	40 + 400

diamond particles with large and small sizes should be >4.5. If the octahedral gap is filled, the particle size ratio of diamond particle PSR > 2.5 is required. Therefore, in the design of diamond particle size ratio, all gaps shall be fully filled as far as possible, and PSR shall be set as 2, 4, 6, 8, and 10. The design of diamond particle size ratio is shown in Table 5.3.

5.4.2 Effect of Secondary Diamond Addition on Properties of Composites

When the diamond particle size ratio is 8, the thermal conductivity of diamond/copper composites with different secondary diamond additions is shown in Figure 5.34. The thermal conductivity increases with the addition of secondary particles and decreases rapidly when the addition of secondary particles exceeds 30 vol%. The thermal conductivity reaches the maximum value of 803 W/mK when the addition of secondary diamond is about 25 vol%. The thermal conductivity of single-particle composite is 684 W/mK. With the introduction of secondary particles, the thermal conductivity is all above 700 W/mK, which is obviously higher than single particle. The results show that the thermal conductivity of the composites can be greatly improved by adding secondary diamond particles and designing the thermal conductivity configuration with appropriate gradation.

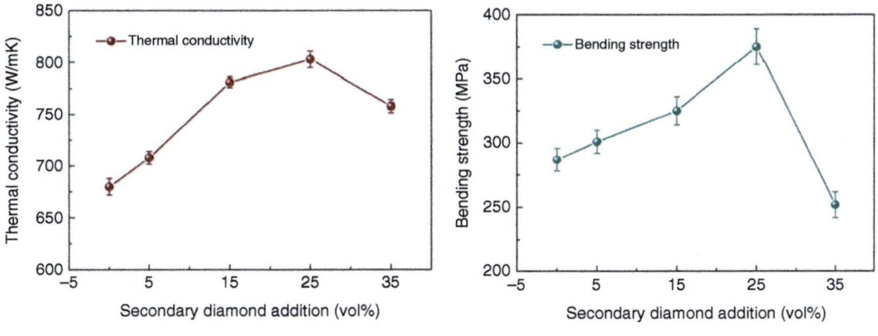

Figure 5.34 Thermal conductivity and bending strength of diamond/copper with different additions of secondary particles. Source: Hong Guo, Zhongnan Xie, DingBang Xiong.

When the diamond particle size ratio is 8, the bending strength of diamond/copper composites with different secondary diamond additions is shown in Figure 5.34. When secondary particles' content is less than 25 vol%, the bending strength increases with the increase of the content of secondary particles. The bending strength of diamond/copper reaches the maximum value of 375 MPa at 25 vol% secondary particles' addition. The bending strength of diamond/copper composite with single-particle size is 284 MPa. With the increase of the addition of secondary diamond, the composites' bending strength continues to increase. However, when the addition of secondary particles reaches 35 vol%, the composites' bending strength decreases rapidly, even lower than that of single-particle size.

5.4.3 Effect of Secondary Particle Size on the Properties of Composites

When SDV = 25 vol%, the thermal conductivity of diamond/copper with different secondary particle sizes is shown in Figure 5.35a. The thermal conductivity of diamond/copper composites increases with the increase of particle size ratio. When PSR = 10, the composite thermal conductivity reaches the maximum value of 853 W/mK. Compared with single-particle diamond/copper composites' thermal conductivity, the thermal conductivity of the composites with thermal conductivity configuration design exceeds 700 W/mK, which is obviously better than that of diamond/copper composites without gradation design.

With the increase of PSR, the total integral number of diamond keeps increasing. The thermal conductivity of diamond/copper composites is 710 W/mK at the PSR = 2. The highest value is 853 W/mK when the PSR = 10. It shows that the bridging secondary diamond particles should not be too large. Excessive secondary diamond particles will widen the distance of the main diamond particles, which will lead to the decrease of heat conduction channel flux, while smaller secondary diamond particles can play the best bridging role and get wider contact, thereby obtaining a longer heat conduction chain and a larger number of heat conduction channels.

When SDV = 25 vol%, the bending strength of diamond/copper composites with different particle sizes of secondary diamond particles is shown in Figure 5.35b. After the design of thermal conductivity configuration with diamond grain

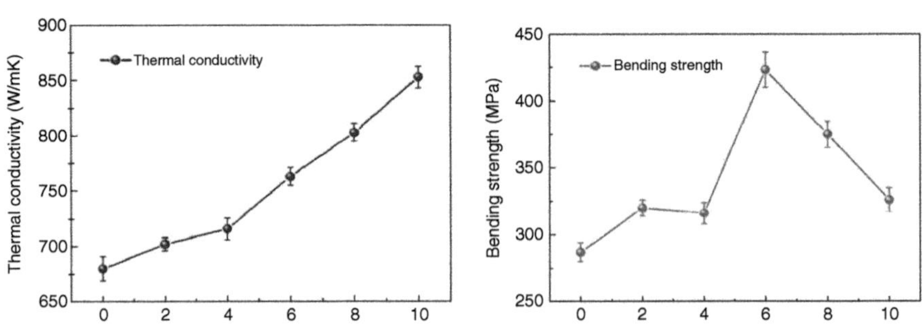

Figure 5.35 Thermal conductivity and bending strength of diamond/copper with different particle size ratios. Source: Hong Guo, Zhongnan Xie, DingBang Xiong.

gradation, the bending strength of diamond/copper composites is above 300 MPa, which can meet the industrial application demand. The composite bending strength reaches the maximum value of 420 MPa at PSR = 6. After PSR exceeds 8, the bending strength of diamond/copper shows a downward trend. However, when PSR = 10, the bending strength of diamond/copper still keeps 325 MPa.

5.4.4 Thermal Expansion Behavior of Composite Materials with Different Thermal Conductive Configurations

The coefficient of thermal expansion (CTE) of ceramic particle-reinforced metal matrix composites decreases with the increase of the inclusion volume fraction. The CTE of diamond/copper is mainly affected by the diamond content. At the same time, it is also reported in the literature that the composites CTE is also affected by the interfacial bonding strength between diamond and copper. But it is obvious only when the interfacial bonding strength is low. At present, Turner model and Kerner model are commonly used to evaluate CTE of particle-reinforced metal matrix composites.

The expression of Turner model is

$$\alpha_{\mathrm{c}} = \frac{\varphi_{\mathrm{m}}\alpha_{\mathrm{m}}K_{\mathrm{m}} + \varphi_{\mathrm{p}}\alpha_{\mathrm{p}}K_{\mathrm{p}}}{\varphi_{\mathrm{m}}K_{\mathrm{m}} + \varphi_{\mathrm{p}}K_{\mathrm{p}}} \tag{5.11}$$

The expression of Kerner model is

$$\alpha_{\mathrm{c}} = \alpha_{\mathrm{m}}V_{\mathrm{m}} + \alpha_{\mathrm{p}}V_{\mathrm{p}} + \frac{V_{\mathrm{m}}V_{\mathrm{p}}(\alpha_{\mathrm{p}} - \alpha_{\mathrm{m}})(K_{\mathrm{p}} - K_{\mathrm{m}})}{V_mK_m + V_{\mathrm{p}}K_{\mathrm{p}} + \dfrac{3K_{\mathrm{p}}K_{\mathrm{m}}}{4G_{\mathrm{m}}}} \tag{5.12}$$

The parameters used to predict the CTE of diamond/copper composites are shown in Table 5.4.

The selected research objects in this section are diamond/copper composites with single-particle size and optimized gradation. The CTE and the change of CTE with temperature were studied. The parameter information of selected diamond/copper samples is shown in Table 5.5.

The thermal expansion is the result of atomic spacing increase, and it is closely related to the bonding force between atoms. With the increase in temperature, the bonding force between atoms decreases, resulting in the increase in thermal expansion coefficient. This is the intrinsic reason why the coefficient of thermal expansion of materials increases with temperature. Figure 5.36 shows the variation

Table 5.4 Parameters used to predict thermal expansion coefficient.

Materials	α (ppm/K)	K(GPa)	G(GPa)
Diamond	1.86	441	360
Cu	17.7	134.8	45.5

Table 5.5 Parameters of graded diamond/copper composites.

Type	PSR	SDV (vol%)	Density (g/cm^3)	Diamond content (vol%)	Thermal conductivity (W/mK)
A	—	—	5.50	63	695
B	4	10	5.22	68.1	716
C	6	20	4.99	72.5	757
D	10	30	4.41	81.8	853

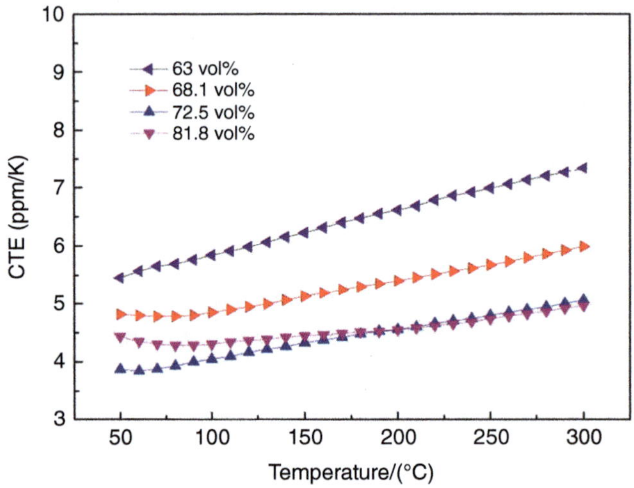

Figure 5.36 The CTE of diamond/copper composites with different diamond content changes with temperature. Source: Hong Guo, Zhongnan Xie, DingBang Xiong.

of thermal expansion coefficient of composites with different diamond contents with temperature. From 50 °C, the CTE increases linearly with the increase of temperature. It is worth noting that in the temperature range of 50–300 °C, the CTE of 68.1–81.8 vol% diamond/copper composites is all between 4 and 6 ppm/K. The thermal expansion coefficient of diamond/copper composite is well matched with that of GaN and other semiconductors.

References

1 Sun, Z.Z., Yan, Z., Yao, J. et al. (2010). Growth of graphene from solid carbon sources. *Nature* 468: 549–552.

2 Cao, M., Xiong, D.B., Tan, Z.Q. et al. (2017). Aligning graphene in bulk copper: Nacre-inspired nanolaminated architecture coupled with in-situ processing for enhanced mechanical properties and high electrical conductivity. *Carbon* 117: 65–74.

3 Cao, H.J., Xiong, D.B., Tan, Z.Q. et al. (2019). Thermal properties of in situ grown graphene reinforced copper matrix laminated composites. *Journal of Alloys and Compounds* 771: 228–237.

4 Wejrzanowski, T., Grybczuk, M., Chmielewski, M. et al. (2016). Thermal conductivity of metal-graphene composites. *Materials & Design* 99: 163–173.

5 Dreyer, D.R., Park, S., Bielawski, C.W. et al. (2010). The chemistry of graphene oxide. *Chemical Society Reviews* 39: 228–240.

6 Cancado, L.G., Jorio, A., Ferreira, E.H.M. et al. (2011). Quantifying defects in graphene via Raman spectroscopy at different excitation energies. *Nano Letters* 11: 3190–3196.

7 Zhang, D.D. and Zhan, Z.J. (2016). Experimental investigation of interfaces in graphene materials/copper composites from a new perspective. *RSC Advances* 6: 52219–52226.

8 Chen, F.Y., Ying, J.M., Wang, Y.F. et al. (2016). Effects of graphene content on the microstructure and properties of copper matrix composites. *Carbon* 96: 836–842.

9 Nan, C.W., Birringer, R., Clarke, D.R. et al. (1997). Effective thermal conductivity of particulate composites with interfacial thermal resistance. *Journal of Applied Physics* 81: 6692–6699.

10 Jagannadham, K. (2012). Thermal conductivity of copper-graphene composite films synthesized by electrochemical deposition with exfoliated graphene platelets. *Metallurgical and Materials Transactions B* 43: 316–324.

11 Hasselman, D.P.H. and Johnson, L.F. (1987). Effective thermal conductivity of composites with interfacial thermal barrier resistance. *Journal of Composite Materials* 21: 508–515.

12 Wang, Z.Q., Xie, R.G., Bui, C.T. et al. (2011). Thermal transport in suspended and supported few-layer graphene. *Nano Letters* 11: 113–118.

13 Vaidya, R.U. and Chawla, K.K. (1994). Thermal expansion of metal-matrix composites. *Composites Science and Technology* 50: 13–22.

14 Firkowska, I., Boden, A., Boerner, B. et al. (2015). The origin of high thermal conductivity and ultralow thermal expansion in copper-graphite composites. *Nano Letters* 15: 4745–4751.

15 Yoon, D., Son, Y.W., and Cheong, H. (2011). Negative thermal expansion coefficient of graphene measured by Raman spectroscopy. *Nano Letters* 11: 3227–3231.

16 Oddone, V., Wimpory, R.C., and Reich, S. (2019). Understanding the negative thermal expansion in planar graphite-metal composites. *Journal of Materials Science* 54: 1267–1274.

17 Cao, H.J., Tan, Z.Q., Fan, G.L. et al. (2020). Wide and fine alignment control and interface modification for high-performance thermally conductive graphite/copper composite. *Composites Part B: Engineering* 191: 107965.

18 Cao, H.J., Tan, Z.Q., Lu, M.H. et al. (2019). Graphene interlayer for enhanced interface thermal conductance in metal matrix composites: An approach beyond surface metallization and matrix alloying. *Carbon* 150: 60–68.

6

Preparation and Performance Study of Silicon Nitride Ceramic Substrate with High Thermal Conductivity

Yao Dongxu[1], *Wang Weide*[2], *and Zeng Yu-ping*[1]

[1] *Chinese Academy of Sciences, Shanghai Institute of Ceramics, State Key Laboratory of High Performance Ceramics and Superfine Microstructure, 1295 Dingxi Road, Shanghai 200050, China*
[2] *National University of Defense Technology, College of Aerospace and Engineering, Science and Technology on Advanced Ceramic Fibers & Composites Laboratory, 109 De Ya Rd, Changsha, Hunan 410073, China*

6.1 Rapid Nitridation of Silicon Compact

Silicon nitride (Si_3N_4) ceramic is an ideal material for the high-power density-integrated circuit heat dissipation substrates due to its outstanding mechanical properties and high theoretical thermal conductivity. The preparation of high thermal conductivity Si_3N_4 ceramics by sintered reaction-bonded silicon nitride (SRBSN) using silicon powder as raw material has become research hotspot in recent years. However, the long nitridation period of silicon (Si) adds to the cost. In this chapter, a novel sintering aid system (Y_2O_3:Eu[x mol%]-MgO) was designed to simultaneously promote the nitridation of silicon powder and improve the thermal conductivity of sintered Si_3N_4 at the same time, and its effects on nitridation rate, grain growth, microstructure, thermal conductivity, and mechanical properties were researched in detail.

6.1.1 Rapid Nitridation of Silicon Compact

In this experiment, the sintering additives were MgO (purity ≥99.9 wt%, Sinopharm Group Holdings Co., Ltd.) and $(YEu)_2O_3$ (purity ≥ 99.9 wt%, Fujian Changting Jinlong Rare Earth Co., Ltd.). The additive ratio was calculated by generating Si_3N_4 by complete nitride of Si, and samples were prepared in the following molar ratio, Si_3N_4: $(YEu)_2O_3$: MgO $= 93:x:y$, labeled xYEyM, where $x+y = 7$, $x = 0, 1, 2, 3, 4, 5, 6, 7$, correspondingly $y = 7, 6, 5, 4, 3, 2, 1, 0$. The proportionally weighed raw materials are ball-milled in a nylon ball milling jars with Si_3N_4 balls as the ball milling medium for 4 hours. The speed was 300 rpm, and the solvent was ethanol. The ball-milled slurry was dried in a vacuum drying chamber at 70 °C for 4 hours and sieved through a 60-mesh. Die press molding was taken out by pressing 1.5 and 12 g of mixed powder in Ø20 mm and 40 mm × 40 mm stainless steel molds,

Thermal Management Materials for Electronic Packaging: Preparation, Characterization, and Devices, First Edition. Edited by Xingyou Tian.

Table 6.1 The different compositions and theoretical densities of samples with different Y_2O_3: Eu (4 mol%)/MgO ratios.

Sample	Si_3N_4 (mol%)	Si (mol%)	Y_2O_3:Eu (4 mol%) (mol%)	MgO (mol%)	ρ_{th}
0YE7M	93	279	0	7	3.197
1YE6M	93	279	1	6	3.217
2YE5M	93	279	2	5	3.237
3YE4M	93	279	3	4	3.256
4YE3M	93	279	4	3	3.275
5YE2M	93	279	5	2	3.306
6YE1M	93	279	6	1	3.318
7YE0M	93	279	7	0	3.329

respectively, then by cold isostatic pressing at 250 MPa. The properties of the raw materials used in the experiment are listed in Table 6.1. In this experiment, the highly thermal conductivity silicon nitride sample was done by two sintering methods, and the compacts were nitrided in a graphite furnace at 1400 °C, N_2 (5 vol% H_2) for 4 h, and then sintered in the same furnace afterward at 1900 °C with 1 MPa N_2.

Table 6.1 lists the theoretical density of the raw material, which is calculated according to the mixing law, as calculated by the following formulas:

$$\rho_{th} = \sum V_i \rho_i$$

The density of the sample was determined using the Archimedes method. The mass loss after gas pressure sintering is calculated by mass change after sintering. The oxygen atoms content is determined by an O/N analyzer (TC-600C, LECO Co., USA). A scanning electron microscope (SEM, Magellan 400, FEI, USA) observes the polished surface. Phase analysis was performed by XRD (XRD, D8 ADVANCE, Bruker, Germany) Cu-target Kα radiometry. The sintered sample is processed into a 3 mm × 4 mm × 36 mm test bar, and then the bending strength is measured on a universal mechanical test machine (Model 5566, Instron Co., High Wycombe, UK) by the three-point bending method. Polishing strips with finished bending strength to test toughness and hardness. Sintered samples were processed into Ø10 mm × 2.5 mm circles to measure the thermal diffusion coefficient (α). α is measured using the laser flash method (LFA 467 Hyper flash, NETZSCH Instruments Co. Ltd., Selb, Germany). Thermal conductivity (κ) is calculated by the following formula:

$$k = \rho C_p \alpha$$

where ρ is the bulk density, C_p is the specific heat capacity, and α is the thermal diffusivity. A constant value of 0.68 J/(g K) was used for the specific heat (C_p) of Si_3N_4 [1, 2].

The nitridation degree (ND) of the samples was calculated based on the weight change as follows:

$$NR = \frac{3 \times M_{Si} \times \Delta m}{4 \times M_N \times m_{Si}}$$

In the above equation, M_{Si} and M_N represent the atomic masses of *Si* and *N*, respectively; and Δm and m_{Si} indicate the weight change of the sample during the nitridation process and the mass of the Si powder in the raw material, respectively.

6.1.1.1 Optimization $(YEu)_2O_3$/MgO Sintering Additive

The XRD patterns of samples with different $(YEu)_2O_3$/MgO ratios after nitridation are shown in Figure 6.1. After the nitriding was completed, the diffraction peaks of α-Si_3N_4 and β-Si_3N_4 appeared in all samples, and no diffraction peaks of Si were detected, indicating that the Si powder had been completely nitrided at this process and was converted to Si_3N_4. The diffraction peaks of the $YSiO_2N$ and $Y_2Si_3O_3N_4$ intermediate phases appeared in the sample 2YE5M, and with the increase of the content of rare earth sintering additives in the sample, the diffraction peaks of the intermediate phase increased, indicating that the content of the intermediate phase gradually increased. The intermediate phase is formed by the cooling crystallization of the liquid phase generated by the reaction of rare earth, MgO, SiO_2, and Si_3N_4; therefore, the content of the intermediate phase gradually increases with the increase of rare earth content.

The nitriding degree (ND) and β-Si_3N_4 phase content of the samples are listed in Table 6.2. From Table 6.2, it can be seen that the nitriding degree of sample 0YE7M is only 83%, and after the addition of rare earth sintering additives, the nitriding rate of the sample is increased to more than 90%, and with the increase of rare earth addition, the nitriding rate of the sample gradually increases. In previous

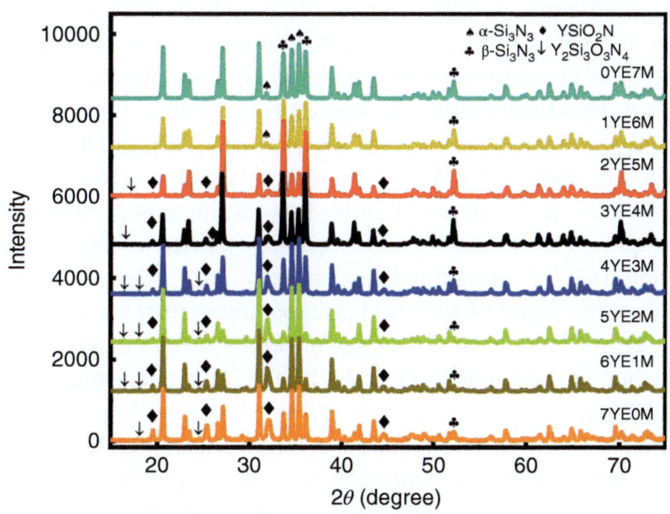

Figure 6.1 XRD of nitrided samples with different Y_2O_3:Eu(4 mol%)/MgO ratios. Source: Yao Dongxu, Wang Weide, Zeng Yu-ping.

Table 6.2 Properties of samples with different Y_2O_3:Eu(4 mol%)/MgO ratios after nitridation.

	0YE7M	1YE6M	2YE5M	3YE4M	4YE3M	5YE2M	6YE1M	7YE0M
β	48.1	58.5	77	68.9	40.5	12.7	18.2	33.8
ND(%)	83	90.23	95.5	97.7	99.2	98.4	100	98.5

studies, the introduction of both Y_2O_3 and Eu_2O_3 can increase the nitriding rate and shorten the nitriding cycle of Si powder compared with pure Si powder [3–5]. Since Si_3N_4-SiO_2-Y_2O_3-MgO forms a liquid phase (formation temperature above 1200 °C) during nitridation, the viscosity decreases when the liquid phase contains Eu and Mg. Y and Eu in the liquid phase will not only affect the composition and viscosity of the liquid phase but also affect the nucleation and phase transformation of Si_3N_4. With the increase of rare earth elements in the liquid phase, the activation energy of nucleation is reduced and the rare earth atoms are adsorbed on the prism of Si_3N_4 to limit the binding of Si atoms and N atoms on side ridge surface so that the grains can only grow in the direction of (001). An abundance of rare earth atoms in the liquid phase and an abundance of low-viscosity liquid phases facilitated the transition from α-Si_3N_4 to β-Si_3N_4 and the growth of β-Si_3N_4. Therefore, as shown in Table 6.2, the content of β-Si_3N_4 in the sample gradually increases as the $(YEu)_2O_3$/MgO ratio increases. As the $(YEu)_2O_3$/MgO ratio increased further, the content of β-Si_3N_4 in the sample decreased. Due to Y^{3+}, Eu^{3+} replaces Si^{4+} in the liquid phase, and as Y^{3+} increases, the liquid phase must dissolve more N to maintain the electrical neutrality requirements, this increases the N/O ratio of the liquid phase. The coordination number of N in the liquid phase (3) is higher than the coordination number of O (2), and the substitution of O by N leads to an increase in the degree of crosslinking of the nitrox-glass phase, resulting in an increase in the glass transition temperature and viscosity of the liquid phase [6–9]. The high-viscosity liquid phase is not conducive to the growth of β-Si_3N_4. Therefore, when the ratio of $(YEu)_2O_3$/MgO is too high, the content of β-Si_3N_4 in the sample is reduced. In sample 0YE7M, the content of β-Si_3N_4 was also increased. Although Si_3N_4-MgO formed a liquid phase at 1500 °C and MgO-SiO_2 formed a liquid phase at 1600 °C [10], the liquid-phase formation temperature of Si_3N_4-MgO-SiO_2 was 1390 °C, and the low-viscosity liquid phase could promote the occurrence of phase transition. However, due to the absence of rare earth elements, the phase change activation energy is higher, so the content of β-Si_3N_4 in 0YE7M is not as high as in 1YE6M, 2YE5M and 3YE4M.

Figure 6.2 shows the SEM with different $(YEu)_2O_3$/MgO ratios after Si powder nitride. It can be seen that the number of long-rod β-Si_3N_4 in sample 2YE5M is significantly more than that of other samples, and its length is 1–2 μm. These long-rod β-Si_3N_4 act as crystalline nuclei during post-sintering, in which β-Si_3N_4 continues to grow and develop to form large-sized grains. Most of the isoaxial α-Si_3N_4 are found

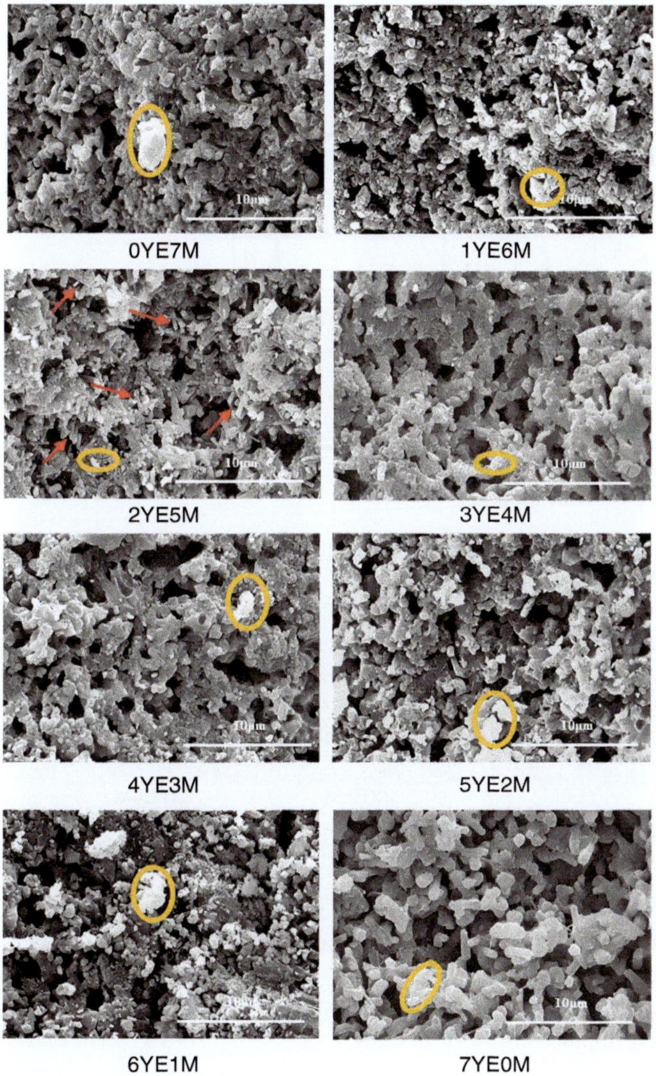

Figure 6.2 SEM of samples with different $(YEu)_2O_3$ / MgO ratios after nitridation. Source: Yao Dongxu, Wang Weide, Zeng Yu-ping.

in 5YE2M and 6YE1M, which favors the compaction process during post-sintering, because α-Si_3N_4 has higher sintering activity than β-Si_3N_4.

Figure 6.3 shows the XRD pattern of the SRBSN samples with different $(YEu)_2O_3$/MgO. The sample is all β-Si_3N_4, indicating that the phase transformation in the post-sintering process has been fully completed, and the different $(YEu)_2O_3$/MgO have little impact on the phase transition in the post-sintering process, and the temperature is the most important reason affecting the phase transition rather than the type and content of the additives. However, with the increase in $(YEu)_2O_3$/MgO ratio, the second crystalline phase appeared in the sample and the composition of

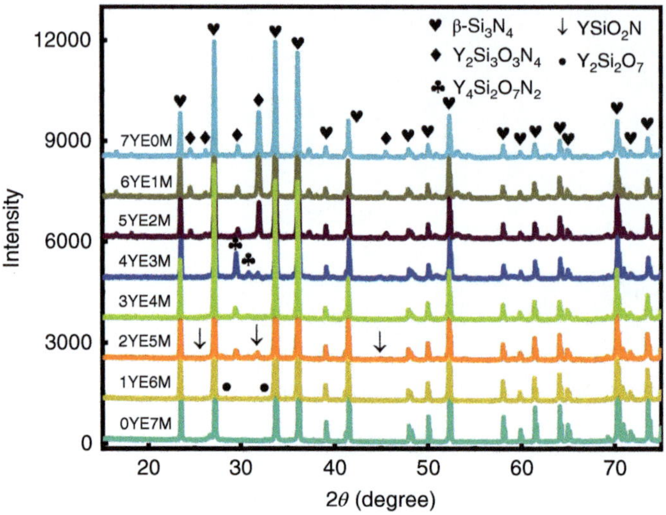

Figure 6.3 XRD pattern of post-sintered samples with different $(YEu)_2O_3$/MgO ratios. Source: Yao Dongxu, Wang Weide, Zeng Yu-ping.

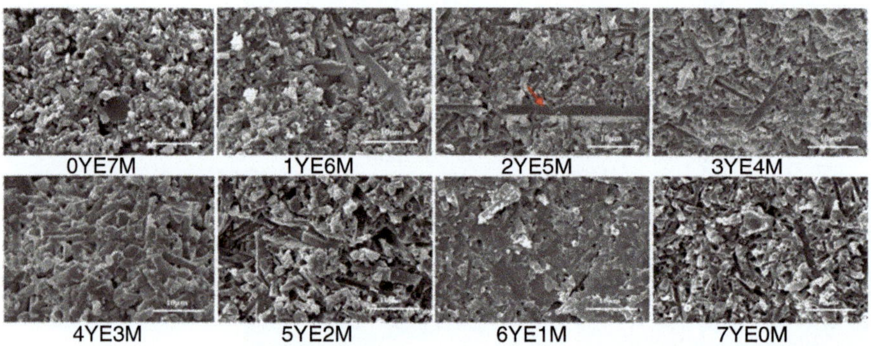

Figure 6.4 SEM of sintered samples with different $(YEu)_2O_3$/MgO ratios. Source: Yao Dongxu, Wang Weide, Zeng Yu-ping.

the second crystalline phase also changed. There is no second phase present except β-Si_3N_4 in 0YE7M and 1YE6M. With the increase in $(YEu)_2O_3$ content, the second crystalline phase of $Y_4Si_2O_7N_2$ appeared in samples 2YE5M, 3YE4M, and 4YE3M, and with the further increase in $(YEu)_2O_3$ content, the $Y_2Si_3O_3N_4$ second phase appeared in the sample. The SEM of the post-sintering sample is shown in Figure 6.4. It can be seen that a small number of elongated rod Si_3N_4 grain appears in 0YE7M, because the sintering additive is only MgO, which forms a liquid phase with SiO_2 at low temperatures. But the liquid phase is volatile at high temperatures, resulting in an insufficient liquid phase which is not conducive to the occurrence of phase transition. With the increase of $(YEu)_2O_3$/MgO, the number of elongated rod Si_3N_4 grains in the sample began to increase and then decrease. The elongated Si_3N_4 in samples 2YE5M and 3YE4M are distributed in the fine round Si_3N_4 grains, forming a typical bimodal morphology, which advantageously improves the bending strength. On the one hand, rare earth ions can absorb on the prism of the β-Si_3N_4 nuclei in the liquid phase, which promotes their growth and development along the direction of (0001). On the other hand, the existing presence of β-Si_3N_4 grains in the nitrided samples as a crystal nucleus favors the growth and development of the grains during post-sintering. However, the number of elongated Si_3N_4 in samples 4YE3M, 5YE2M, 6YE1M, and 7YE0M decreased again. This is because, with the decrease of MgO content and the increase of $(YEu)_2O_3$, the increase of liquid-phase viscosity at high temperatures is negative to grain development. And the field strength and coordination number of Y^{3+} in the liquid phase are higher than that of Mg^{2+}, which leads to an increased degree of crosslinking of the liquid phase and leads to increased viscosity [6, 8].

The density, relative density, and weight change of the SRBSN samples with different $(YEu)_2O_3$/MgO are shown in Figure 6.5. Figure 6.5 shows that with the increase of $(YEu)_2O_3$/MgO, the MgO is volatile at high temperatures [11], so the weight loss of the samples gradually decreases after sintering with the decrease of MgO content. When the MgO content decreases to a certain extent, the weight loss of the sample gradually decreases, while the weight of samples 6YE1M and 7YE0M is slightly increased. Except for the error caused by weighing, the weight increase of this part may be due to the mass increase caused by the liquid-phase formation at high temperatures, and the entry of N into the liquid phase in 1 MPa N_2. MgO at high temperatures is volatile, which also resulted in the low relative density of sample 0YE7M being less than 80%. When 1 mol $(YEu)_2O_3$ is added to the sample, the density rapidly increases to 3.147, and the relative density increases to 97.82%. With the continuous increase of $(YEu)_2O_3$ content, the relative densities of 2YE5M, 3YE4M, and 4YE4M are above 99%, which can be considered to be completely dense. However, the relative densities of 5YE2M, 6YE1M, and 7YE0M decreased slightly, but the

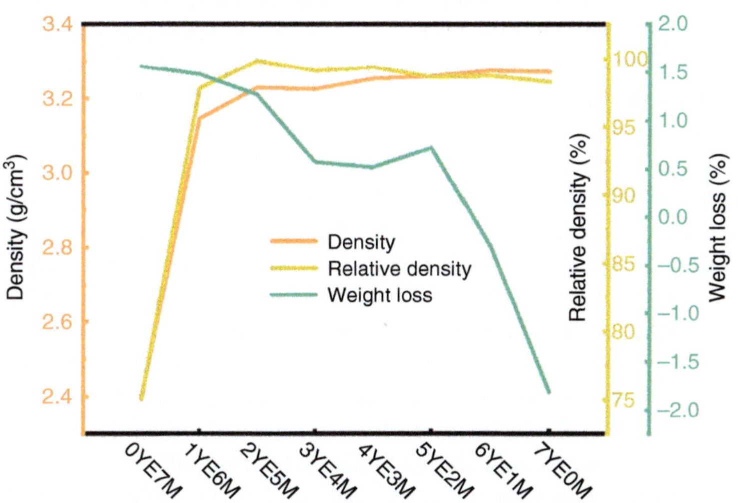

Figure 6.5 Density, relative density, and weight loss of SRBSN samples with different $(YEu)_2O_3$/MgO ratios. Source: Yao Dongxu, Wang Weide, Zeng Yu-ping.

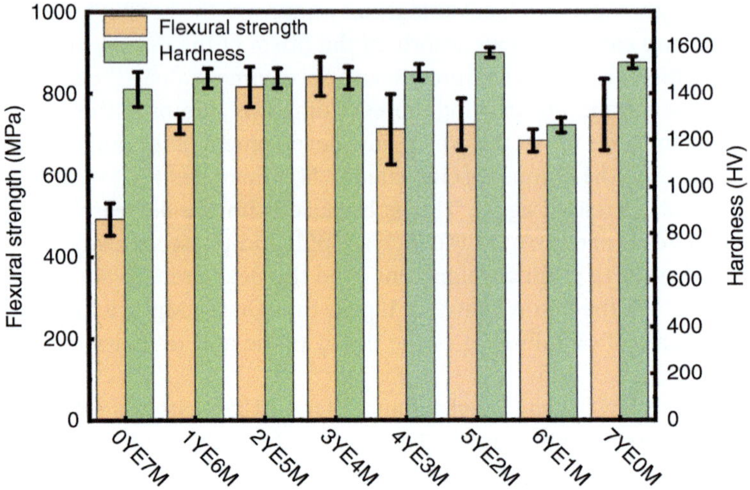

Figure 6.6 Bending strength and hardness of SRBSN samples with different ratios of Y_2O_3: Eu (4 mol%)/MgO. Source: Yao Dongxu, Wang Weide, Zeng Yu-ping.

whole was higher than 98%. This is due to the too-small MgO content in the sample, resulting in insufficient liquid-phase content at high temperatures. However, the relative density of 7YE0M can also reach more than 98%, due to the promoting densification effect of Y_2O_3 on Si_3N_4.

The flexural strength and hardness of SRBSN samples with the different $(YEu)_2O_3$/MgO are shown in Figure 6.6. Sample 0YE7M has the lowest flexural strength because of the poor density after sintering, fewer elongated grains, and no bimodal morphology. With the increase of the $(YEu)_2O_3$/MgO ratio, the bending strength and hardness of the sample first increased and then decreased. Samples 2YE5M and 3YE4M have higher bending strength and hardness, due to the typical bimodal morphology formed after the posterior sintering, which is consistent with the above analysis results. The bending strengths of 4YE3M, 5YE2M, 6YE1M, and 7YE0M are greatly reduced compared with 2YE5M and 3YE4M, which is greatly related to its microstructure and slightly decreased density.

Thermal diffusion and thermal conductivity of SRBSN samples are shown in Figure 6.7. With the increase of $(YEu)_2O_3$/MgO, thermal diffusion and thermal conductivity decrease after the increase. The 2YE5M has the highest thermal diffusion and thermal conductivity (80 W m^{-1} K^{-1}). Although 0YE7M has a higher thermal diffusion, its poor density leads to its lowest thermal conductivity.1YE6M and 2YE5M have high thermal diffusion and thermal conductivity because their samples are completely dense, and the large size of Si_3N_4 grains in the microstructure, which facilitates thermal conductivity. Although 3YE4M also has a higher density, the grain size of Si_3N_4 in its microstructure is small, which leads to the increase of the grain-boundary and the enhanced scattering effect on phonons, which is negative to the improvement of thermal conductivity. 4YE3M, 5YE2M, 6YE1M, and 7YE0M have a low thermal conductivity due to small grains and slightly decreased density.

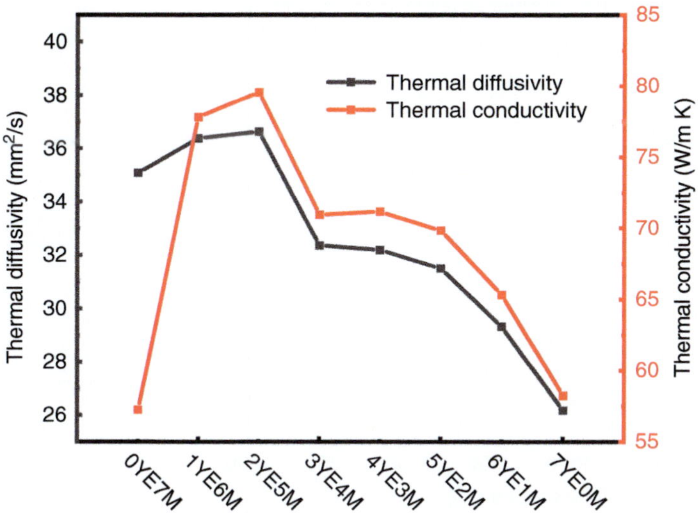

Figure 6.7 Thermal diffusion and thermal conductivity of SRBSN samples with different Y_2O_3: Eu (4 mol%)/MgO ratios. Source: Yao Dongxu, Wang Weide, Zeng Yu-ping.

Considering the resultant density, bending strength, hardness, and thermal conductivity of the sample after sintering, 2YE5M has the best comprehensive performance. After subsequent sintering, the relative density is greater than 99%, which is considered completely dense. Flexural strength and hardness were 810.06 MPa and 1476 HV, respectively, with thermal conductivity of nearly 80 W m^{-1} K^{-1}. Therefore, the optimized $(YEu)_2O_3$/MgO is 2 : 5.

6.1.1.2 Further Optimization of the SRBSN with 2YE5M as Sintering Additive

The optimal $(YEu)_2O_3$/MgO of 2 : 5 (mole ratio) was selected in 6.1, and the samples were prepared with the same raw material as follows: Si_3N_4: $(YEu)_2O_3$/MgO = 93 : 2 : 5, Si_3N_4:Y_2O_3:MgO = 93 : 2 : 5, Si_3N_4:Eu_2O_3:MgO = 93 : 2 : 5, and Si_3N_4:Y_2O_3:Eu_2O_3:MgO = 93 : 1.91 : 0.09 : 5, according to Si complete nitrogen to generate Si_3N_4. The above samples were labeled YEu, Y, Eu, and Y + Eu. The additional raw materials are Y_2O_3 (purity≥99.99 wt%, Shanghai Yuelong Non-ferrous Metals Co., LTD.) and Eu_2O_3 (purity≥99.9 wt%, Shanghai Aladdin Biochemical Technology Co., LTD.). The raw material will be weighed in a nylon ball grinding jar with Si_3N_4 ball as ball grinding medium for 4 hours at 300 rpm and solvent with ethanol. The slurry was dried in a 70 °C vacuum drying tank for 4 hours and screened over 60 meshes. Take 1.5 and 12 g mixed powder to die press molding in Ø 20 and 40 mm stainless steel molds, respectively, followed by cold isostatic pressure at 250 MPa. Characterization methods are performed as described in Section 6.1.

Thermogravimetric (TG) curves of the four compacts are shown in Figure 6.8. As can be seen from the curve, the trends of thermal weight gains were generally consistent across the four samples during the low-temperature section (before the 1200 °C). Initial nitriding of Si powder was started at around 950 °C so that the four different additive species had little effect on the initial nitriding of Si powder. But

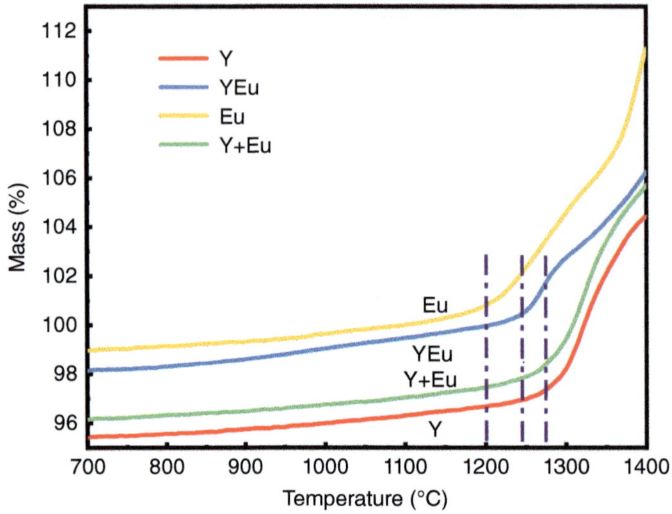

Figure 6.8 Thermogravimetric curves of the green body. Source: Reproduced with permission from Ref. [2]; © 2023 John Wiley & Sons. Source: Yao Dongxu, Wang Weide, Zeng Yu-ping.

there were noticeable differences in nitride rates of different samples after 1200 °C. The sample with Eu started a rapid nitride at 1200 °C, followed by a slightly slower rate until a second rapid nitride phase occurred at 1350 °C. This result matches the data from a previous study [3]. The temperature point of rapid nitride in the sample with YEu is around 1250 °C, which is related to the content of Eu_2O_3. The sample with YEu also showed a second rapid nitride peak at 1300 °C. The rapid nitride of both the Y + Eu and Y samples, at roughly the same temperature, was 1280 °C, and both nitride rates began to slow after 1350 °C, while the rapid nitride of pure Si powder was set at 1380 °C [3]. Therefore, both Y_2O_3 and Eu_2O_3 can promote the nitride of Si powder, but the catalytic nitride effect of Y_2O_3 is not obvious, while a small amount of Eu_2O_3 in YEu can greatly improve the catalytic nitride effect. However, the catalytic nitride effect of the sample with Y + Eu obtained by the mechanical mixture of Y_2O_3 and Eu_2O_3 is not obvious, which may be related to the uneven distribution of Eu_2O_3 in the sample.

The XRD patterns after nitridation of the four samples with Y, YEu, Y + Eu, or Eu are shown in Figure 6.9. The diffraction peak of Si was not detected in the XRD spectra of the whole samples, which indicates that the silicon powder had been completely monitored. Besides the sample with Eu, the second crystalline phase of $Y_2Si_3O_3N_4$ appeared in all three samples, and it was more abundant in the Y sample. Both α-Si_3N_4 and β-Si_3N_4 were present in the Eu samples, and other crystalline phases were not detected. The content of the β phase varied after the completion of the nitride in the four samples. The β phase content after nitride is listed in Table 6.3. In the sample with YEu, the highest content of β-Si_3N_4 is 72.6%, which is basically consistent with that discussed in Section 2.1. The β-Si_3N_4 content in the sample with Y was 51.1%, which was lower than the 61.3% content in the sample

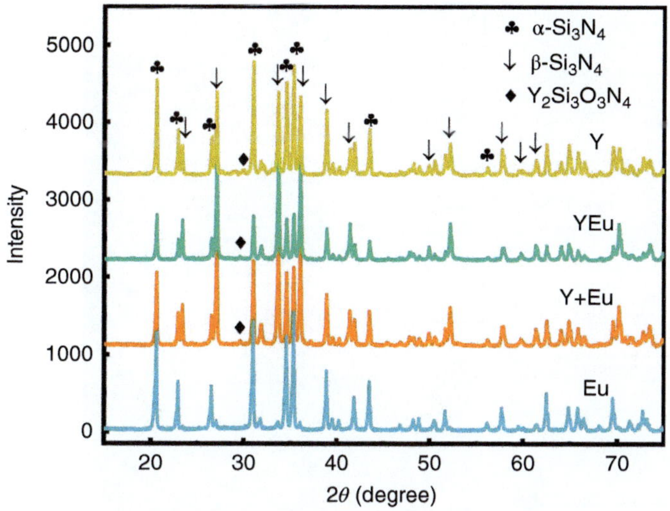

Figure 6.9 XRD of four samples of Y, YEu, Y + Eu, and Eu after nitridation. Source: Reproduced with permission from Ref. [2]; © 2023 John Wiley & Sons.

Table 6.3 Content of β-Si_3N_4 after nitridation of four samples of Y, YEu, Y + Eu, and Eu.

Sample	Y	YEu	Y + Eu	Eu
β-Si_3N_4	51.1	72.6	61.3	6.5

with Y + Eu. This is because the addition of Eu_2O_3 catalyzes the nitride reaction, and the Si powder nitride is a strongly exothermic reaction that leads to local silicon melting, and the molten Si during the nitride favors β-Si_3N_4 generation. On the other hand, the presence of the liquid phase during the nitride favors the growth and development of β-Si_3N_4 due to the low melting point of the glass phase containing Eu_2O_3.

Figure 6.10 shows the results of SEM photos after the nitriding of the four samples. As can be seen in Figure 6.10, all long-rod β-Si_3N_4 were generated in (a), (b), and (c), while both isoaxial Si_3N_4 were found in (d). This indicates that Eu_2O_3 does not favor the α-Si_3N_4 to β-Si_3N_4 transition, but a small amount of Eu_2O_3 has little effect in the presence of little Y_2O_3 [12]. Nitrided long-rod β-Si_3N_4 can act as a seed in the posterior sintering process, and during the dissolution–precipitation and Ostwald ripening process, the Si_3N_4 continues to grow up to generate abnormally large grains.

The nitriding degree, expansion degree (ED), and relative density (RD) of the four samples are listed in Table 6.4. As can be seen from Table 6.4, the nitride rates of samples with YEu and Y + Eu are 96.7% and 97%, respectively, which is greatly improved compared with the nitride rate of 93.8% of the sample with Y, indicating that the addition of Eu_2O_3 can catalyze the nitride of Si powder and improve the nitride rate of Si powder. The expansion rate of the sample Eu is 0.39%, which is too low

Figure 6.10 Microstructures of samples after nitridation at 1400 °C for 4 h. (a) Y, (b) YEu, (c) Y + Eu, (d) Eu. Source: Reproduced with permission from Ref. [2]; © 2023 John Wiley & Sons.

Table 6.4 Nitridation rate, expansion rate, and relative density of samples after nitridation at 1400 °C for 4 h.

	Y	YEu	Y + Eu	Eu
ND(%)	93.8	96.7	97	96.3
ED(%)	1.36	1.49	1.24	0.39
RD(%)	70.7	70.7	71.5	73.1

compared with the other three samples, due to the low melting point of the glass phase containing Eu_2O_3. The liquid phase has been generated at the nitride temperature (1400 °C). The presence of the liquid phase shrinks the sample, resulting in a lower expansion rate. The relative density of the sample with Eu was the highest among the four samples, which was also due to the presence of a liquid phase during the nitride. The relative density of the remaining three samples is more than 70%, which is greater than the relative density of the sample after 250 MPa cold isostatic pressure before nitride.

The nitride-completed samples were sintered. The XRD spectra of the samples after 1900 °C insulation for 4 h are shown in Figure 6.11. Figure 6.11 shows that all the samples are β-Si_3N_4, indicating that the phase transition of the samples is completely completed after post-sintering. The density and relative density of the samples after sintering at different posterior sintering temperatures and high-temperature insulation times are shown in Figure 6.12. As can be seen from Figure 6.12, the relative density of the sample with Y approached 100% under all

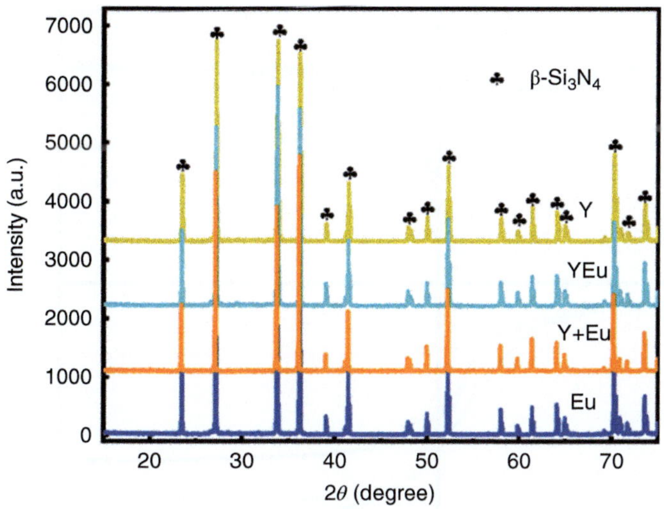

Figure 6.11 XRD of the sample after holding at 1900 °C for 4 h. Source: Reproduced with permission from Ref. [2]; © 2023 John Wiley & Sons.

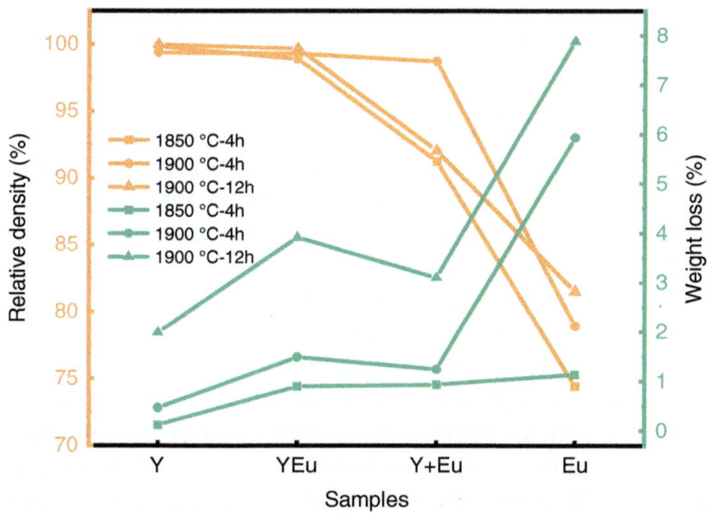

Figure 6.12 Density and relative density of samples at different sintering systems Source: Reproduced with permission from Ref. [2]; © 2023 John Wiley & Sons.

three sintering systems, and it was completely dense. However, the sample with Eu was not dense under the three sintering systems, and the relative density was below 85%. Samples with YEu and Y + Eu had a relative density of greater than 99% after 4 h at 1900 °C, whereas the YEu sample was slightly reduced after 1850 °C for 4 h and 1900 °C for 4 h, but the relative density of a sample with Y + Eu was between 90% and 95%. This is due to excessive thermal insulation at 1900 °C, which is negative to compaction; and at 1850 °C for 4 h, the viscosity of the liquid phase is too large, which is negative to the compaction process.

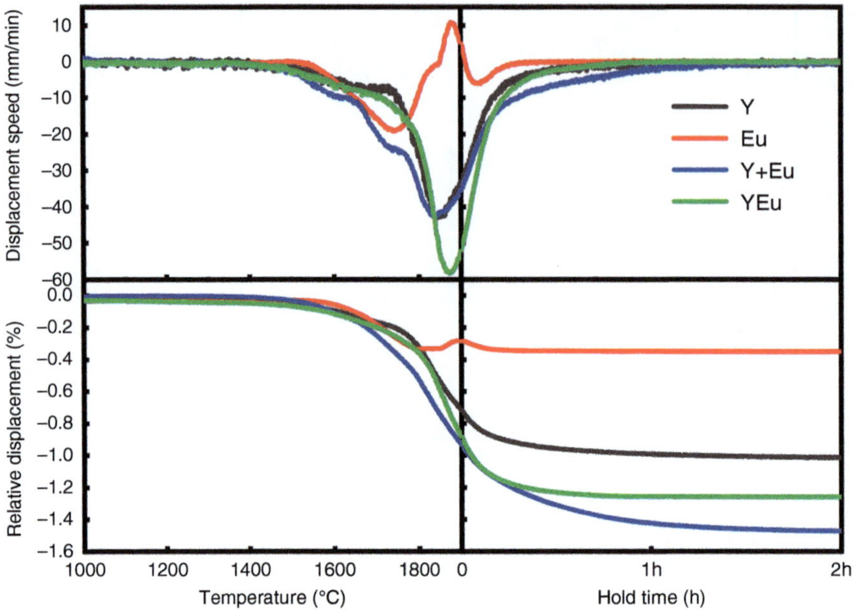

Figure 6.13 The shrinkage curves of four samples of Y, YEu, Y + Eu, and Eu. Source: Reproduced with permission from Ref. [2]; © 2023 John Wiley & Sons.

To study the effect of different sintering additives on densification behavior, the shrinkage curve of the four samples was analyzed, and the results are shown in Figure 6.13. It can be seen that the maximum shrinkage rate of sample Eu corresponds to a temperature below 1800 °C, lower than the maximum shrinkage rate of the other three samples, and the starting shrinkage temperature of the sample with Y is the highest among the four samples. This further illustrates that the glass phase containing Eu_2O_3 has a lower melting point than the glass phase containing Y_2O_3. The analysis in this matches the results discussed in Table 6.4. At 1900 °C, the sample with Eu had a distinct expansion peak, and its final shrinkage degree was smaller than that in the other three samples. There is an expansion peak because the glass phase containing Eu_2O_3 is volatile serious under high temperatures. While the sample large size grain basically developed, and the large grains got into the solid-phase diffusion stage, large grains overlap each other to form a hole. And volatilization of the liquid phase causes insufficient liquid phase, and the sample expansion rate in the solid-phase diffusion phase is higher compared to the sample shrinkage rate, thus presenting the phenomenon of sample expansion. Moreover, both samples with YEu and Y + Eu finally showed more shrinkage than that of the sample with Y, which indicates that the addition of a small amount of Eu_2O_3 is conducive to densification compared with the rare earth oxide Y_2O_3 alone.

Figure 6.14 shows the SEM photos of the sample after sintering at 1900 °C for 4 h. Large-size grains in samples with Y, YEu, and Y + Eu are shown in Figure 6.14. The grain size in the sample with YEu is significantly larger than that in Y, and the grain in the intercrystalline phase is significantly well developed, but there are a few

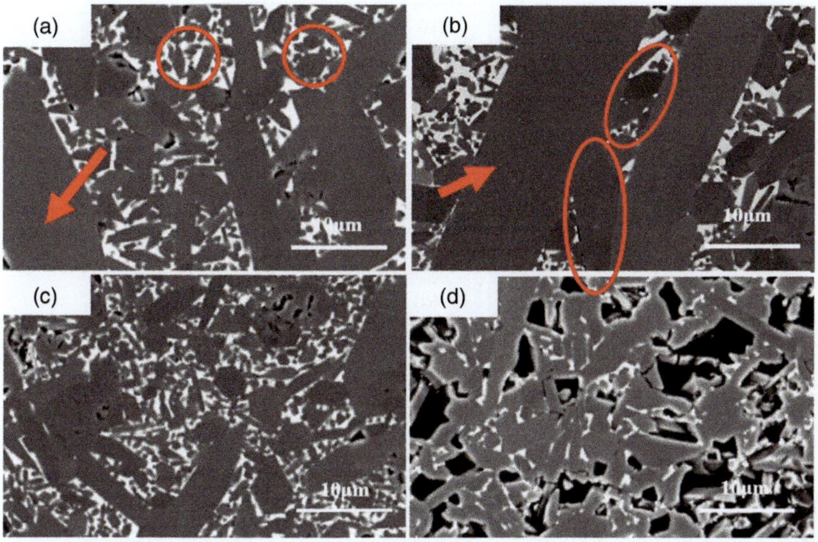

Figure 6.14 SEM of the nitrided sample after sintering at 1900 °C for 4 h. (a) Y, (b) YEu, (c) Y + Eu, (d) Eu. Source: Reproduced with permission from Ref. [2]; © 2023 John Wiley & Sons.

pores, due to the volatilization of the liquid phase containing Eu_2O_3. In the sample with Eu, a large number of holes appeared, and the grain size is significantly smaller compared to the other three samples. The appearance of a large number of holes is negative to the improvement of thermal conductivity and the optimization of their mechanical properties.

The mechanical properties of the samples after sintering at 1900 °C for 4 h are shown in Figure 6.15. Figure 6.15a shows that the sample with YEu has the highest hardness, up to 1639 HV and high flexural strength (776.8 MPa). However, the bending strength of sample Y was only 735 MPa and hardness of only 1534 HV. Sample Eu had the lowest bending strength and hardness, with 476 MPa and 654 HV, respectively. Because the relative density of sample Eu

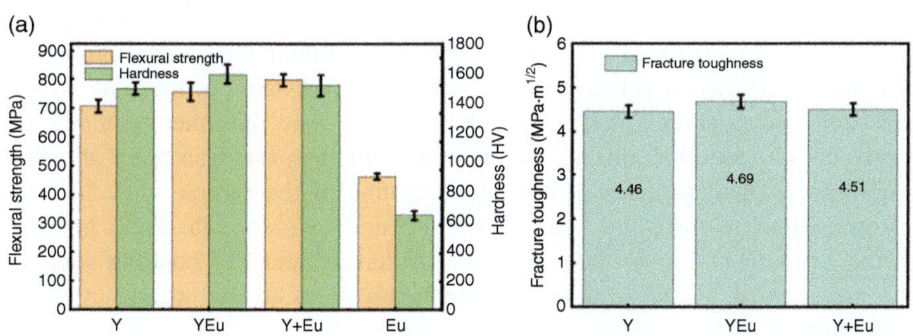

Figure 6.15 Mechanical properties of samples after sintering at 1900 °C for 4 h. (a) Flexural strength and hardness. (b) Fracture toughness. Source: Reproduced with permission from Ref. [2]; © 2023 John Wiley & Sons.

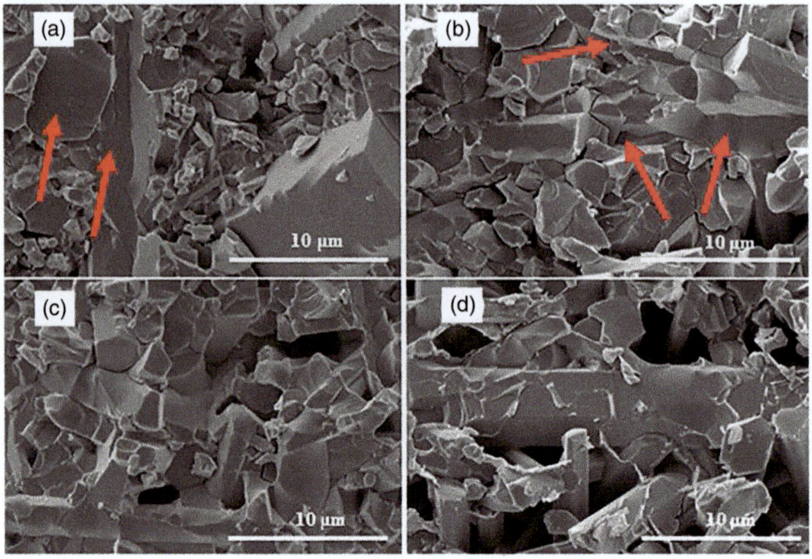

Figure 6.16 SEM of the fracture surface of (a) Y, (b) YEu, (c) Y + Eu, (d) Eu. Source: Reproduced with permission from Ref. [2]; © 2023 John Wiley & Sons.

is low, it presents a porous state, and it has few abnormally large-sized grains, which is negative to the improvement of mechanical properties. Figure 6.15b shows the toughness test results of the sample, in which the toughness of the sample Eu is not listed, because there are too many holes in the sample Eu, and the toughness is tested by the indentation method, and the line and crack length of the crack extension cannot be determined. It can be seen that the sample with YEu has the highest toughness, reaching 4.69 MPa $m^{1/2}$. The mechanical properties are closely related to the microstructure of the sample. Figure 6.16 shows the scanning electron microscope photo of the sample section. There are too many holes in the sample with Eu, which is not conducive to improving its mechanical properties. There are large-size β-Si_3N_4 grains distributed in small-size β-Si_3N_4 grains, forming a typical bimodal morphology which is conducive to the improvement of their mechanical properties. Moreover, from (a) to (c), we can see some grooves left by the grain extraction, with no obvious fracture, which indicates that the fracture along the crystal mainly occurred during the fracture, which is the reason for the high toughness of these samples. However, the grains in the sample with Eu overlap each other to form the holes, which do not have enough liquid phase to fill the holes, resulting in their low relative density and low bending strength. The presence of a large number of pores also leads to the weak binding of each grain.

Figure 6.17 shows the thermal diffusion and thermal conductivity of the four samples at different temperatures and holding times. As can be seen from Figure 6.17a,b, the thermal diffusion and thermal conductivity increase as the

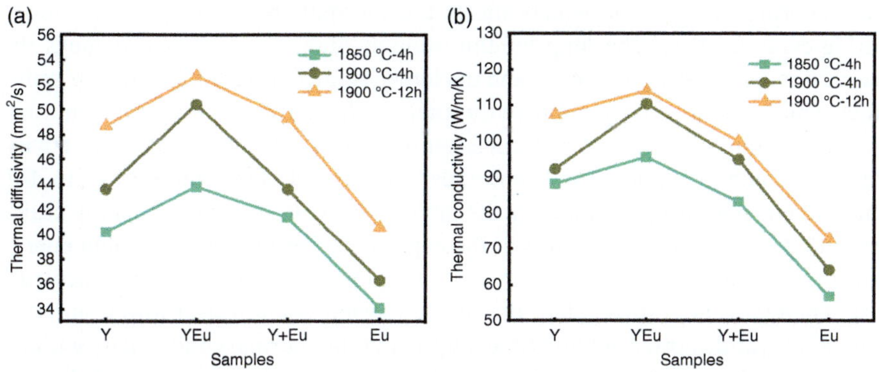

Figure 6.17 (a) Thermal diffusion and (b) thermal conductivity of four samples at different post-sintering temperatures and holding times. Source: Yao Dongxu, Wang Weide, Zeng Yu-ping.

sintering temperature and holding time increase. Under 1900 °C sintering temperature, the thermal conductivity is relatively high, but with the extension of the holding time under 1900 °C, the holding time is extended from 4 to 12 h, and the thermal conductivity increase is not obvious. After 4 h of 1900 °C insulation, the thermal conductivity of the sample with YEu increased by 16.5% compared to that of sample Y, above 110 W/(m K), while the thermal conductivity of samples with Y, Y + Eu, and Eu were 94.7 W/(m K), 94.9 W/(m K), and 64.14 W/(m K), respectively.

Considering the resultant density, mechanical properties, and thermal conductivity of the sample comprehensively, we found that $(YEu)_2O_3$ is an effective sintering additive that can improve both mechanical properties and thermal conductivity. When the molar ratio of $(YEu)_2O_3$ and MgO is 2 : 5, the optimal sintering system is 1900 °C and insulated for 4 h, and the thermal conductivity increases significantly under this system.

6.2 Optimization of Sintering Aids for High Thermal Conductivity Si_3N_4 Ceramics

The SiO_2 on the surface of the silicon nitride powder and the added oxide sintering aid introduce oxygen into the sintering system, which can result in high oxygen content in the liquid phase. Due to the $\alpha \rightarrow \beta$ phase transformation (dissolution of α-Si_3N_4 in the liquid phase and precipitation of β-Si_3N_4) and Ostwald ripening (dissolution of small β-Si_3N_4 grains in the liquid phase and precipitation growth on large β-Si_3N_4 grains) during the sintering process, the formation of lattice oxygen in the β-Si_3N_4 lattice during the dissolution–precipitation is a kinetic process, which is mainly controlled by the oxygen content in the liquid phase [13]. Therefore, a liquid phase with high oxygen content generates more lattice oxygen in the newly formed β-Si_3N_4 lattice, leading to an increase in silicon vacancies and enhancing the scattering of phonons, which affects the thermal conductivity of the silicon nitride ceramics [14]. The long holding time at high temperatures prolongs the dissolution–precipitation process, and the lattice is further purified while the grains grow, which can effectively improve thermal conductivity. However, the abnormal growth of grains alone does not consistently improve the thermal conductivity, which gradually slows down and converges to a constant as the grains grow [15, 16]. This value is mainly determined by the lattice defects in the grains (mainly lattice oxygen). And this improvement is at the expense of flexural strength, high energy consumption, long cycle time, high loss of equipment, and high manufacturing cost. Therefore, how to regulate the lattice oxygen content at a lower sintering temperature and shorter holding time, improving the thermal conductivity without deteriorating its strength, and achieving the harmonious development of thermal conductivity and strength is the key point [17].

Non-oxide additives such as nitrides, fluorides, carbides, etc. instead of oxide sintering additives were used to reduce the introduction of oxygen, which can reduce the lattice oxygen content. However, some of the non-oxide additives also have some shortcomings, such as the reaction of fluoride sintering additives with SiO_2 can produce biotoxic SiF_4(g) [18]; the use of $Y_2Si_4N_6C$ instead of Y_2O_3 can also effectively enhance the thermal conductivity of silicon nitride ceramics, but the preparation process of $Y_2Si_4N_{6C}$ is complicated, and the preparation of this additive in large quantities has not been reported [19]. Therefore, the search for safe, non-toxic, and more efficient non-oxide sintering aids is still on the way. Si_3N_4 has two elements, Si and N. Metal nitrides such as $MgSiN_2$ (which also contains Si) be an effective aid for improving the thermal conductivity of silicon nitride [20]. So far, no report has been seen on the use of metal silicides as sintering aids to study the thermal conductivity of silicon nitride. The experiments in this chapter envisage the use of $ZrSi_2$ for the preparation of high thermal conductivity silicon nitride. Its effects on the densification, phase composition, microstructure, thermal conductivity, and mechanical properties of silicon nitride are investigated, and its reaction mechanism is explored to lay the foundation for the preparation of high thermal conductivity silicon nitride ceramic materials.

6.2.1 Preparation of High Thermal Conductivity Silicon Nitride Ceramics Using $ZrSi_2$ as a Sintering Aid

6.2.1.1 Reaction Mechanism of $ZrSi_2$

To figure out the reaction mechanism of $ZrSi_2$ during the sintering process, a powder mixture with a composition of $ZrSi_2$: SiO_2 = 1 : 1 at mol ratio was heated at 1000 °C, 1300 °C, 1400 °C, and 1500 °C in nitrogen for 2 h, respectively. The XRD results before and after being heated at different temperatures for 2 h are shown in Figure 6.18.

As can be seen from Figure 6.18a, the main crystalline phases were $ZrSi_2$ and SiO_2 in the mixture without heat treatment; this is consistent with the composition of the starting raw mixture. It was found that the diffraction peak of $ZrSi_2$ and SiO_2 diminished gradually as the temperature increased and vanished completely after

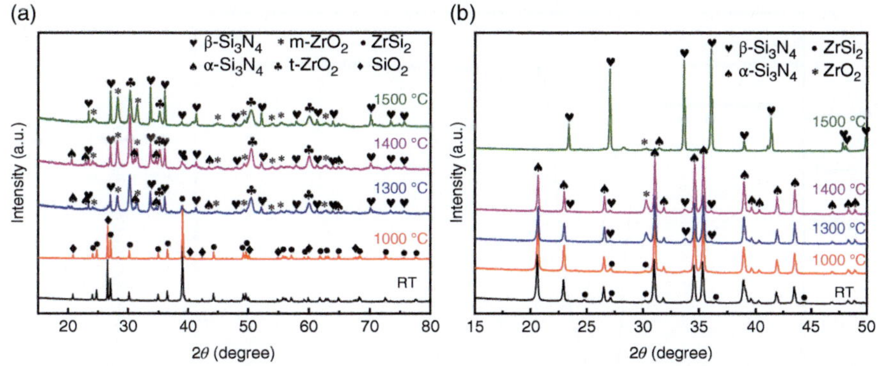

Figure 6.18 The XRD patterns of samples before and after being heated in various temperatures for 2 h: (a) $ZrSi_2$: SiO_2 = 1 : 1. (b) Si_3N_4: $ZrSi_2$: MgO = 93 : 3 : 4. Source: Reproduced with permission from Ref. [21]; © 2019 John Wiley & Sons.

being heated at 1500 °C for 2 h. In the meantime, ZrO_2, α-Si_3N_4, and β-Si_3N_4 came into being at 1300 °C. Besides, α-Si_3N_4 transformed into β-Si_3N_4 gradually as the temperature increased. Obviously, the formation of α-Si_3N_4 and β-Si_3N_4 was mainly attributed to the reaction between SiO_2 and $ZrSi_2$ since there was no β-Si_3N_4 or α-Si_3N_4 in the raw powder mixture.

As shown in Figure 6.18b, weak peaks of ZrO_2 and β-Si_3N_4 were found in sample 3Z4M at 1300 °C which were in accordance with behaviors in the $ZrSi_2$ and SiO_2 powder mixture. However, the diffraction intensity of β-Si_3N_4 remains unchanged even when heated at 1400 °C for 2 h; besides, α-Si_3N_4 was not detected after being heated at 1500 °C for 2 h, which means that $\alpha \rightarrow \beta$ phase transformation of Si_3N_4 in 3Z4M sample starts above 1400 °C and was completed after being heated at 1500 °C for 2 h.

The as-formed ZrO_2 collaborated with MgO and Si_3N_4 to form the Si-Zr-Mg-O-N liquid phase promoting the densification of Si_3N_4. This is consistent with the observation from the thermal dilatometer which was shown in Figure 6.19. It was found that the shrinkage of sample 3Z4M started around 1400 °C, and drastic shrinkage was detected around 1600 °C indicating that a considerable amount of liquid phase had formed in the sample. The identical conclusions were obtained in the HAADF-STEM image of typical sample 3Z4M. As shown in Figure 6.20, β-Si_3N_4 grains and intergranular phases were recognized as dark and bright fields because of atomic composition contrast. Adjacent β-Si_3N_4 grains were typically separated by a thin amorphous grain-boundary film, and the amorphous intergranular phases tended to distribute at the junctions of triple and multiple grains. The marked position in glass pockets was analyzed by EDS, and Zr, Si, Mg, N, and O were contained which confirmed the rationality of the previous deduction. The detailed results are summarized in Figure 6.20b.

Besides, β-Si_3N_4 particles appeared by the reaction between $ZrSi_2$ and SiO_2 at a much lower temperature than the $\alpha \rightarrow \beta$ phase transformation of Si_3N_4. As shown in Figure 6.21, the rod-like grains were generated and grew up preferentially when

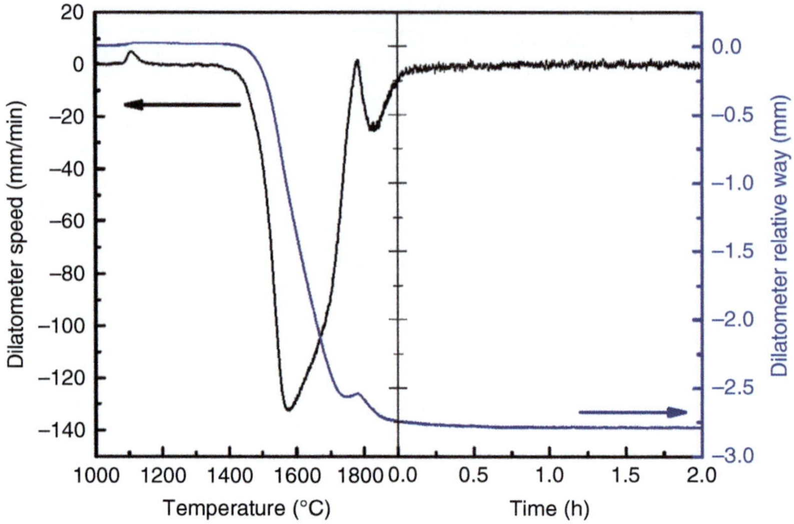

Figure 6.19 The *in situ* sintering shrinkage behavior of 3Z4M. Source: Reproduced with permission from Ref. [21]; © 2019 John Wiley & Sons.

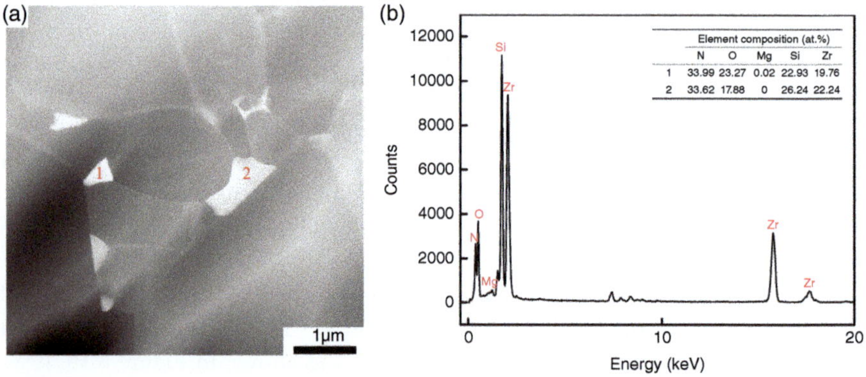

	Element composition (at.%)				
	N	O	Mg	Si	Zr
1	33.99	23.27	0.02	22.93	19.76
2	33.62	17.88	0	26.24	22.24

Figure 6.20 (a) HAADF-STEM image of sample 3Z4M; (b) EDS spectrum and composition of multiple grains junction of sample 3Z4M. Source: Reproduced with permission from Ref. [21]; © 2019 John Wiley & Sons.

$ZrSi_2$ and SiO_2 powder mixture was heated at 1400 °C for 2 h. Likewise, additive $ZrSi_2$ would react with native SiO_2 layer on silicon nitride raw powder in sample 3Z4M to generate β-Si_3N_4 particles which play a dominating role in promoting the abnormal grain growth.

Based on the foregoing analysis, two stages were possibly included in sample 3Z4M. Initially, $ZrSi_2$ reacted with SiO_2 and N_2 to form ZrO_2, α-Si_3N_4 and β-Si_3N_4. Then α-Si_3N_4 transformed into β-Si_3N_4 via the dissolution–precipitation mechanism in the liquid phase, which was generated by the reaction among Si_3N_4, MgO, and newly formed ZrO_2. *In situ*-formed β-Si_3N_4 particles play a dominating role in promoting the abnormal grain growth. The schematic drawing of densification

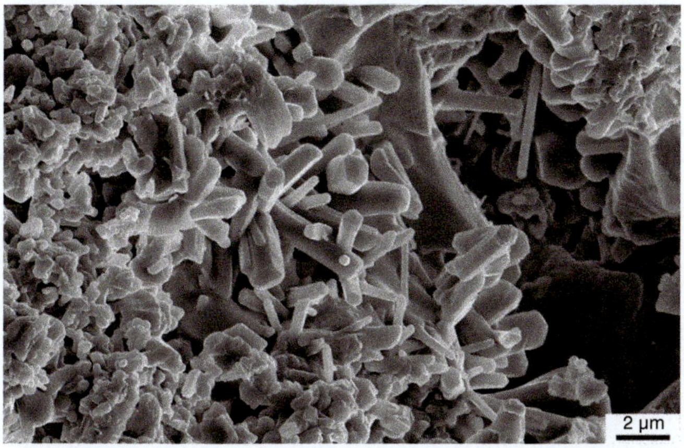

Figure 6.21 The microstructure of $ZrSi_2$ and SiO_2 mixture powder after being heated at 1400 °C for 2 h. Source: Reproduced with permission from Ref. [21]; © 2019 John Wiley & Sons.

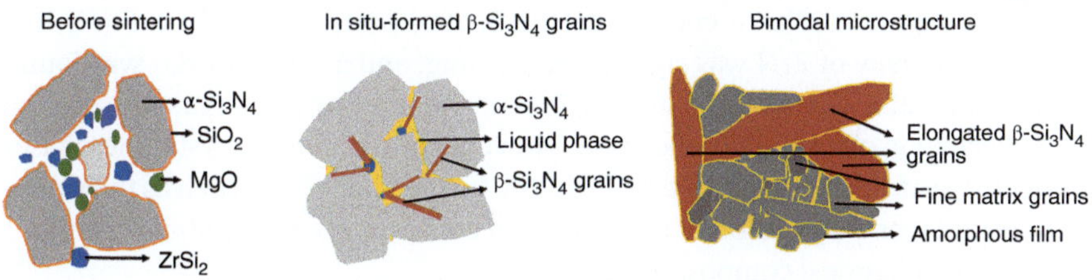

Figure 6.22 Schematic drawing of the densification mechanism of Si_3N_4 ceramics with $ZrSi_2$ and MgO additives during the gas pressure sintering. Source: Reproduced with permission from Ref. [21]; © 2019 John Wiley & Sons.

mechanism of Si_3N_4 ceramics with $ZrSi_2$ and MgO additives during the gas pressure sintering is shown in Figure 6.22

6.2.1.2 Effect of $ZrSi_2$ on the Phase Composition

Figure 6.23a shows the phase identification result of the grinded and polished surfaces (center) of sintered samples at 1900 °C. Evidently, β-Si_3N_4 was the dominant phase, while α-Si_3N_4 peaks were not detected in all cases, indicating the $\alpha \rightarrow \beta$ phase transformation was completed after sintering. The ZrO_2 phase was observed in all samples except for 0Z7M, and the ZrO_2 and ZrN coexisted in the inner parts of samples 5Z2M, 6Z1M, and 7Z0M. Compared to 3O4M, the diffraction intensity of ZrO_2 in 3Z4M was a little bit weak. The ZrN phase was not found in both 3O4M and 3Z4M, probably below the detection limit of the apparatus. The Mg-containing phases were not found in all the XRD patterns, indicating the Mg-containing secondary phases were in an amorphous state or the crystalline amount was below the detection limit of the apparatus.

It is worthwhile mentioning that a golden yellow coating appeared at the surface of all samples, while the inner parts of the samples appear gray. To figure out the

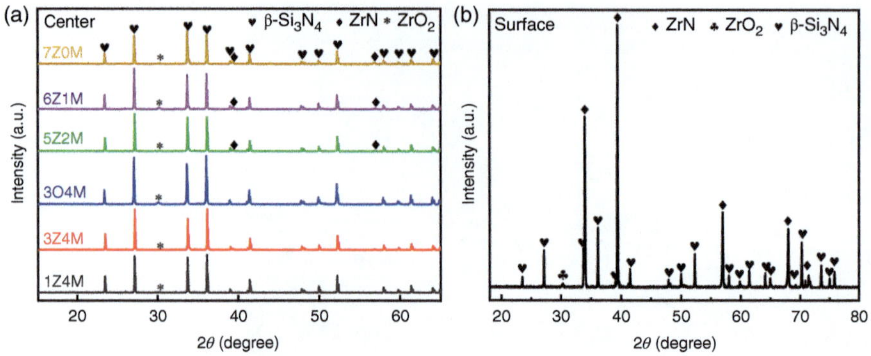

Figure 6.23 The XRD patterns of the Si_3N_4 ceramics samples after sintering at 1900 °C for 12 h: (a) the center, (b) the un-machined surface of sample 3Z4M. Source: Reproduced with permission from Ref. [21]; © 2019 John Wiley & Sons.

possible reasons for the differences between the internal and external parts of the samples, the phase identification of the un-machined surface of typical sample 3Z4M is shown in Figure 6.23b. In contrast to the central parts, except for β-Si_3N_4, the diffraction intensity of ZrN was significantly strong, and a trace of ZrO_2 was found also. The crystallization of ZrO_2 and/or ZrN in the interior parts must be affected by the specimens chemical composition. However, the different crystalline phases between the surface and the inside must be due to the other effects such as the sintering atmosphere or evaporation of the elemental constituents during sintering rather than an initial chemical composition.

The following reactions presumably occurred among Si_3N_4, ZrO_2, N_2, and C (graphite atmosphere inside graphite resistor furnace).

$$4Si_3N_4(s) + 6ZrO_2(s) = 12SiO(g) + 6ZrN(s) + 5N_2(g)$$

$$\Delta_rG^\circ\ (2173\ K) = -76.03\ kJ \cdot mol^{-1}, K_p\ (2173\ K) = 6.72 \times 10^1$$

$$2ZrO_2(s) + 4C(s) + N_2(g) = 2ZrN(s) + 4CO(g)$$

$$\Delta_rG^\circ\ (2173\ K) = -130.41\ kJ \cdot mol^{-1}, K_p\ (2173\ K) = 1.36 \times 10^3$$

6.2.1.3 Effect of $ZrSi_2$ on Microstructure

The low-magnification and high-magnification SEM images of the polished and plasma-etched surfaces of the inner parts of sintered samples are illustrated in Figures 6.24 and 6.25, respectively. It is evident that both 0Z7M and 7Z0M show a porous microstructure due to the low densities. The observations suggest that a bimodal microstructure composed of a fraction of large elongated grains and a majority of fine matrix grains was achieved in densified samples. Besides, compared to 3O4M, the selective grain growth in the $ZrSi_2$-doped samples was so pronounced that the β-Si_3N_4 grains larger than 100 μm in length were observed in low-magnification images, which means the addition of $ZrSi_2$ does stimulate the growth of the elongated β-Si_3N_4 grains.

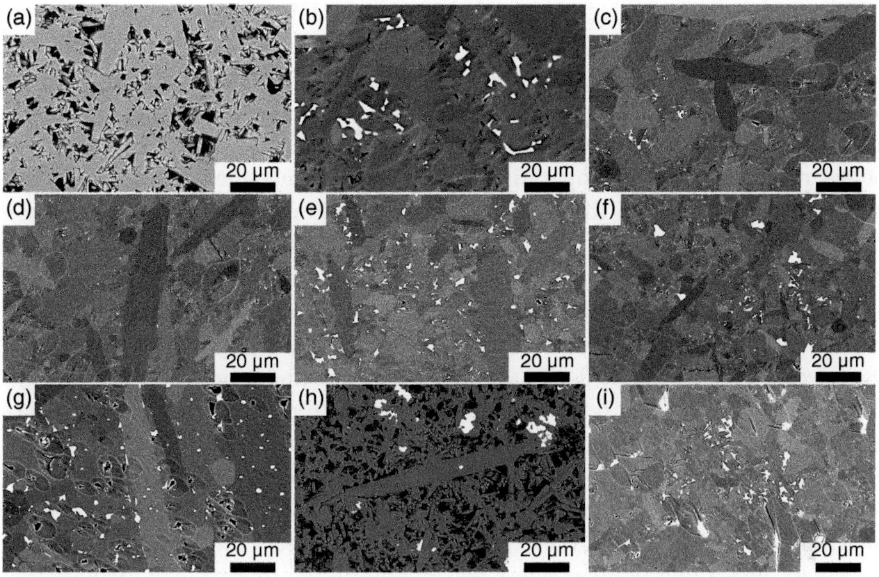

Figure 6.24 The low-magnification images of polished and plasma-etched surfaces of Si_3N_4 ceramics after sintering at 1900 °C for 12 h: (a) 0Z7M, (b) 1Z6M, (c) 2Z5M, (d) 3Z4M, (e) 4Z3M, (f) 5Z2M, (g) 6Z1M, (h) 7Z0M, (i) 3O4M. Source: Reproduced with permission from Ref. [21]; © 2019 John Wiley & Sons.

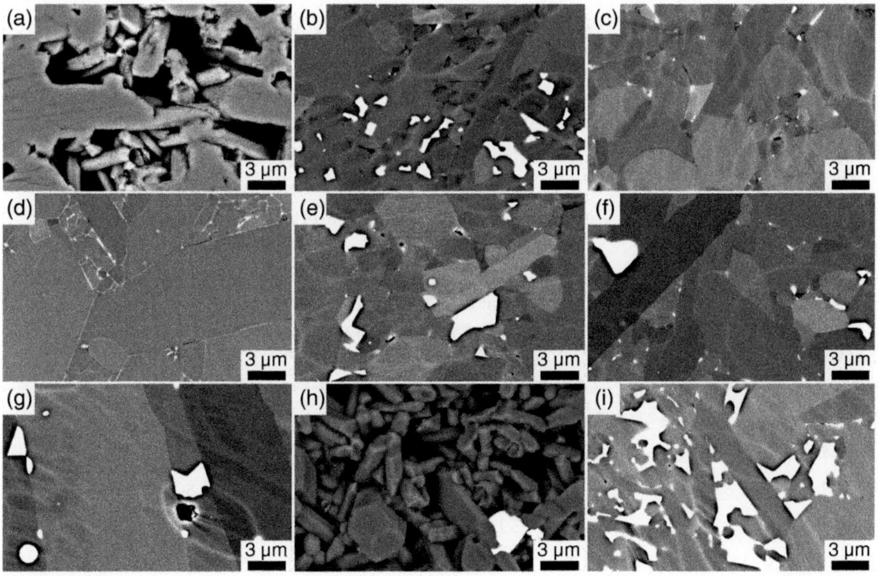

Figure 6.25 The high-magnification images of polished and plasma-etched surfaces of Si_3N_4 ceramics after sintering at 1900 °C for 12 h: (a) 0Z7M, (b) 1Z6M, (c) 2Z5M, (d) 3Z4M, (e) 4Z3M, (f) 5Z2M, (g) 6Z1M, (h) 7Z0M, (i) 3O4M. Source: Reproduced with permission from Ref. [21]; © 2019 John Wiley & Sons.

The experiments revealed that the formation and abnormal growth of elongated β-Si_3N_4 grains depend strongly on the amount, grain-size distribution, and morphology of β-Si_3N_4 grain [22]. As discussed previously, a large number of β-Si_3N_4 crystal seeds were *in situ* generated by the reaction between $ZrSi_2$ and native SiO_2 at low temperatures. However, β-Si_3N_4 grains would not be generated until $\alpha \rightarrow \beta$ phase transformation in comparative sample 3O4M. Thus, the abnormal growth of large elongated grains in the $ZrSi_2$-doped samples is possibly attributed to the *in situ*-formed β-Si_3N_4 grains. The high-magnification images illustrated that 3Z4M possesses a lower volume fraction of glassy phase than 3O4M. The ZrO_2 in 3Z4M was produced by the reaction between SiO_2 and $ZrSi_2$, and thus the glassy phase was reduced due to the consumption of SiO_2 by $ZrSi_2$. Besides, a significant difference in the microstructure of $ZrSi_2$-doped samples with some literature [18, 19, 23, 24], also the most interesting one, is that the distribution state of the grain-boundary phase and the thickness of grain-boundary film according to the high-magnification observations. The secondary phases were enriched at triple and multiple junctions, and the grain-boundary film was extremely thin in densified samples indicating higher Si_3N_4-Si_3N_4 contiguity. Both the experiment and the theory proved that the isolated secondary phase morphology is less detrimental to the thermal conductivity than the interconnected secondary phase [15, 25].

6.2.1.4 Effect of $ZrSi_2$ on Thermal Conductivity

As can be seen from Table 6.5, the thermal conductivity of 0Z7M was only 59.39 W/m K due to poor relative density. The thermal conductivity of sample 1Z6M drastically increased to 89.22 W/m K with only 1 mol% of $ZrSi_2$ being added. The thermal conductivity of samples keeps increasing slightly with the dosage of $ZrSi_2$. When 3 mol% $ZrSi_2$ was added, the thermal conductivity of the sample 3Z4M reached a maximum value of 113.91 W/m K. However, the thermal conductivity of 3O4M was only 84.58 W/m K. With the amount of $ZrSi_2$ kept increasing, thermal

Table 6.5 Relative density, weight loss, and thermal properties of the prepared Si_3N_4 ceramics.

Sample	Bulk density (g/cm^3)	Relative density (%)	Weight loss (%)	Thermal diffusivity (mm^2/s)	Thermal conductivity (W/m K)	Flexural strength (MPa)	Fracture toughness (MPa·m$^{1/2}$)	Electrical resistivity (Ω cm)
0Z7M	2.53	79.25	8.21	34.52	59.39	—	—	—
1Z6M	3.17	98.66	4.34	41.39	89.22	505 ± 50	5.67 ± 0.12	4.40×10^{14}
2Z5M	3.20	99.58	3.93	49.89	108.58	524 ± 20	5.87 ± 0.15	4.33×10^{14}
3Z4M	3.22	99.52	3.21	52.02	113.91	553 ± 39	6.36 ± 0.32	5.65×10^{14}
4Z3M	3.21	98.77	3.12	48.09	104.97	519 ± 58	6.27 ± 0.42	1.40×10^{14}
5Z2M	3.21	98.21	2.43	45.78	99.94	467 ± 32	6.83 ± 0.28	1.32×10^{14}
6Z1M	3.20	98.07	1.27	43.49	94.65	495 ± 60	6.56 ± 0.35	7.89×10^{13}
7Z0M	2.32	70.95	0.80	31.08	49.04	—	—	—
3O4M	3.22	99.69	3.85	38.63	84.58	552 ± 48	7.35 ± 0.25	2.32×10^{14}

conductivity began to decrease. The thermal conductivity of the sample without MgO addition (7Z0M) was quite low because of its porous microstructure. Thus, it is difficult to densify silicon nitride ceramics with the sole addition of $ZrSi_2$ or MgO.

The drastic increase of thermal conductivity with the dosage of $ZrSi_2$ till around 2 mol% was attributed to the remarkably improved densification, which means substantial amounts of liquid phase were formed. After samples were densified completely when the addition of $ZrSi_2$ was 2 mol%, the increase of thermal conductivity with the increase of $ZrSi_2$ addition was probably attributed to the consumption of SiO_2. The process of oxygen dissolved into the lattice was inhibited because the oxygen content in the liquid phase was reduced.

The higher thermal conductivity of 3Z4M than of 3O4M was attributed to three factors. First, as discussed before, SiO_2 would not be consumed in the ZrO_2-dope sample, so sample 3O4M probably possesses a higher oxygen concentration than 3Z4M in the liquid phase during the sintering of Si_3N_4. As a result, oxygen dissolved in the β-Si_3N_4 grain lattice would be higher in 3O4M. Second, $ZrSi_2$ facilitated the elimination of SiO_2 and the reduction of the low thermal conductivity intergranular phase. Third, the amount of large elongated grains in 3Z4M was higher than in 3O4M. Therefore, the thermal conductivity was effectively improved by the addition of $ZrSi_2$ instead of ZrO_2. Once the amount of $ZrSi_2$ was enough, the excess addition of $ZrSi_2$ would have a negative effect on thermal conductivity. The diffraction intensity of ZrO_2 increased with the increment of $ZrSi_2$ addition. A typical thermal conductivity value for ZrO_2 is as low as 2 W/m K [26]. So, the thermal conductivity of sintered silicon nitride was partly weakened by ZrO_2. Moreover, the superfluous $ZrSi_2$ would be nitride to ZrN, which would be detrimental to the densification of silicon nitride as an inert secondary phase. Apart from this, the amount of liquid phase was reduced as the decreased amount of MgO with the increased ratio of $ZrSi_2$/MgO. This is consistent with the high-magnification observation in Figure 6.24 where some pores were found in samples 4Z3M, 5Z2M, and 6Z1M.

6.2.1.5 Effect of $ZrSi_2$ on Mechanical Properties and Electrical Resistivity

Thermal conductivity could be improved by the large grains; however, the reduction of flexural strength was inevitable due to abnormal grain growth. Measured by three-point bending test, samples from 1Z6M to 6Z1M showed a flexural strength of around 500 MPa which was comparable to that of the comparative sample 3O4M, and values have been reported in a similar sintering process [27, 28]. Taking the coarse microstructure into consideration, this value was reasonable. As shown in Figure 6.24, an interlocking microstructure consisting of large-elongated β-Si_3N_4 grains was formed which is a prerequisite for achieving toughening of Si_3N_4 by crack bridging or crack deflection mechanisms. However, the formation of elongated grains is not sufficient for obtaining toughening effects [29, 30]. Toughening by a crack bridging or crack deflection process occurs only when cracks propagate along the prismatic faces of the elongated β-Si_3N_4 grain, so an appropriately weak grain boundary is necessary. When non-oxide additive $ZrSi_2$ was applied, the volume fraction of glassy phase was reduced, and the grain-boundary film was thinned indicating a strong connection of Si_3N_4–Si_3N_4 grains; thus, interfacial debonding

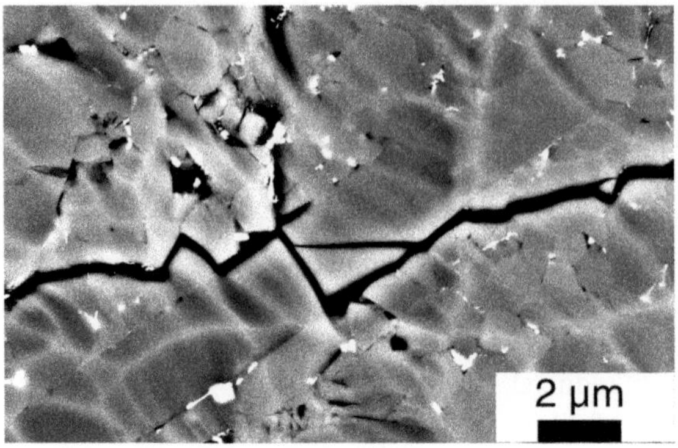

Figure 6.26 SEM view of a crack path in 3Z4M. Source: Yao Dongxu, Wang Weide, Zeng Yu-ping.

was hindered (Figure 6.26). These factors may account for lower fracture toughness than that of the comparative sample 3O4M.

The electrical resistivity was higher than 1×10^{14} Ω·cm except for sample 6Z1M (7.89×10^{13} Ω·cm). Despite the electrically conductive phase ZrN being formed inside the Zr-containing samples, the insulation performance was not severely deteriorated in comparison to the value that has been reported [24, 31]. The electrical resistivity of silicon nitride is governed by the formation of electrically conductive path. It was reported that the electrical resistivity of typical Y-Si-A1 oxynitride glasses was lower than 10^{11} Ω·cm, while the resistivity of 10^{16} Ω cm was reported for Si_3N_4 on n-type silicon [21, 32]. In those oxide additives doped silicon nitride, the liquid phases remain as amorphous and/or crystalline oxynitride glasses at multigrain junction and grain-boundary films after cooling; the electrical resistivity was deteriorated by the lower resistivity oxynitride glasses, especially the grain-boundary films path. In this work, a non-oxide additive was applied, and the volume fraction of the glassy phase was reduced, so deterioration from oxynitride glasses was weakened. Although a tiny amount of ZrN was formed, it is isolated and dispersed in the β-Si_3N_4 matrix, and chains of electroconductive ZrN particles cannot be formed. Therefore, insulation performance would not be deteriorated by the addition of $ZrSi_2$.

6.2.2 High Thermal Conductivity Si_3N_4 Sintered with YH_2 as Sintering Aid

In the powder metallurgy industry, oxygen in raw material powders can also adversely affect the properties of metal products, for example, high oxygen content in raw Ti powder can make titanium alloys less ductile [33]. Due to the demand for oxygen removal in low-cost, high-performance titanium alloys, researchers have introduced rare earth metal monomers in the form of rare earth metal hydrides [34, 35]. The rare earth metal hydrides decompose into rare earth metal monomers and H_2. The highly reactive rare earth metal monomers act as oxygen absorbers, and H_2 also has a cleaning effect on the oxide layer on the surface of the metal powder particles [36]. As a potential sintering additive for Si_3N_4 ceramics, rare earth hydride YH_2 may react with native SiO_2 to generate Y_2O_3, and thereby the Y_2O_3/SiO_2 ratio

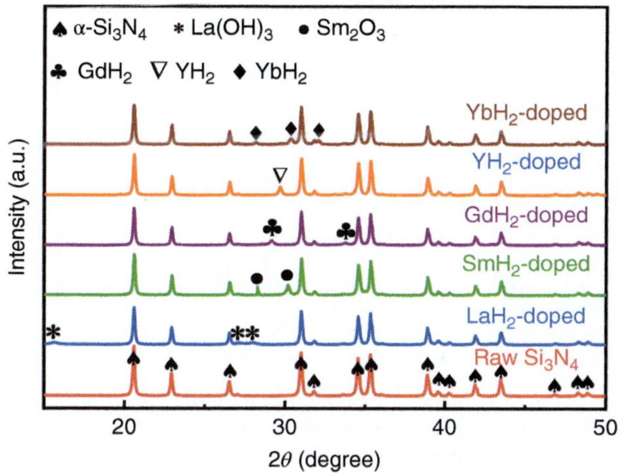

Figure 6.27 The phase composition of mixtures doped with REH_2 after ball milling. Source: Yao Dongxu, Wang Weide, Zeng Yu-ping.

can be tailored by the addition of YH_2. Therefore, it was expected that YH_2 could be an efficient sintering additive for improving the thermal conductivity of Si_3N_4 ceramics.

All rare earth hydrides can react with water to produce the corresponding hydroxide and release H_2. The prerequisite for carrying out the next experiment is that the added metal hydrides are not oxidized or react with water vapor during the ball milling process. Therefore, a vacuum ball milling tank was used for this experiment. First, a material with a mass ratio of Si_3N_4: REH_2: MgO = 94.7 : 2 : 1.5 was prepared to verify the stable presence of REH_2 during the mixing process.

The phase composition of the materials after ball milling is shown in Figure 6.27. GdH_2, YH_2, and YbH_2 can be well preserved, while LaH_2 and SmH_2 are no longer present. LaH_2 was transformed to $La(OH)_3$, and SmH_2 was transformed to Sm_2O_3. This indicates that LaH_2 and SmH_2 are less stable and react with oxygen or water vapor in the air during the batching or ball milling process, so only the effect of adding GdH_2, YH_2, and YbH_2 as sintering aids for Si_3N_4 was studied.

In this work, the amount of MgO was fixed at 1.5 wt% while the amount of YH_2 was set as 0, 1, 2, 3, and 4 wt%, and labeled as 0YM,1YM, 2YM, 3YM, and 4YM, respectively. In addition, a comparative experiment (Si_3N_4: Y_2O_3: MgO = 94.7 : 3.8 : 1.5 at a mass ratio, abbreviated as YOM) was conducted.

6.2.2.1 Pre-sintering of the Compact

The planetary ball milling is a mechanical crushing process containing considerably high energy. So, the powder mixtures were ball-milled within a sealed polymer container. The XRD patterns of the Si_3N_4 samples after ball milling are shown in Figure 6.28 (a). YH_2 phase was observed in all samples, and the diffraction intensity of YH_2 increased with the increase of YH_2 content, indicating that YH_2 was not oxidized during the ball milling process. Then, specimens were heated at 1200 °C for

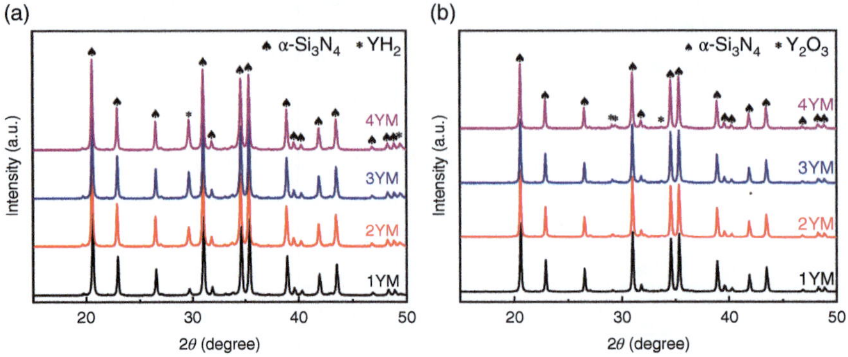

Figure 6.28 The XRD patterns of the Si_3N_4 ceramics: (a) after ball milling, (b) after pre-sintering at 1200 °C for 4 h. Source: Reproduced with permission from Ref. [37]; © 2020 Elsevier Ltd.

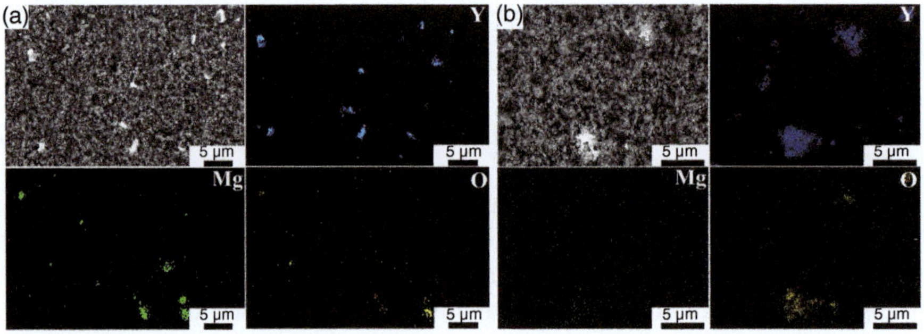

Figure 6.29 The EDS mapping of the Si_3N_4 ceramics: (a) after ball milling, (b) after pre-sintering at 1200 °C for 4 h. Source: Yao Dongxu, Wang Weide, Zeng Yu-ping.

4 h under an N_2 pressure of 0.1 MPa. Figure 6.28b shows the XRD results of the Si_3N_4 samples after heat treatment. It was found that YH_2 vanished, and Y_2O_3 was formed in all YH_2-doped samples. Besides, the diffraction intensity of Y_2O_3 increased with the increase of YH_2 content. Identical conclusions can be retrieved from the elements distribution of sample 3YM after heat treatment in Figure 6.29. Obviously, the Y element coexisted with the O element in the same position. The compound was Y_2O_3. Table 6.6 gives the oxygen contents of samples before and after heat treatment. The oxygen content in all samples decreased after heat treatment, indicating Y_2O_3 was generated by the reactions inside the Si_3N_4 samples instead of the oxidation of YH_2 by external factors.

To figure out the reaction mechanism of YH_2 during the sintering process, the powder mixture with a composition of YH_2: SiO_2 = 4 : 3 at mol ratio was heated at 800, 1000, and 1200 °C under vacuum conditions, respectively. As shown in Figure 6.30, the diffraction intensities of YH_2 and SiO_2 phases remain unchanged even after being heated at 1000 °C, which means YH_2 decomposes at >1000 °C in the powder mixture of YH_2 and SiO_2. Besides, the YH_2 and SiO_2 phases vanished,

Table 6.6 Oxygen contents of the Si_3N_4 samples before and after pre-sintering at 1200 °C for 4 h.

Sample	Oxygen content (wt%)		Y_2O_3/SiO_2 ratio
	Before heat treatment	After heat treatment	
YOM	2.24	2.16	0.34
1 YM	1.75	1.66	0.17
2 YM	1.80	1.61	0.45
3 YM	1.74	1.54	1.01
4 YM	1.72	1.50	2.70

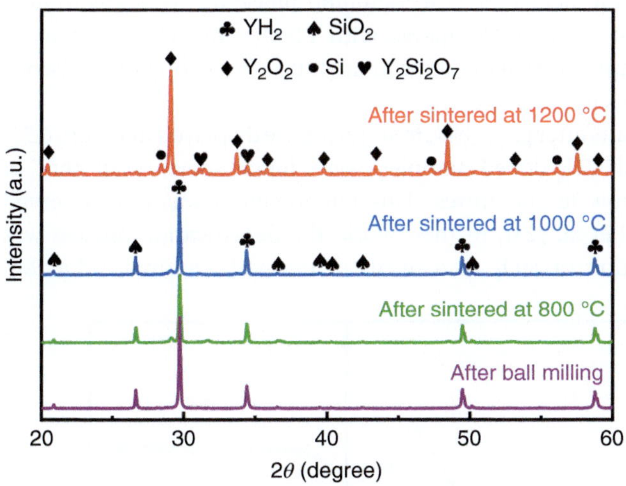

Figure 6.30 The XRD patterns of powder mixture of YH_2 and SiO_2 before and after being heated at various temperatures. Source: Reproduced with permission from Ref. [9]; © 2020 John Wiley & Sons.

and Si, Y_2O_3, and $Y_2Si_2O_7$ came into being after heated at 1200 °C. The following chemical reactions probably involved during heat treatment:

$$4Y(s) + 3SiO_2(s) = 3Si(s) + 2Y_2O_3(s)$$

$$\Delta_r G^{\circ}\ (1473\ K) = -1016.00\ kJ \cdot mol^{-1}, K_p\ (1473\ K) = 1.07 \times 10^{36}$$

$$2Y(s) + 3SiO_2(s) = 3SiO(g) + Y_2O_3(s)$$

$$\Delta_r G^{\circ}\ (1473\ K) = -206.15\ kJ \cdot mol^{-1}, K_p\ (1473\ K) = 2.04 \times 10^{7}$$

$$Y_2O_3(s) + 2SiO_2(s) = Y_2Si_2O_7$$

$$\Delta_r G^{\circ}\ (1473\ K) = -75.63\ kJ \cdot mol^{-1}, K_p\ (1473\ K) = 4.807 \times 10^{2}$$

As reported previously, the YH_2 phase is unstable [38]. YH_2 decomposed into metallic Y particles and atomic H during pre-sintering. The atomic H resulting from the decomposition of YH_2 probably induced a surface-cleaning effect on silicon nitride powders [39]. The scavenging of oxygen by *in situ*-formed metallic Y was expected due to the high oxygen affinity of Y. The native SiO_2 film on the raw silicon nitride powder was consumed, and Y_2O_3 was *in situ* generated. What is more, in this work, all the oxygen was assumed to exist as SiO_2 film enfolding the Si_3N_4 particles. Thus, an approximate value for the Y_2O_3/SiO_2 ratio can be calculated, and the result is shown in Table 6.6. It is evident that the Y_2O_3/SiO_2 ratios in samples keep increasing with the increase of YH_2 content due to the elimination of SiO_2 and the generation of Y_2O_3.

6.2.2.2 Effect of YH_2 on the Densification and Weight Loss

As shown in Figure 6.31, the shrinkage of YOM and YHM started around 1300 and 1400 °C, respectively, which means the eutectic liquid phase in YHM was formed at 100 °C higher than that in YOM. Compared with YHM, a higher displacement speed was detected in sample YOM at the same temperature indicating a less viscous oxynitride liquid phase in sample YOM.

The relative density results after gas pressure sintering are displayed in Figure 6.32. The relative densities of YH_2-doped samples were much lower than that of Y_2O_3-doped samples at low temperatures. This phenomenon was also observed by Kim et al. [40] and Li et al. [23]. In their work, the densification process was retarded by the elimination of SiO_2 as a result of the carbothermal reduction.

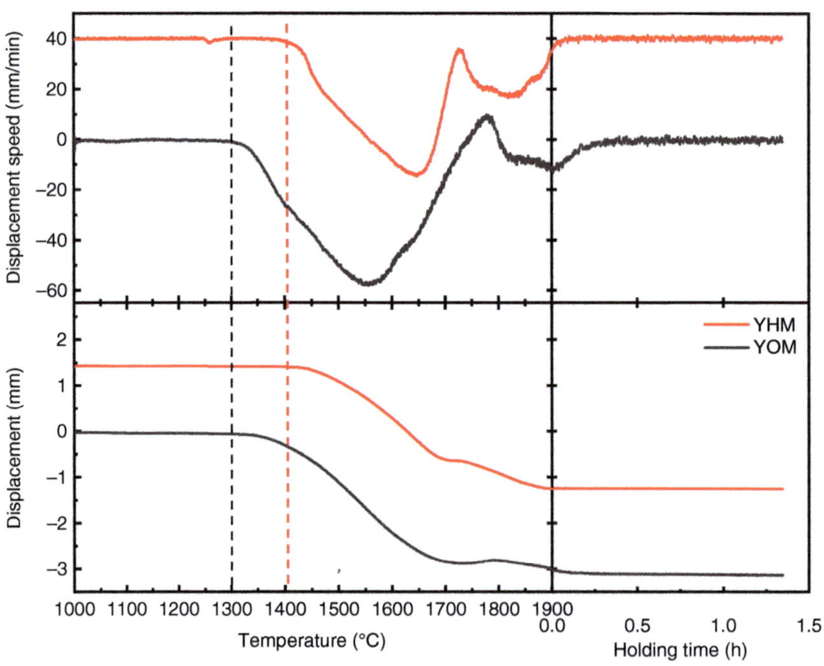

Figure 6.31 The shrinkage curves of the Si_3N_4 ceramics. Source: Reproduced with permission from Ref. [9]; © 2020 John Wiley & Sons.

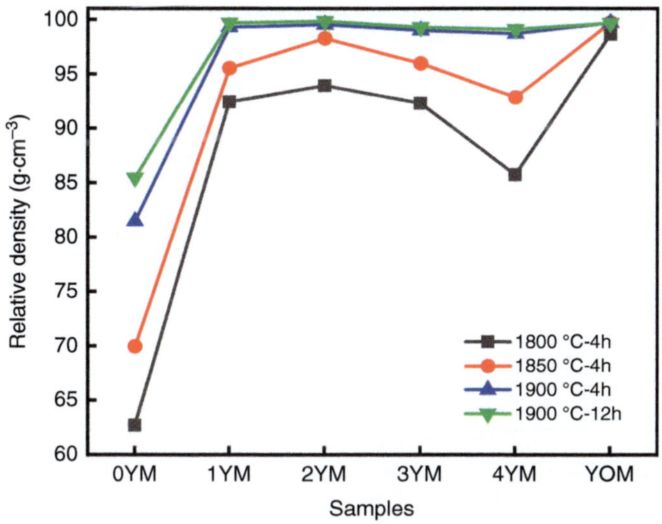

Figure 6.32 The relative density of Si_3N_4 samples after gas pressure sintering. Source: Reproduced with permission from Ref. [37]; © 2020 Elsevier Ltd.

N/O atomic ratio in the liquid phase was increased by the substitution of Y_2O_3 with YH_2. It is generally accepted that the shrinkage behavior is controlled by the glass transition temperature and the viscosity of the liquid phase [40, 41]. As elucidated previously, both of which increased with the increasing N/O atomic ratio because an increased N/O ratio makes the glass structure more rigid resulting from the replacement of a twofold-coordinated oxygen atom by a threefold-coordinated nitrogen atom [42, 43]. So, the densification in sample YHM was retarded due to a higher glass transition temperature and a more viscous liquid phase. The relative density of silicon nitride ceramics doped with unitary MgO was only 85.5%, even sintered at 1900 °C for 12 h. Sinterability was improved when YH_2 was added since the Y^{3+} ions act as a modifier of the oxynitride glass network by enabling the fast diffusion of Si and N ions in the liquid phase. Full density was achieved in all of the YH_2-doped samples after sintered at 1900 °C for 4 h, indicating YH_2-MgO is efficient to densify Si_3N_4 ceramics.

Inversed V-shaped density curves against the YH_2 content were found after sintered at 1800 and 1850 °C for 4 h. The relative density reached a maximum when 2 wt% YH_2 was added. The addition of 2 wt% initially promoted the densification, but further addition inhibited the densification. The discrepancy of relative densities tends to decrease with the increased sintering temperature as well as the prolonged dwelling time. The differences in relative densities were hard to distinguish after sintered at 1900 °C for 12 h due to nearly full densification for all samples. The densification in YH_2-doped samples was controlled by two contradictory factors: the number of Y^{3+} cations and the Y_2O_3/SiO_2 ratio. Densification was promoted with the increasing number of Y^{3+} cations, which enabled the fast diffusion of Si and N ions in the oxynitride liquid phase. However, the increased Y_2O_3/SiO_2 ratio leads to a higher N/O ratio due to the electric neutrality requirement in the liquid phase [42, 44]. Numerous works of oxynitride glasses have confirmed that the

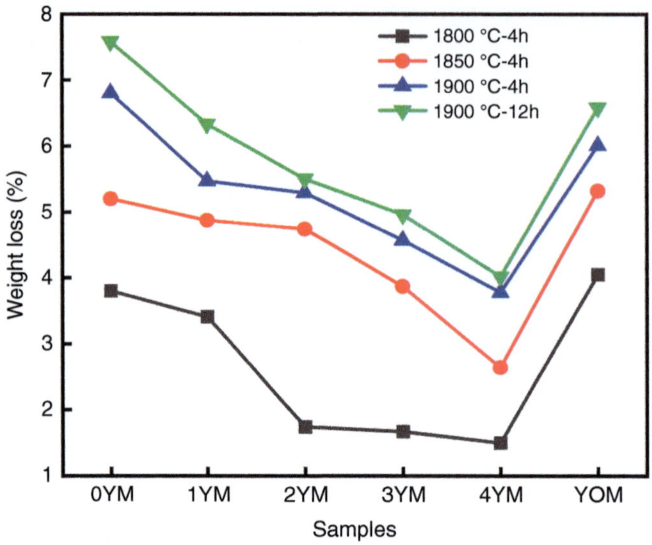

Figure 6.33 The weight loss of Si_3N_4 samples after gas pressure sintering. Source: Reproduced with permission from Ref. [37]; © 2020 Elsevier Ltd.

glass transition temperature and viscosity increase as the N/O atomic ratio increases. Thus, densification was inhibited with the increasing Y_2O_3/SiO_2 ratio. Therefore, when YH_2 was introduced and further increased to 2 wt%, the increased number of Y^{3+} plays a dominant role in promoting the densification of silicon nitride ceramics, but the increased Y_2O_3/SiO_2 ratio by further addition of YH_2 tends to retard the densification.

As shown in Figure 6.33, the weight losses of all the samples increased with the increased sintering temperature as well as the prolonged dwelling time. It was well recognized that weight loss of silicon nitrides during sintering results from many factors, including the vaporization loss of liquid phase, the decomposition of Si_3N_4 itself, and the interaction between Si_3N_4 and SiO_2. Among them, the last factor where volatile SiO(g) and N_2(g) were released by the reaction between SiO_2 and Si_3N_4 would be responsible for the discrepancy of weight loss among samples [45, 46]. The native SiO_2 was consumed by the addition of YH_2, and the amount of residual SiO_2 decreased with the increased addition of YH_2. Therefore, the amount of SiO_2 involved in the following reaction was reduced, resulting in decreased weight loss.

$$3SiO_2(l) + Si_3N_4\,(s) = 2N_2(g) + 6SiO(g)$$

6.2.2.3 Effect of YH_2 on Elements Distribution and Phase Composition

The elements distribution of sample 3YM after sintering at 1900 °C for 12 h is shown in Figure 6.34. It was found that the elements of Y, Mg, and O coexist as intergranular phases. The intergranular phases tended to be enriched at the triple and multiple junctions of Si_3N_4 grains.

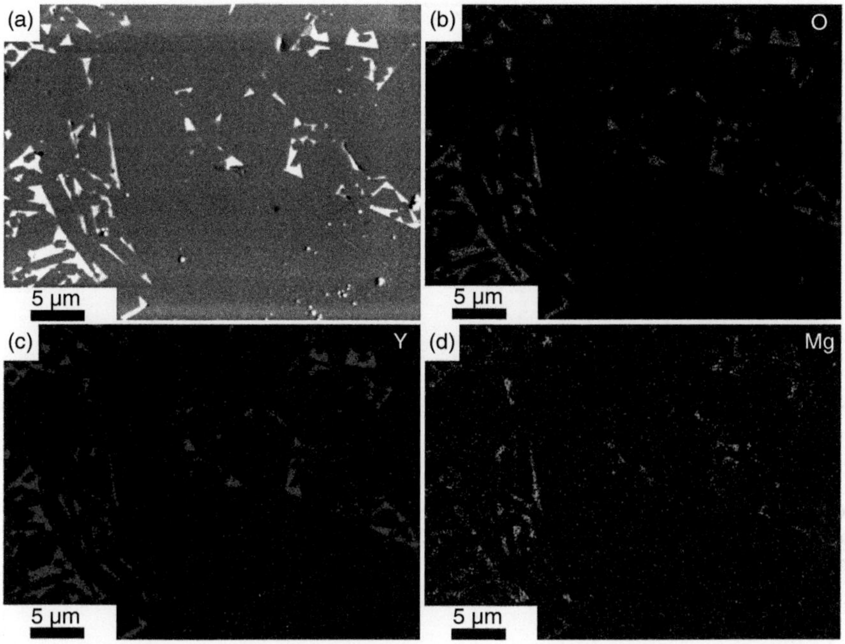

Figure 6.34 (a) The SEM micrograph of the polished surface of sample 3YM after sintering at 1900 °C for 12 h; EDS mapping analysis: O element (b), Y element (c), and Mg element (d). Source: Reproduced with permission from Ref. [37]; © 2020 Elsevier Ltd.

No crystalline secondary phases were detected in Y_2O_3-doped sample, suggesting an amorphous state in the intergranular phases. Different secondary phases were detected in YH_2-doped samples. $Y_2Si_2O_7$ and $YSiO_2N$ were detected in samples 1YM and 2YM, respectively. Both $Y_4Si_2O_2N_7$ and $Y_2Si_3O_3N_4$ phases were found in samples 3YM and 4YM, while the diffraction intensity of the secondary phases in 4YM was stronger than in 3YM (Figure 6.35).

The Y_2O_3/SiO_2 ratios in the samples keep increasing with the increase of YH_2 content. Thus, the XRD results indicate that the composition of secondary phases shifts from $Y_2Si_2O_7$ ($Y_2O_3{\cdot}2SiO_2$, $Y_2O_3/SiO_2 = 1/2$) to $YSiO_2N$ ($2Y_2O_3{\cdot}SiO_2{\cdot}Si_3N_4$, $Y_2O_3/SiO_2 = 2/1$) and finally to $Y_4Si_2O_2N_7$ ($4Y_2O_3{\cdot}SiO_2{\cdot}Si_3N_4$, $Y_2O_3/SiO_2 = 4/1$) and $Y_2Si_3O_3N_4$ ($Y_2O_3{\cdot}Si_3N_4$, $Y_2O_3/SiO_2 = 1/0$) as the Y_2O_3/SiO_2 ratios increase. As shown in Figure 6.36, a Y_2O_3-SiO_2-Si_3N_4 abridged diagram can be constructed from the XRD results in this study and previously published works [47–50]. It was assumed that the phases detected via XRD were in equilibrium with the liquid phase at the annealing temperature. Therefore, the green and blue regions correspond to the phase composition of samples 1YM and 2YM, respectively, while the orange region corresponds to the phase composition of samples 3YM and 4YM.

6.2.2.4 Effect of YH_2 on Microstructure

Figures 6.37 and 6.38 show the low and high-magnification SEM micrographs of the polished surfaces of Si_3N_4 ceramics after sintered at 1900 °C for 12 h, respectively.

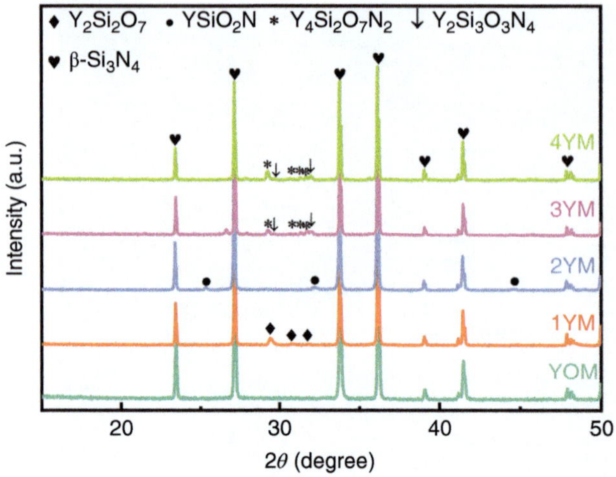

Figure 6.35 The XRD patterns of Si_3N_4 samples after sintered at 1900 °C for 12 h. Source: Reproduced with permission from Ref. [37]; © 2020 Elsevier Ltd.

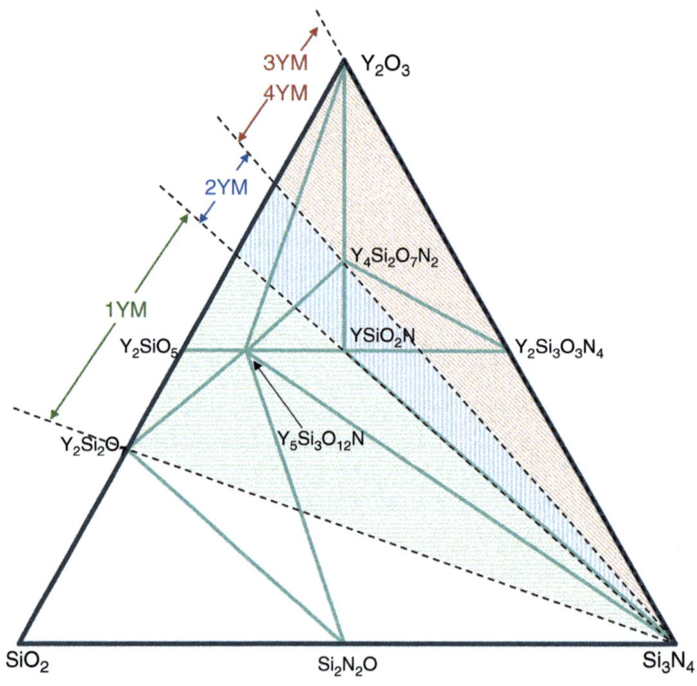

Figure 6.36 Schematic illustration of the Si_3N_4–Y_2O_3–SiO_2 phase diagram at 1900 °C. Source: Reproduced with permission from Ref. [37]; © 2020 Elsevier Ltd.

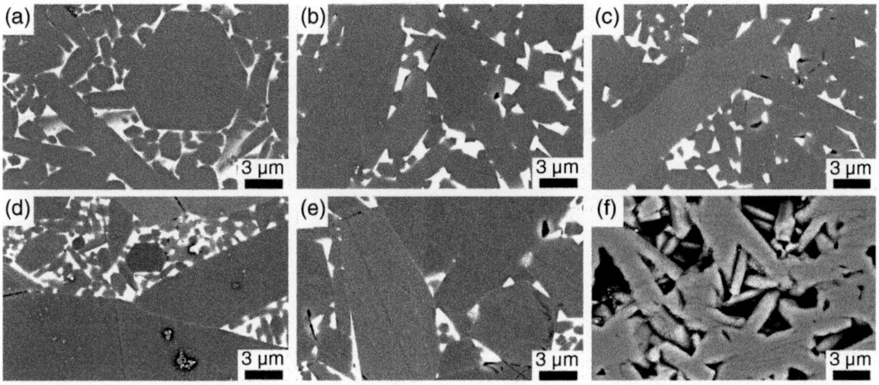

Figure 6.37 The low-magnification SEM micrographs of the polished surfaces of Si_3N_4 ceramics after sintered at 1900 °C for 12 h: (a) YOM, (b) 1YM, (c) 2YM, (d) 3YM, (e) 4YM, (f) 0YM. Source: Reproduced with permission from Ref. [37]; © 2020 Elsevier Ltd.

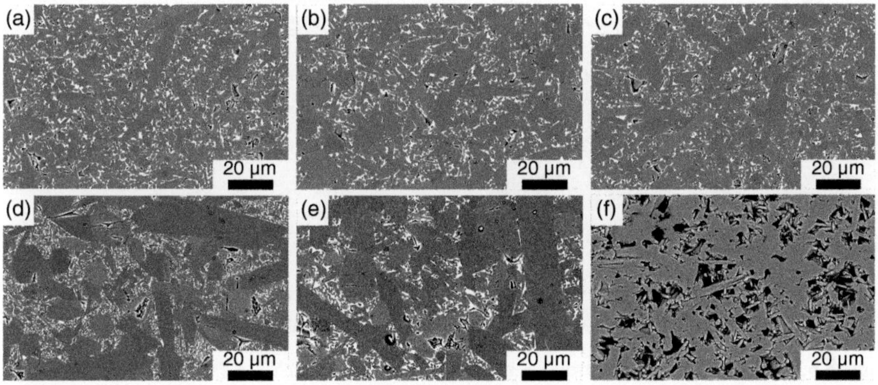

Figure 6.38 The high-magnification SEM micrographs of the polished surfaces of Si_3N_4 ceramics after sintered at 1900 °C for 12 h: (a) YOM, (b) 1YM, (c) 2YM, (d) 3YM, (e) 4YM, (f) 0YM. Source: Reproduced with permission from Ref. [37]; © 2020 Elsevier Ltd.

Table 6.7 summarizes the grain sizes, secondary phases volume fractions, and thermal conductivities of the materials after sintered at 1900 °C for 12 h. A porous structure was found in the 0YM sample due to poor densification. The grain size increases as the Y_2O_3/SiO_2 ratio increases from sample 1YM to sample 4YM. A salient increase in grain size occurs at 3YM, with a Y_2O_3/SiO_2 ratio of 1.01. The microstructure discrepancy was probably attributed to the Y_2O_3/SiO_2 ratio [42]. More dissolved nitrogen was needed with the increased number of Y^{3+} to enhance the negative valence of silicate chain according to the electric neutrality principle [42, 43]. Therefore, the increase in Y_2O_3/SiO_2 ratio enhanced the solubility of Si_3N_4.

The HAADF images of Si_3N_4 samples, as well as the corresponding EDS analyses of the pocket phases, are shown in Figure 6.39. Y, Si, Mg, O, and N were detected in the multiple junctions indicating the intergranular phases were Y-Si-Mg-O-N phases. It is evident that the N/O atomic ratio in the multiple junctions of sample

Table 6.7 Grain sizes, secondary phases' volume fractions, and thermal conductivities of the Si_3N_4 samples after sintering at 1900 °C for 12 h.

Sample	YOM	1 YM	2 YM	3 YM	4 YM
Grain size (μm)	2.46	3.05	3.13	4.88	5.14
Secondary phases	/	$Y_2Si_2O_7$	$YSiO_2N$	$Y_4Si_2O_2N_7$ and $Y_2Si_3O_3N_4$	$Y_4Si_2O_2N_7$ and $Y_2Si_3O_3N_4$
Secondary phases volume fraction (vol%)	6.91	5.80	5.40	5.36	5.68
Thermal conductivity (W/m K)	95.30	97.92	106.74	123.00	117.51

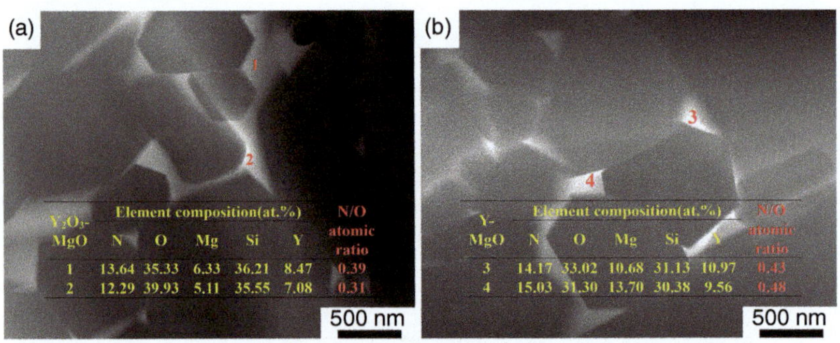

Figure 6.39 HAADF images and EDS analyses of samples (a) YOM, (b) YHM. Source: Reproduced with permission from Ref. [9]; © 2020 John Wiley & Sons.

YHM was higher than that of sample YOM. It means that N/O atomic ratio in the liquid phase was increased by the substitution of Y_2O_3 with YH_2. The substitution of Y_2O_3 by YH_2 led to an increase in the N/O ratio, which benefits the grain growth of β-Si_3N_4. Besides, the volume fraction of secondary phases was reduced, and the Si_3N_4-Si_3N_4 contiguity was enhanced due to the consumption of SiO_2 by the addition of YH_2 in YH_2-doped samples, which is favorable to the thermal conductivity.

6.2.2.5 Effect of YH_2 on Thermal Conductivity

Figure 6.40 shows the thermal conductivity values of Si_3N_4 samples after gas pressure sintering. The thermal conductivity increased with the increased sintering temperature as well as the prolonged holding time due to the enhanced densification and promoted growth of β-Si_3N_4 grains. Sample 0YM possesses the lowest thermal conductivity due to the lowest density in all sintering conditions. As for those samples sintered at 1900 °C, the thermal conductivity improved drastically with only 1 wt% YH_2 being added and keeps increasing gradually with further

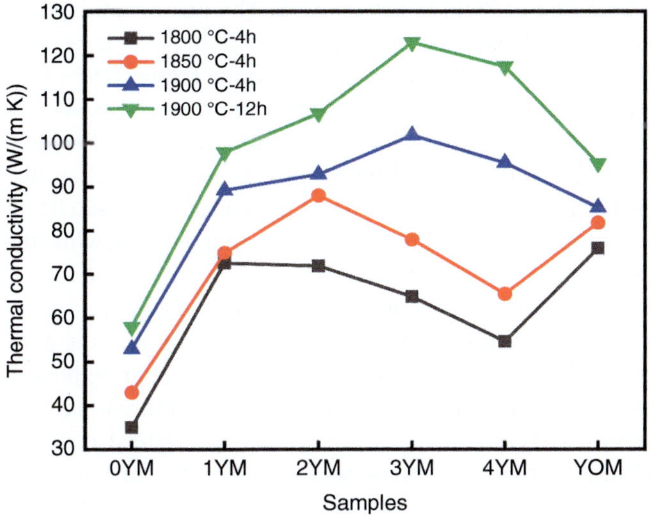

Figure 6.40 The thermal conductivity of Si_3N_4 samples after gas pressure sintering. Source: Reproduced with permission from Ref. [37]; © 2020 Elsevier Ltd.

addition to 3 wt%. A slight decrease was found when the addition of YH_2 increased from 3 to 4 wt%. Ultimately, the highest thermal conductivity value of 123 W/m K was obtained using 3 wt% YH_2 and 1.5 wt% MgO after sintering at 1900 °C for 12 h.

6.2.2.6 Effect of YH_2 on Mechanical Properties

The flexural strength of the silicon nitride ceramics is shown in Figure 6.41a; with the increase of the sintering temperature and the extension of the holding time, the grain size of silicon nitride gradually increases, so the flexural strength gradually decreases. 2YM has a flexural strength of 1018 ± 42 MPa after sintering at 1800 °C

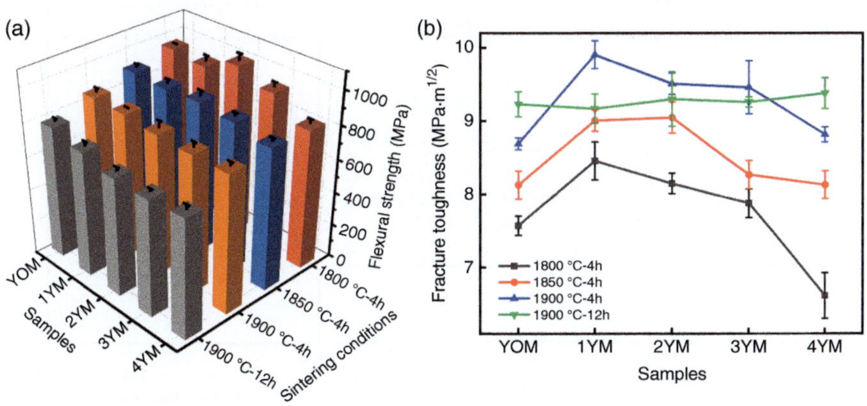

Figure 6.41 The mechanical properties of Si_3N_4 ceramics: (a) flexural strength, (b) fracture toughness. Source: Yao Dongxu, Wang Weide, Zeng Yu-ping.

for 4 h, which is higher than that of YOM at 959 ± 5 MPa. The flexural strength of the other samples doped with YH_2 was lower than that of YOM. This is because the YH_2-doped samples have more developed grains due to the higher N/O in the liquid phase, which affects the flexural strength to some extent. Nevertheless, the flexural strength of the YH_2-doped samples was above 700 MPa after 12 h of holding at 1900 °C. The fracture toughness of the YH_2-doped samples was lower than that of YOM after sintering at 1800 °C and 1850 °C for 4 h. After sintering at 1900 °C for 4 h, the fracture toughness of 1YM and 2YM was better than that of YOM, while the fracture toughness of 3YM and 4YM was comparable to that of YOM. The factors affecting the fracture toughness are complex, not only related to the grain size but also the composition of the grain-boundary film between the grain and the glass phase. It has been suggested that an increase in the O/N ratio at the grain boundaries is beneficial for interfacial exfoliation, thus improving the fracture toughness [51, 52]. In this experiment, YH_2 was used as a sintering aid instead of Y_2O_3, and the increase of N/O ratio in the liquid phase was not favorable for interfacial debonding.

The reaction mechanism of YH_2 is shown in Figure 6.42. Y_2O_3 was *in situ*-formed by a metal hydride reduction reaction between YH_2 and native SiO_2 during the first-step sintering procedure. Thus, the N/O atomic ratio in the Y-Si-Mg-O-N liquid phase was increased, resulting in a higher viscosity of the liquid phase. This inhibits densification to some extent, but the low SiO_2 activity in the liquid phase

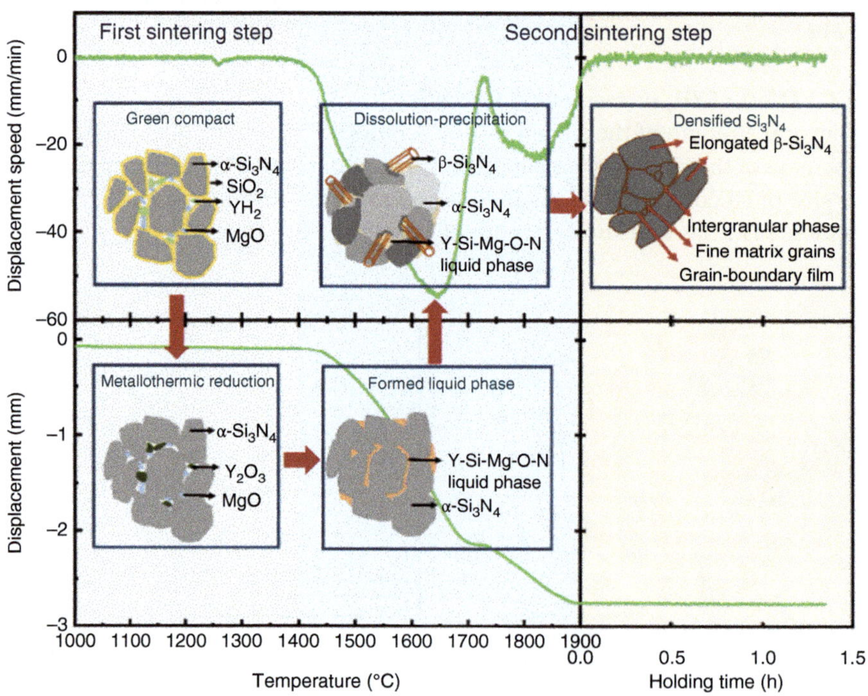

Figure 6.42 Schematic diagram of the action mechanism of YH_2 as a sintering additive in the sintering of Si_3N_4 ceramics. Source: Yao Dongxu, Wang Weide, Zeng Yu-ping.

helps to reduce the lattice oxygen content in the silicon nitride lattice, which in turn improves the thermal conductivity. The high N/O atomic ratio in the liquid phase facilitates the development of silicon nitride grains and the formation of prominent bimodal morphology, which affects the mechanical properties of silicon nitride ceramics.

6.2.2.7 Differences in the Effect of Different REH_2 on the Thermal Conductivity of Silicon Nitride

The use of YH_2 instead of Y_2O_3 as a sintering aid effectively improved the thermal conductivity. In addition to YH_2, the variability of the effect of other REH_2 on the thermal conductivity of silicon nitride was investigated. The thermal conductivity of silicon nitride prepared by adding YbH_2, YH_2, and GdH_2 as sintering aids at 1900 °C with different holding times is shown in Table 6.8. After holding at 1900 °C for 4 h, the thermal conductivity of silicon nitride prepared by adding different aids was $YH_2 > YbH_2 > GdH_2$, $YH_2 > GdH_2 > YbH_2$ after holding at 1900 °C for 12 h, $GdH_2 > YH_2 > YbH_2$ after holding at 1900 °C for 24 h.

The values of thermal conductivity obtained under the same sintering conditions were similar and did not show a consistent regularity with the change of ionic radius of rare earth elements in the additives. The analysis suggests that when metal hydrides, which are non-oxides, are used as sintering aids, the glass transition temperature and viscosity increase due to the high N/O ratio in the liquid phase, resulting in more diffusion-controlled grain growth of silicon nitride, and the increase in liquid-phase viscosity due to the higher N/O atomic ratio in the diffusion-controlled mode masks the effect of cation field strength on viscosity in the liquid phase. In addition, the addition of REH_2 as a sintering aid removes SiO_2 sufficiently, reduces SiO_2 activity, and reduces lattice defects, and the increase in thermal conductivity brought about by the replacement of RE_2O_3 by REH_2 as aid is much greater than that caused by the different radii of rare earth ions in the liquid phase.

Table 6.8 The thermal conductivity of Si_3N_4 doped with different REH_2.

Additive	Ionic radius (nm)	Annealing time at 1900 °C (h)	Thermal conductivity (W/m K)
GdH_2 [52]	0.094	4	98.07
		12	119.07
		24	134.90
YH_2 [37]	0.089	4	101.80
		12	123.00
		24	131.60
YbH_2	0.086	4	100.20
		12	118.90
		24	131.15

6.3 Investigation of Cu-Metalized Si_3N_4 Substrates Via Active Metal Brazing (AMB) Method

The Cu-metalized Si_3N_4 substrates were brazed with Ag–Cu–Ti-based brazing filler. On the one hand, it is easy for the copper foil to be wetted by molten brazing alloy at high temperatures due to the similar nature of metals. Moreover, the high bond strength of the brazing alloy/copper foil interface can be obtained by atom thermal diffusion. On the other hand, it has been shown that the wetting angle decreases sharply with the Ti content increased in the filler. For example, when the Ti reaches a certain content, the wetting angle between the brazing filler and Si_3N_4 is only about 30° at 880 °C. It is favorable for the spread of the brazing filler on the surface of the Si_3N_4. The active element Ti in the brazing material can react with silicon nitride at high temperatures to drive the wetting, and an interfacial reaction layer consisting of TiN and Ti_5Si_3 is generated, thus realizing the connection of ceramic–metal heterogeneous materials. The reaction between Si_3N_4 and Ti is as follows:

$$Si_3N_4(s) + 9Ti(l) = 4TiN(s) + Ti_5Si_3(s)$$

Ag–Cu–Ti solder paste is widely used in the preparation of high-voltage and high-power IGBT power devices due to the ease of coating and the amount that can be precisely controlled. Ag–Cu–Ti brazing paste is mainly composed of Ag–Cu–Ti metal powder and an organic vehicle. The metal powder content (solid content) and the organic vehicle content generally account for 85–90 and 10–15 wt%, respectively.

Compared with Ti powder, TiH_2 powder has a lower cost and less oxygen content, and the dehydrogenation reaction that occurs during brazing can clean the particle surface and improve the reactivity, so TiH_2 powder was used to replace Ti powder. Meanwhile, in order to investigate more deeply the effects of brazing components on the microstructure and mechanical properties of Cu-metalized Si_3N_4 substrates, it was decided to use Ag powder, Cu powder, and TiH_2 powder alone to prepare brazing paste. The preparations of active solder paste and brazing process are the key factors affecting the connection properties of Cu-metalized Si_3N_4 substrates. The effects of brazing temperature and holding time, ball milling time of Ag powder, Cu powder and TiH_2 powder, the ratio of Ag powder and Cu powder, the ratio of solid powder to an organic carrier and the content of TiH_2 powder on the microstructure, and peeling strength of Cu-metalized Si_3N_4 substrates were investigated.

6.3.1 Effect of Brazing Temperature on the Peeling Strength of Cu-Metalized Si_3N_4 Substrates

Brazing temperature can directly affect the interfacial reaction between brazing material and base material, thus affecting the interfacial bonding strength between ceramic and copper. The effect of brazing temperature on the peeling strength of Cu-metalized Si_3N_4 substrates was analyzed by testing the peeling strength at 850 °C–900 °C for 30 min.

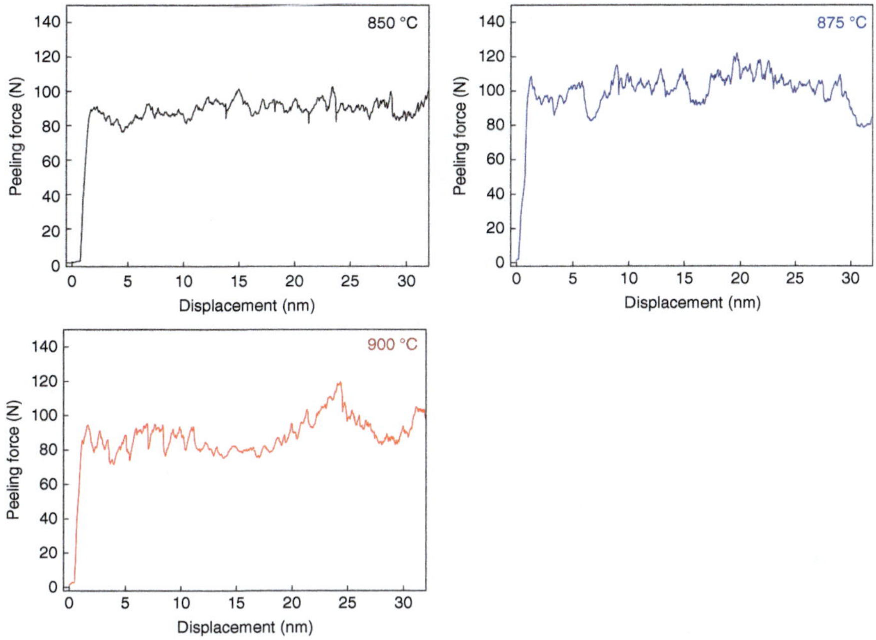

Figure 6.43 Peeling curves of Cu-metalized ceramic substrates prepared at different brazing temperatures. Source: Yao Dongxu, Wang Weide, Zeng Yu-ping.

Figure 6.43 shows the peeling strengths of the Cu-metalized Si_3N_4 substrates obtained at different brazing temperatures. The average peeling strength of the Cu-metalized ceramic substrates reached 179 N /cm at a holding time of 30 min at 850 °C; when the brazing temperature was increased to 875 °C, the peeling strength increased to 195 N/cm; with further increase in brazing temperature, the peeling strength of the Cu-metalized ceramic substrates obtained at 900 °C decreased to 180 N/cm.

6.3.2 Effect of Holding Time on the Peeling Strength of Cu-Metalized Ceramic Substrates

At a certain brazing temperature, the molten brazing material fully wets the base material and requires a certain amount of holding time. When the holding time is insufficient, the contact time between the brazing material and the base material is short and the reaction is not sufficient, which may cause welding defects such as false welding and reduce the interfacial bond strength; when the holding time is too long, the interfacial reaction layer is too thick and the residual stress in the Cu-metalized ceramic substrates increases, which easily produces microcracks in the reaction layer and reduces the strength of the Cu-metalized ceramic substrates. Holding times of 10, 30, and 60 minutes were set at 875 °C to study the corresponding changes in the peeling strength of Cu-metalized Si_3N_4 substrates.

Figure 6.44 shows the peeling strengths of Cu-metalized Si_3N_4 substrates obtained at different holding times, from which it can be seen that the peeling

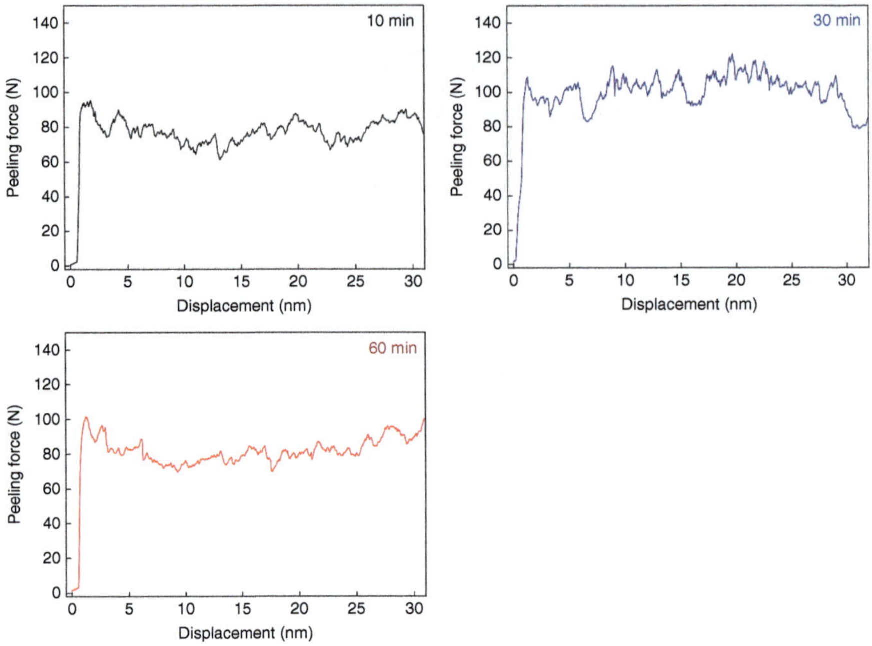

Figure 6.44 Peeling curves of Cu-metalized ceramic substrates prepared at different holding times. Source: Yao Dongxu, Wang Weide, Zeng Yu-ping.

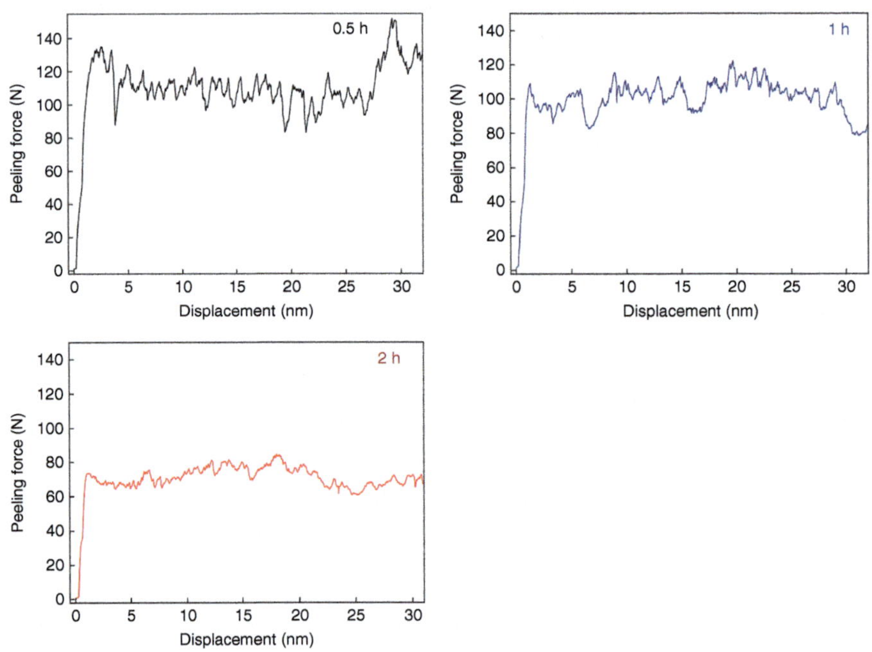

Figure 6.45 Peeling curves of Cu-metalized ceramic substrates prepared at different brazing ball milling times. Source: Yao Dongxu, Wang Weide, Zeng Yu-ping.

strength increases and then decreases with increasing holding time, reaching a maximum value of 195 N/cm at 30 minutes of holding time.

6.3.3 Effect of Brazing Ball Milling Time on the Peeling Strength of Cu-Metalized Ceramic Substrates

The mixed powder was ball-milled for 0.5, 1, and 2 h to make an active solder paste and brazed at 875 °C for 30 min to study the effect of different brazing material ball milling times on the peeling strength of Cu-metalized ceramic substrates.

Figure 6.45 shows the peeling strength of Cu-metalized Si_3N_4 substrates brazed with active solder paste prepared by using different ball milling times, and it can be seen from the figure that the peeling strength was maximum 225 N/cm at 0.5 h of ball milling time; with the extension of the ball milling time to 1 and 2 h, the peeling strength decreased by 30 and 70 N/cm, respectively.

References

1 Zhu, X. and Sakka, Y. (2006). The effect of embedding conditions on the thermal conductivity of β-Si_3N_4. *Journal of the Ceramic Society of Japan* 1093–1096.

2 Li, S. and Yao, D. (2023). Enhanced thermal conductivity and flexural strength of sintered reaction-bonded silicon nitride with addition of $(Y_{0.96}Eu_{0.04})_2O_3$. *International Journal of Applied Ceramic Technology* 1: 465–474.

3 Hyuga, H. and Zhou, Y. (2011). Nitridation behaviors of silicon powder doped with various rare earth oxides. *Journal of the Ceramic Society of Japan* 1387: 251–253.

4 Jin, X. and Xing, P. (2019). Function mechanism of Y_2O_3 additive in silicon powder nitridation. *International Journal of Applied Ceramic Technology* 4: 1407–1415.

5 Wu, L.-X. and Zhu, L.-L. (2021). Fabrication and properties of pressure-sintered reaction-bonded Si_3N_4 ceramics with addition of Eu_2O_3–MgO–Y_2O_3. *Ceramics International* 1: 935–942.

6 Hampshire, S. (2003). Oxynitride glasses, their properties and crystallisation – a review. *Journal of Non-Crystalline Solids* 1: 64–73.

7 Li, Y. and Kim, H.-N. (2018). Enhanced thermal conductivity in Si_3N_4 ceramic with the addition of $Y_2Si_4N_6C$. *Journal of the American Ceramic Society* 9: 4128–4136.

8 Ohashi, M. and Nakamura, K. (1995). Formation and properties of Ln-Si-O-N glasses (Ln = lanthanides or Y). *Journal of the American Ceramic Society* 71: 1136.

9 Wang, W. and Yao, D. (2020). Improved thermal conductivity of β-Si_3N_4 ceramics by lowering SiO_2/Y_2O_3 ratio using YH_2 as sintering additive. *Journal of the American Ceramic Society* 10: 5567–5572.

10 Moulson, A.J. (1979). Reaction-bonded silicon nitride: its formation and properties. *Journal of Materials Science* 5: 1017–1051.

11 Li, Y. and Kim, H.-N. (2021). Microstructure and thermal conductivity of gas-pressure-sintered Si_3N_4 ceramic: the effects of Y_2O_3 additive content. *Journal of the European Ceramic Society* 1: 274–283.

12 Dai, J. and Li, J. (2003). Effects of rare-earth oxide additions and heat-treatment temperature on the transformation and microstructure of silicon nitride. *Physica Status Solidi (A)* 1: 91–98.

13 Zhu, X. and Zhou, Y. (2007). Processing and thermal conductivity of sintered reaction-bonded silicon nitride: (II) effects of magnesium compound and yttria additives. *Journal of the American Ceramic Society* 6: 1684–1692.

14 Zhou, Y. and Hyuga, H. (2011). A tough silicon nitride ceramic with high thermal conductivity. *Advanced Materials* 39: 4563–4567.

15 Kitayama, M. and Hirao, K. (1999). Thermal conductivity of β-Si_3N_4: I, effects of various microstructural factors. *Journal of the American Ceramic Society* 11: 3105–3112.

16 Kitayama, M. and Hirao, K. (2000). Thermal conductivity of β-Si_3N_4: II, effect of lattice oxygen. *Journal of the American Ceramic Society* 8: 1985–1992.

17 Zhou, Y. and Hyuga, H. (2019). Effects of yttria and magnesia on densification and thermal conductivity of sintered reaction-bonded silicon nitrides. *Journal of the American Ceramic Society* 4: 1579–1588.

18 Liang, H. and Zeng, Y. (2016). Mechanical properties and thermal conductivity of Si_3N_4 ceramics with YF_3 and MgO as sintering additives. *Ceramics International* 14: 15679–15686.

19 Hayashi, H. and Hirao, K. (2001). $MgSiN_2$ addition as a means of increasing the thermal conductivity of β-silicon nitride. *Journal of the American Ceramic Society* 12: 3060–3062.

20 Dressler, W. and Kleebe, H.-J. (1996). Model experiments concerning abnormal grain growth in silicon nitride. *Journal of the European Ceramic Society* 1: 3–14.

21 Wang, W. and Yao, D. (2020). $ZrSi_2$–MgO as novel additives for high thermal conductivity of β-Si_3N_4 ceramics. *Journal of the American Ceramic Society* 3: 2090–2100.

22 Li, Y. and Kim, H.-N. (2019). Enhanced thermal conductivity in Si_3N_4 ceramic by addition of a small amount of carbon. *Journal of the European Ceramic Society* 2: 157–164.

23 Li, Y. and Kim, H.-N. (2017). Improved thermal conductivity of sintered reaction-bonded silicon nitride using a BN/graphite powder bed. *Journal of the European Ceramic Society* 15: 4483–4490.

24 Kim, W.-J. and Kim, D.K. (1996). Morphological effect of second phase on the thermal conductivity of AIN ceramics. *Journal of the American Ceramic Society* 4: 1066–1072.

25 Schlichting, K.W. and Padture, N.P. (2001). Thermal conductivity of dense and porous yttria-stabilized zirconia. *Journal of Materials Science* 12: 3003–3010.

26 Zhu, X. and Hayashi, H. (2004). Influence of additive composition on thermal and mechanical properties of β–Si_3N_4 ceramics. *Journal of Materials Research* 11: 3270–3278.

27 Zhou, Y. and Zhu, X. (2008). Sintered reaction-bonded silicon nitride with high thermal conductivity and high strength. *International Journal of Applied Ceramic Technology* 2: 119–126.

28 Park, H. and Kim, H.E. (1997). Microstructural evolution and mechanical properties of Si_3N_4 with Yb_2O_3 as a sintering additive. *Journal of the American Ceramic Society* 750–756.

29 Šajgalik, P. and Dusza, J. (2005). Relationship between microstructure, toughening mechanisms, and fracture toughness of reinforced silicon nitride ceramics. *Journal of the American Ceramic Society* 2619–2624.

30 Jie, Z. and Xiaoshan, N. (2008). Effect of rare-earth additives on thermal conductivity, mechanical and electrical properties of silicon nitride ceramics. *Rare Metal Materials and Engineering* 693–696.

31 Leedecke, C.J. (1980). Electrical properties of some Y-Si-Al Oxynitride glass-ceramics. *Journal of the American Ceramic Society* 479–480.

32 Hugon, M.C. and Delmotte, F. (1997). Electrical properties of metal–insulator–semiconductor structures with silicon nitride dielectrics deposited by low temperature plasma enhanced chemical vapor deposition distributed electron cyclotron resonance. *Journal of Vacuum Science & Technology A* 6: 3143–3153.

33 Liu, Y. and Liu, Y. (2010). Microstructures evolution and mechanical properties of a powder metallurgical titanium alloy with yttrium addition. *Materials and Manufacturing Processes* 735–739.

34 Liu, Y. and Liu, Y.B. (2012). Rare earth element: is it a necessity for PM Ti alloys? *Key Engineering Materials* 41–48.

35 Yan, M. and Liu, Y. (2012). Simultaneous gettering of oxygen and chlorine and homogenization of the β phase by rare earth hydride additions to a powder metallurgy Ti–2.25Mo–1.5Fe alloy. *Scripta Materialia* 5: 491–494.

36 Khatamian, D. and Manchester, F.D. (1988). The H–Y (hydrogen-yttrium) system. *Bulletin of Alloy Phase Diagrams* 3: 252–260.

37 Wang, W. and Yao, D. (2020). Effect of in-situ formed Y_2O_3 by metal hydride reduction reaction on thermal conductivity of β-Si_3N_4 ceramics. *Journal of the European Ceramic Society* 15: 5316–5323.

38 Robertson, I.M. and Schaffer, G.B. (2010). Comparison of sintering of titanium and titanium hydride powders. *Powder Metallurgy* 12–19.

39 Kim, H.-D. and Han, B.-D. (2002). Novel two-step sintering process to obtain a bimodal microstructure in silicon nitride. *Journal of the American Ceramic Society* 1: 245–252.

40 Riley, F.L. (2004). Silicon nitride and related materials. *Journal of the American Ceramic Society* 245–265.

41 Hakeem, A.S. and Daucé, R. (2005). Silicate glasses with unprecedented high nitrogen and electropositive metal contents obtained by using metals as precursors. *Advanced Materials* 18: 2214–2216.

42 Lemercier, H. and Rouxel, T. (1996). Yttrium SiAlON glasses: structure and mechanical properties — elasticity and viscosity. *Journal of Non-Crystalline Solids* 1: 128–145.

43 Kitayama, M. and Hirao, K. (1999). Control of β-Si_3N_4 crystal morphology and its mechanism (part 1). *Journal of the Ceramic Society of Japan* 930–934.

44 Wada, S. (2001). Control of instability of Si_3N_4 during pressureless sintering. *Journal of the Ceramic Society of Japan* 803–808.

45 Greskovich, C.D. and Prochazka, S. (1981). Stability of Si_3N_4 and liquid phase(s) during sintering. *Journal of the American Ceramic Society*.

46 Kolitsch, U. and Seifert, H.J. (1999). Phase equilibria and crystal chemistry in the Y_2O_3–Al_2O_3–SiO_2 system. *Journal of Materials Research* 2: 447–455.

47 Lin, Y. and Ning, X. (2002). Study on thermal conductivity of Si_3N_4 sintered with Y_2O_3-MgO additive system. *Key Engineering Materials* 813–818.

48 Gauckler, L.J. and Hohnke, H. (1980). The system Si_3N_4-SiO_2-Y_2O_3. *Journal of the American Ceramic Society* 1–2: 35–37.

49 Zhu, X. and Sakka, Y. (2007). Processing and properties of sintered reaction-bonded silicon nitride with Y_2O_3–$MgSiN_2$: effects of Si powder and Li_2O addition. *Acta Materialia* 16: 5581–5591.

50 Becher, P.F. and Painter, G.S. (2000). The importance of amorphous intergranular films in self-reinforced Si_3N_4 ceramics. *Acta Materialia* 18: 4493–4499.

51 Becher, P.F. and Sun, E.Y. (1996). Debonding of interfaces between beta-silicon nitride whiskers and Si Al Y oxynitride glasses. *Acta Materialia* 10: 3881–3893.

52 Wang, W. and Yao, D. (2021). Improved thermal conductivity of β-Si_3N_4 ceramics through the modification of the liquid phase by using GdH_2 as a sintering additive. *Ceramics International* 4: 5631–5638.

7

Preparation and Properties of Thermal Interface Materials

Xiaoliang Zeng, Linlin Ren, and Rong Sun

Chinese Academy of Sciences, Shenzhen Institute of Advanced Technology, Shenzhen Institute of Advanced Electronic Materials, Shenzhen 518055, China

7.1 Conception of Thermal Interface Materials

Miniaturization and integration have emerged as electronic device development trends in recent years. The amount of heat produced is sharply increasing as the power of electronic devices continues to rise [1]. When two solid surfaces are joined, only a small portion of the apparent surface area will actually contact due to the morphology's roughness [2]. Air's thermal conductivity of 0.026 W/m K, four orders of magnitude lower than that of metals, means that heat transfer across the interface through air is negligible. The remaining area will be separated by a gap filled with air. Even for substrates with very low roughness, there will be a significant thermal bottleneck because the majority of the heat flux will pass through the actual contacting points. As can be seen in Figure 7.1, this manifests itself as a drop in temperature across the interface. To increase the interface's thermal conductivity, thermal interface materials (TIMs) are positioned between two mating surfaces.

TIMs are typically utilized at all interfaces between two solid materials in an electronics package. These interfaces typically occur between a heat-generating die and an integrated heat spreader (IHS) or between the IHS and a heat sink. Figure 7.2 depicts these two TIM applications, which are referred to as TIM1 and TIM2, respectively. Additionally, there are packages that connect a die directly to a heat sink and have more than two instances of TIM between the chip and heat sink [4]. The thermal interface resistance (R_{TIM}) of a TIM, which is a measure of how difficult it is for heat to escape the interface, is typically used as the basis for determining a TIM's thermal performance. According to Fourier's law, it is related to the temperature drop (ΔT) over the interface as $\Delta T = R_{\text{TIM}} Q$, where Q is the heat flux. An overarching objective of TIM's development is to minimize R_{TIM}. The resistive components of the thermal interface resistance for a TIM typically include

$$R_{\text{TIM}} = \frac{BLT}{\lambda_{\text{TIM}}} + R_{c1} + R_{c2} \tag{7.1}$$

Thermal Management Materials for Electronic Packaging: Preparation, Characterization, and Devices, First Edition. Edited by Xingyou Tian.

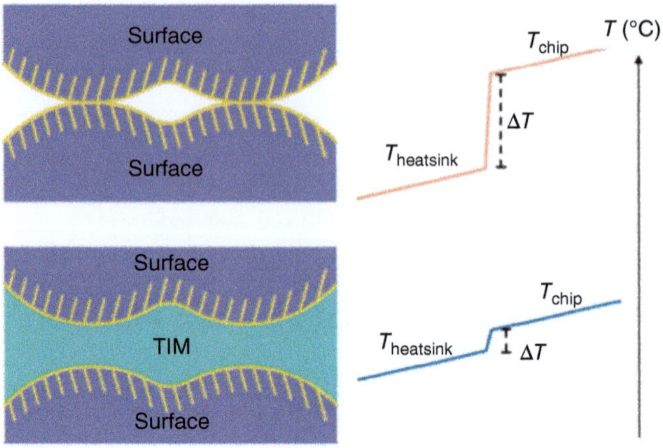

Figure 7.1 Schematic representation of working principle of a TIM. An interface lacking a TIM will have a very concentrated heat flux, causing a large temperature drop at the interface. Source: Reproduced with permission from Ref. [3]; © 2012 Wiley Online Library.

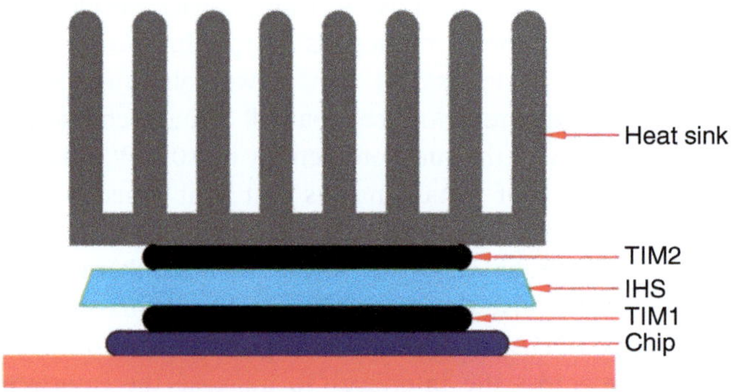

Figure 7.2 Schematic illustration of a typical ball grid array (BGA) electronics package with two TIMs. The heat is conducted through the backside of a BGA chip. There are TIMs between chip and IHS and between IHS and heat sink. The different TIM applications are called TIM1 and TIM2, respectively. Source: Reproduced with permission from Ref. [3]; © 2012 Wiley Online Library.

where R_{TIM} is the total thermal interface resistance, R_{c1} and R_{c2} are the contact resistances at the interface between the TIM and the two substrates, λ_{TIM} is the thermal conductivity of the TIM, and BLT is the bond line thickness, that is, the thickness of the TIM. Depending on the type of TIM and application, the different terms of the equation will have more or less impact on the overall performance, and when choosing or designing a TIM, it is important to identify which parameters to optimize.

Depending on the thermal contact, there is a contact resistance between the substrate and TIM. The degree to which the TIM conforms to the substrate and fills in voids determines this contact resistance. It depends heavily on the applied pressure for some TIMs and is influenced by the TIM's compliance as well as how well it adheres to the substrates. Due to phonon or electron scattering at the interface, there is also a thermal boundary resistance between two distinct materials, even if they are atomically smooth. The term for this effect is Kapitza resistance [5]. A measure of how far apart the two surfaces are, or how thick the TIM is, is the bond line thickness (BLT). It is desirable to reduce the BLT because the TIM typically still has a significantly lower thermal conductivity than either of the substrates. Additionally, a lower BLT typically implies a lower cost as well as less material used. Ideally, the TIM only fills in the original interface's gaps. However, in practice, a mismatch in the coefficient of thermal expansion (CTE) and voiding as a result of uneven TIM coverage can cause reliability issues during thermal cycling. The BLT of TIMs in liquid form is affected not only by the quantity of material but also by the material's viscosity and clamping pressure. In addition, the thickness of the BLT that can be achieved with TIM loaded with solid filler particles may be constrained by the size of the fillers. The BLT is constrained by handling and mechanical requirements when used as a pad for a TIM. The TIM's thermal conductivity is a measure of the material's internal heat conductivity. Thermal conductivity becomes a crucial parameter; for very thin bond lines, contact resistance takes precedence over thermal conductivity, but for thick TIMs, the importance of a high thermal conductivity is proportional to the BLT.

In recent years, there has been a great drive in the industry in reducing R_{TIM}. The heat flux from the chip is non-uniform [6, 7], because both the core and cache are on the same die. Majority of power is dissipated from the core, that is, from a much smaller area of the chip. Even within the core, heat flux (q) is non-uniform. Mahajan et al. discussed the issue of the non-uniform heat fluxes from the die and indicated that cooling solutions were not only needed to maintain the average chip temperature below a design point; it was also important to maintain the temperature of the hottest spot below a certain design point. Therefore, the thermal problem near the chip is very severe. The total thermal resistance for non-uniform heating can be written as [8]

$$\Psi_{j-a} = DF \times R_{jc} + \Psi_{cs} + \Psi_{sa} \tag{7.2}$$

where ψ_{j-a} is the junction to ambient thermal resistance, R_{jc} is the junction to case thermal impedance for a uniformly heated die, ψ_{cs} is the case to sink resistance, and ψ_{sa} is the sink to ambient thermal resistance. DF in Eq. (2) is called "Density Factor" that accounts for the non-uniformity of q and the die size. The unit of DF is cm^{-2}. For a 1 cm^2 uniformly heated chip, the DF is 1. DF is typically larger than 1 for most microprocessors due to high non-uniformity of q and small die sizes, but theoretically DF can approach zero for very large die. According to Eq. (2), a reduction in both R_{jc} and DF leads to a reduction in ψ_{j-a}. Note that the reduction of R_{jc} leads to a greater reduction in ψ_{j-a} if DF is larger than 1. Since R_{jc} is primarily due to TIM thermal resistance, this has led to a great drive in the electronics cooling industry to develop better TIMs. A material with a low thermal contact resistance, a high thermal conductivity, and a low bond line thickness is the ideal TIM. However, when selecting or designing a TIM, a number of additional parameters must be taken into account in addition to the thermal interface resistance. Whether the TIM is electrically insulating or not can be relevant, as other specifications are related to the mechanical properties of the TIM, which affect the handling capability as well as reliability issues. A TIM can be in either liquid or solid form, and not necessarily be the same during handling as during operation. TIMs that are liquid during handling can lead to a somewhat messy process and uncertain BLT but are generally able to infiltrate voids well if the viscosity is low enough, while applying as a pad is a much more simple process but might not be able to conform to the substrates as well. The importance of the conformability parameter depends on the substrate's roughness and flatness.

During operation, liquid TIMs may be susceptible to effects that negatively affect the reliability. During thermal cycling, the package will undergo mechanical deformations, which tends to spread out the TIM further, eventually pumping out material from between the surfaces. This pump-out effect can be a significant problem for liquid TIMs. Another aspect to consider is whether the TIM can act as an adhesive or if it requires external pressure. TIMs that are liquid during operation will need an applied pressure to hold the system together, also depending on the viscosity of the TIM. TIMs that are solid during both handling and operation will generally require a significant pressure to make the TIM conform to the substrates. On the other hand, there are TIMs that also function as adhesives which will bond the two substrates together and do not require any external pressure during operation, although it can cause delamination issues which can be a reliability concern.

7.2 Polymer-Based Thermal Interface Materials

Polymer-based thermal interface materials are the most widely used TIMs because of their good flexibility and deformation recovery ability. The bonding force between polymer and fillers has been of interest for a long time. Most polymers and fillers are poorly bound with each other due to their low affinity and compatibility, which deteriorates the performance of polymer composites such as mechanical properties [9], thermal properties [10], and electrical properties [11]. For example, the interface thermal resistance (ITR) of silicon–polymer interface decreases with the increase of bond strength of interface [12]. Therefore, enhancing the bonding force between filler and polymer or filler and filler can effectively reduce the scattering of phonons at the interface, thereby increasing the thermal conductivity.

7.2.1 Filler Surface Functionalization

Filler surface functionalization is an effective method to enhance the bonding force between filler and polymer [9, 13]. Guo et al. [14] prepared triethoxyvinylsilane (YDH-151)-functionalized multi-walled carbon nanotubes (s-MWCNTs) and combined them with poly(vinylidene fluoride) (PVDF) to fabricate TIMs. Functionalization not only improves the dispersibility of s-MWCNTs in PVDF but also improves the bonding between CNTs. The enhanced interface bonding effectively reduces the ITR and improves the overall thermal conductivity of TIMs. When the filler content of s-MWCNTs is 10.0 wt%, the thermal conductivity is 1.55 W m^{-1} K^{-1}, which is three times larger than that of pure MWCNTs/PVDF composite. In addition, recent studies have shown that wrapping an unfunctionalized filler with a functionalized filler is also an effective way to improve thermal conductivity. Ji et al. [15] found that amino-functionalized MoS_2 with good wettability can effectively enhance the bonding force between the polymer and the filler by wrapping the carbon nanotubes (CNTs), thus reducing the phonon scattering and ITR. The in-plane thermal conductivity of the prepared CNT/MoS_2/graphene/epoxy composite with a filler content of 20.0 wt% is 4.60 W m^{-1} K^{-1}. This value is 2.3 times larger compared to that of the CNT/epoxy composite under the same filler content.

7.2.2 Covalent Bonding Among Fillers

Covalent bonding is considered to be another possible way to improve the thermal transfer capability between filler and filler or between filler and polymer due to its stronger bonding force than the hydrogen bond and the van der Waals force [16–25]. You et al. [17] used plasma-assisted mechanochemistry to form a covalent bond (N–B and C–N) between polyamide 66 (PA66) and hexagonal boron nitride (h-BN), which reduces the phonon scattering at the interface and improves the thermal transfer efficiency. When the h-BN content increases from 1.0 to 20.0 wt%, the in-plane thermal conductivity of PA66/h-BN composites increases from 9.10 to 26.13 W/m K, which is more than four times higher than that of composites with the same h-BN content. To build covalent connection between fillers, usually, the thermal conductivity enhanced by increasing the bonding force between polymer matrix and fillers is not always satisfactory. The main reason is that most polymers are normally bad electrical and thermal conductors. Therefore, improving the bonding force between fillers and fillers, when a thermally conductive pathway is formed, is considered to be another way to enhance the thermal transport of electrons and phonons in TIMs.

7.2.3 Construction of Thermally Conductive Pathways

A well-arranged filler structure can provide efficient channels for the transmission of electrons and phonons to enhance their thermal transport capabilities [18]. There are already many strategies for constructing thermal transfer pathways, such as vacuum-assisted filtration [19], hot-pressing [20], directional stretching [21], ice-template method [22], magnetic field-assisted orientation [23], electric field-assisted orientation [24], and chemical vapor deposition process [25].

7.2.3.1 In-Plane Thermally Conductive Pathways

Vacuum-assisted filtration has been considered a facile method for preparing layered composite in various research fields [26]. During the vacuum-assisted filtration process, fillers are arranged in the in-plane direction under the force of the polymer fluid, which gives the TIMs satisfactory in-plane thermal conductivity due to the formation of electrons and phonons thermal transfer pathways. Cheng et al. [19] assembled hydroxyapatite (HAP) nanowires (NWs) and BN NS into a flexible film by vacuum-assisted filtration. The well-oriented HAP NWs and BN NS lead to an in-plane thermal conductivity of 6.40 W/m K at room temperature and 5.50 W/m K at 100 °C.

Hot-pressing is able to orient the fillers by applying a mechanical force. First, the fillers and the polymer matrix are mixed uniformly, and then the mixture is placed under high temperature and high pressure. The anisotropic filler tends to be aligned along the in-plane direction under pressure [27]. Hu et al. [20] reported an epoxy resin/oriented BN composite prepared by hot-pressing processes. The in-plane thermal conductivity of hot-pressed composite is 6.09 W/m K when the BN content is 50.0 wt%, while the thermal conductivity of epoxy resin/random BN composites is only 2.44 W/m K. Sun et al. [28] studied the hot-pressed BN board/polycarbonate composites, and the in-plane thermal conductivity of the composites with the volume fraction of 18.5 vol% is 3.09 W/m K. Recently, Li et al. [29] proposed a graphene oxide (GO)-assisted gelation method, combined with hot-pressing to prepare reduced graphene oxide/BN/natural rubber (rGO/BN/NR) composites. Due to the high orientation of the filler in the plane [30], the in-plane thermal conductivity of the prepared composites under a BN load of 250 parts per hundred rubber (phr) is 16.0 W/m K. Moreover, Kwon et al. [21] fabricated a film with a high in-plane thermal conductivity by stretching the polyvinyl alcohol (PVA)/h-BN composites as shown in Figure 7.4a. When the stretching rate is 100%, the in-plane thermal conductivity of the PVA/h-BN composite film is 6.40 W/m K, which is 32 times larger compared to the value before stretching. Han et al. [32] constructed a nacre-mimetic three-dimensional (3D) network structure by using bidirectional freezing technology to prepare BN NS/epoxy composite materials as shown in Figure 7.3c. The in-plane thermal conductivity of the BN NS/epoxy composite is 6.07 W/m K when the BN NS loading is 15.0 wt% (Figure 7.3d). Casting mold is a technology that allows fillers to form along a horizontal orientation and is suitable for large-scale production. The mixture containing the filler and polymer is poured on the substrate to make the mixture flow freely under the action of gravity, and the weak shear stress generated can make the filler orientate. Chang et al. [35] created a ball-milled exfoliated graphite NS/silver NWs/polydimethylsiloxane (BMEG NS/Ag NWs/PDMS) TIM with fillers arranged horizontally by casting molding method. 1D Ag NWs and 2D BMEG NS form thermal pathways in PDMS, and the sample with silver NW content of 2.0 mg/ml can thus obtain an in-plane thermal conductivity of 29.2 W/m K (Figure 7.3a,b). Recently, Li et al. [33] constructed a 3D hybridized structure carbon film, graphitized graphene oxide on polyimide (g-GO/PI), by solution dip coating graphene oxide (GO) on polyimide (PI) skeletons, followed by high-temperature annealing (Figure 7.3e). The high in-plane (1428.00 ± 64 W/m K) thermal conductivity is attributed to the favorable orientation of carbon nanofibers and graphene NS in the 3D architecture and the extremely high crystalline structure after the graphene-induced graphitization, which promotes the thermal conduction of electrons and phonons.

7.2.3.2 Out-of-Plane Thermally Conductive Pathways

In recent years, researchers have started to use the ice-template method, a common method for preparing porous ceramics, to prepare 3D filler networks in TIMs [18, 22, 36]. The mechanism of the ice-template method is directional freezing, which uses ice crystals formed in the process of water freezing as a template

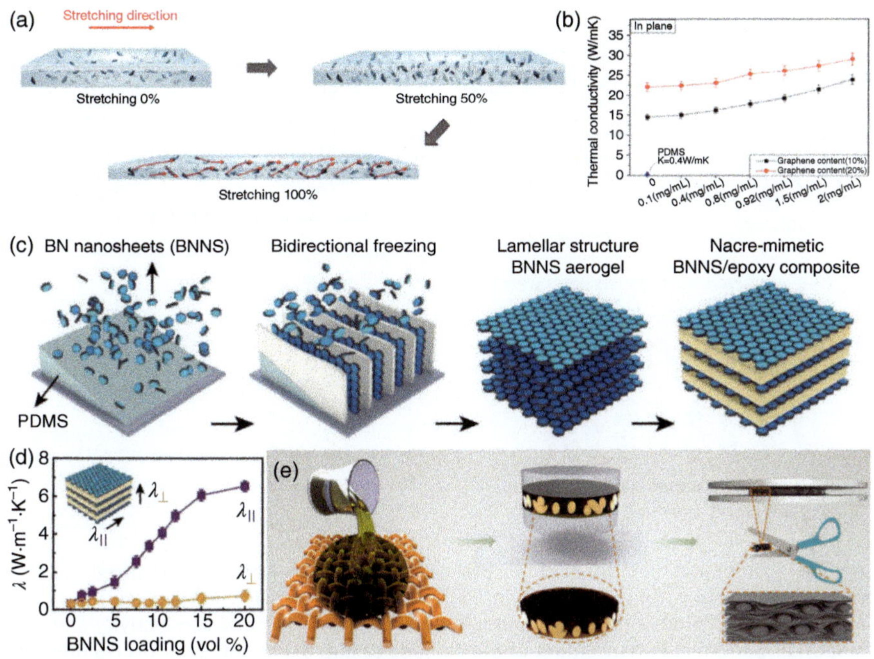

Figure 7.3 Schematic of (a) a PVA/h-BN composite film stretched in the transverse direction, (b) in-plane thermal conductivity of ball-milled exfoliated graphite NS/silver NWs/polydimethylsiloxane (BMEG NS/Ag NWs/PDMS) composite, (c) the manufacturing route using a bidirectional freezing technique and epoxy resin penetration, (d) thermal conductivity of BN NS/epoxy composites in the parallel ($\lambda_{||}$) and vertical ($\lambda_{\perp}$) directions, (e) the graphitized graphene oxide on polyimide (g-GO/PI) film preparation process. Source: Reprinted from Refs. [21, 32–34]; © 2023 American Chemical Society, © 1999–2023 John Wiley & Sons, Inc. and © 2020 Elsevier Ltd.

to prepare an oriented filler network. The specific process is to place the water dispersion containing the fillers under a temperature gradient, and the ice crystals will grow from low temperature to high temperature, while the fillers will be squeezed out during the growth process, thus forming a 3D network along the growth direction. After freeze-drying to remove ice, an oriented filler skeleton is obtained [37]. Yao et al. [22] combined ice-template and infiltrating methods to prepare a 3D phonon skeleton structure consisting of stacked BN platelets reinforced with reduced graphene oxide (rGO) (Figure 7.4a). The formation of a phonon-matched 3D BN-rGO network results in a better phonon transport, and this 3D BN-rGO/epoxy resin composite exhibits an out-of-plane thermal conductivity of 5.05 W/m K at a volume fraction of 13.2 vol%. Fu et al. [38] reported an Ag NPs-BN NS/Ag NWs/epoxy resin composite prepared by the ice-template method, and an out-of-plane thermal conductivity of 0.80 W/m K was obtained.

Magnetic field orientation provides another method to realize the orientation of the filler in a specific direction [39]. Kang et al. [23] synthesized a rod-shaped liquid crystal monomer PNP-6MA to prepare TIMs (Figure 7.4b). The out-of-plane thermal conductivity can be enhanced to be 2.5 W/m K, which is much higher than that of

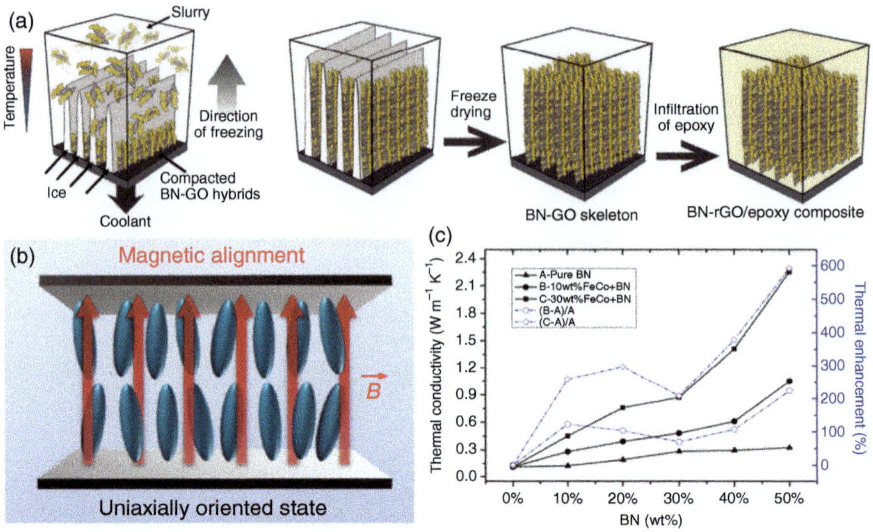

Figure 7.4 Schematic diagram of (a) 3D BN-reduced graphene oxide (rGO)/epoxy composite preparation, (b) rod-shaped liquid crystal monomer PNP-6MA molecules arranged unidirectionally under a magnetic field (B = 2T), (c) thermal conductivity and corresponding thermal enhancement of the composite film using only pure BN NS filler or FeCo-BN NS filler. Source: Reprinted from Refs. [22, 23, 31]; © 2023 American Chemical Society and © 1999–2023 John Wiley & Sons, Inc.

traditional epoxy resin. Yuan et al. [31] fabricated FeCo/hexagonal-BN NS/PDMS (FeCo/h-BN NS/PDMS) composites, and FeCo nanocubes were used to help BN NS to be arranged vertically in a magnetic field to form thermal conductive pathways. This structure facilitates the thermal transport in the vertical direction. The out-of-plane thermal conductivity of the composite film is 2.25 W/m K when the content of FeCo is 30 wt% and BN is 50 wt%, which is seven times that of the film containing 50 wt% BN (random arrangement) (Figure 7.4c).

In addition, some researchers have prepared 3D networks using other methods. An et al. [40] prepared graphene hybrid foam with vertical network arrangement in graphene oxide/graphene nanoplatelets (GO/GNP) suspension by KOH-assisted hydrothermal method and air-drying method (Figure 7.5a). Figure 7.5b shows the image of freeze-dried rGO/GNP hybrid foam. Song et al. [42] prepared a BN/polyvinylidene fluoride (x-BN/PVDF) composite film using a general stepwise orientation method without chemical modification or special treatment. Specifically, it involves several processes: (i) preparation of BN films with in-plane orientation by bar coating (*x*–*y* plane); (ii) hot-pressing of multilayer film stacks to form 3D BN/PVDF composite materials; (iii) cut the 3D BN/PVDF composite into BN/PVDF film along the vertical direction (*x*–*z* plane). Therefore, for the *x*–*y* plane, BN platelets are oriented along an in-plane direction, but for the *x*–*z* plane, they are oriented along out-of-plane. Vu et al. [25] prepared thin SiC sheets from GO sheets by a facile in situ vapor–solid reaction using the template-assisted CVD technique. The vertical arrangement of SiC (Figure 7.5c) gives the composite a high out-of-plane thermal conductivity of 14.32 W/m K at a volume fraction of 3.7 vol%.

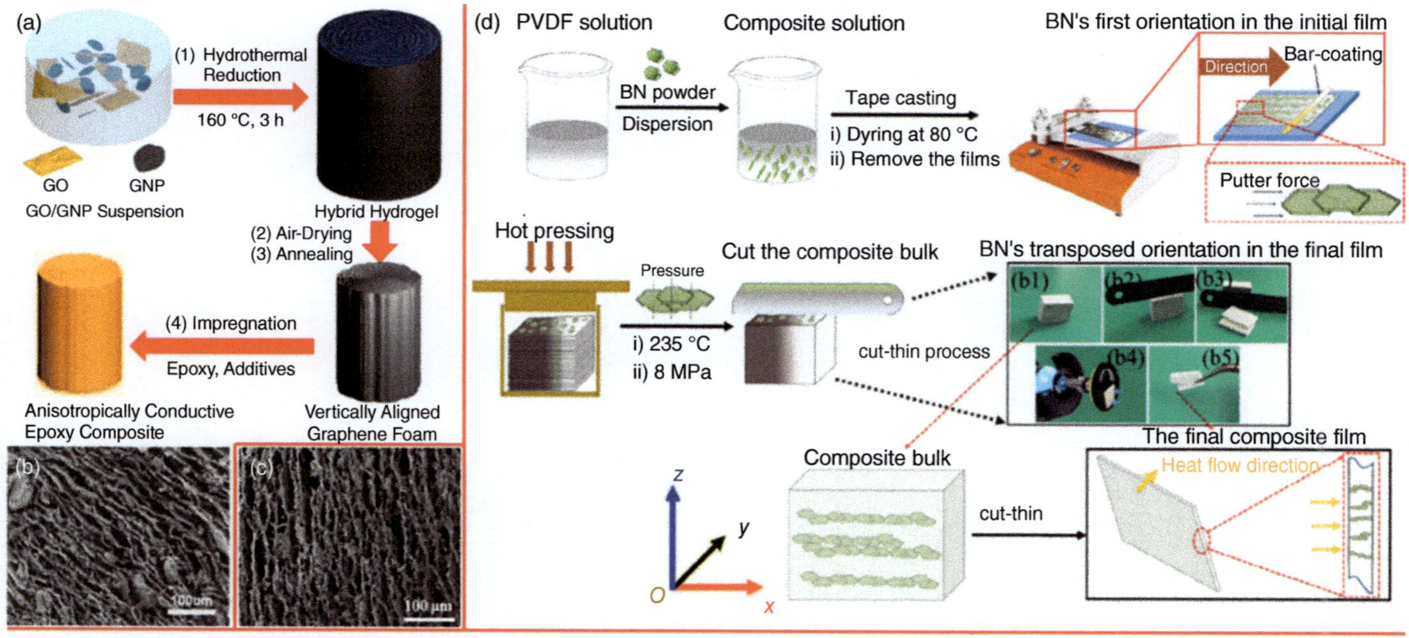

Figure 7.5 Schematic diagram of (a) preparation of graphene oxide/graphene nanoplatelets/epoxy (GO/GNP/epoxy) resin composite with vertically aligned structure, (b) SEM image of freeze-dried rGO/GNP hybrid foam, (c) SEM image of top view of SiC3D structure, (d) the stepwise orientation method that used to fabricate BN/PVDF composite films. Source: Reprinted from Refs. [5, 40, 41]; © 2023 The American Society of Mechanical Engineers, © 2023 American Chemical Society, and © 2019 Elsevier Ltd.

7.2.3.3 Isotropic Thermally Conductive Pathways

Randomly distributing thermally conductive fillers in the polymer matrix is the simplest way to prepare TIMs [43]. Solution blending is the most common method for preparing randomly distributed filler–polymer composites. The thermally conductive filler is mixed with liquid polymer or solid polymer dispersed in a solvent uniformly, and then placed in a mold for curing to obtain TIMs [44, 45]. However, TIMs prepared by this method always exhibit unsatisfactory thermal conductivity, because the fillers tend to be separated by the polymer, which significantly affects the transport of electrons and phonons. The bridging effect formed by connecting low-dimensional materials between fillers can address the problem. Feng et al. [46] functionalized graphene with flame retardant molecules (GP-DOPO) and mixed them with Ag NWs and epoxy resin to prepare TIMs. The addition of GP-DOPO not only improves the dispersion of Ag NWs but also causes bridging effect to make the material form thermal conduction pathways (Figure 7.6a). When the volume fraction of Ag NWs is 4 vol% and filler content of GP-DOPO is 2 wt%, the thermal conductivity of the composite is 1.41 W/m K.

There are other methods of constructing isotropic thermally conductive networks [43, 48]. Tian et al. [47] reported a direct foaming method for preparing 3D interconnected layered porous BN/epoxy composite. It has been shown that this method can produce isotropic porous structures with low filler content, and micro-BN can better align along the 3D network than submicron BN (Figure 7.6b). Therefore, when the filler content of micron BN is 24.4 wt%, the in-plane and out-of-plane

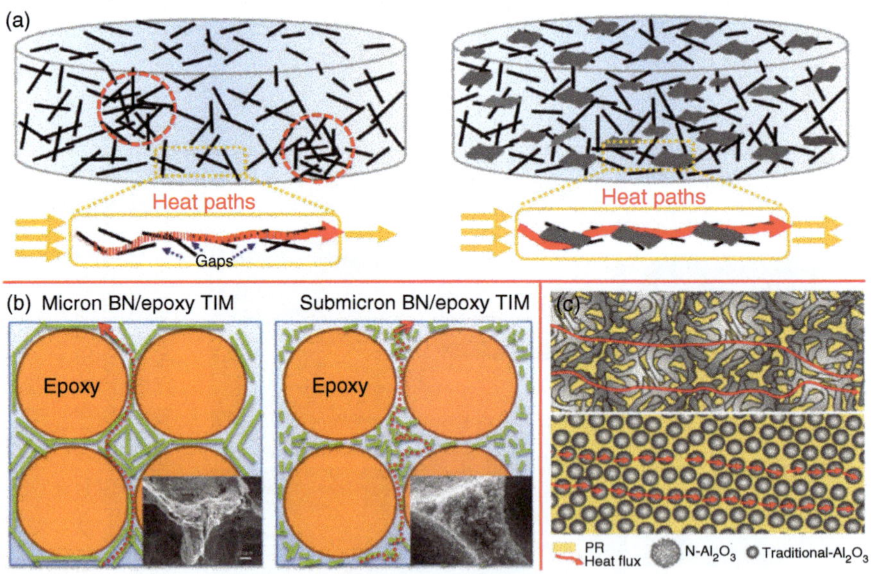

Figure 7.6 (a) Thermal conductive models of epoxy/Ag NW (left) and epoxy/Ag NW/functionalized graphene with flame retardant molecules (GP-DOPO) (right). (b) Heat flow model of micron and submicron BN/epoxy composites. (c) Heat flow model of N-Al_2O_3/PR composites and traditional composites. Source: Reprinted from Refs. [46, 47]; © Royal Society of Chemistry 2023 and © 2023 American Chemical Society.

thermal conductivity of the prepared BN/epoxy composite are 5.19 and 3.48 W/m K, respectively.

7.2.4 Enhance the Bonding Force and Construct Thermally Conductive Pathways

7.2.4.1 Non-Covalent Bonds and Thermally Conductive Pathways

Jiang et al. [49] prepared polyamide-imide/3D-BN (PAI/3D-BN) composites by adjusting the hydrogen bond between PAI and BN. The formation of 3D-BN structure can be controlled as shown in Figure 7.7a. Wang et al. [50] reported a fluorinated CNT/nanofibrillated cellulose (FCNT/NFC) composite. The prepared film has a good layered structure arranged in the in-plane direction and strong hydrophobic-to-hydrophobic interaction between NFCs and FCNTs (Figure 7.7b), which leads to efficient in-plane thermal conduction pathways in the composite film. Yang et al. [51] fabricated the BN/nitrile butadiene rubber (BN/NBR) composite by mixing poly(dopamine) (PDA) and γ-(2,3-epoxy propoxy) propyltrimethoxysilane (KH560) functionalized BN platelets with NBR (Figure 7.7c,d). Functionalized BN enhances the interface interaction between BN and NBR, which not only effectively reduces the ITR at interface but also enables the filler to be uniformly dispersed to form a thermally conductive network. Wang et al. [54] obtained few-layered BN NS via biomolecular stripping technology and then fabricated layered BN NS/ethylene-vinyl acetate copolymer (EVA) composite film through vacuum-assisted filtration. Ma et al. [45] assembled the vertically aligned carbon fibers (CFs) skeleton by ice-template method, and then mixed it with epoxy resin to produce a 3D-CFs/epoxy composite. Surface functionalization of CFs improves the interaction between CFs and epoxy, while the 3D-CFs network can form thermal conduction channels for electrons and phonons. Therefore, when the volume fraction of CFs is 13.0 vol%, the out-of-plane thermal conductivity of the composite is 2.84 W/m K compared with 0.19 W/m K of pure epoxy resin.

7.2.4.2 Covalent Bonds and Thermally Conductive Pathways

The construction of covalent bonding and thermal conduction pathways has also been extensively studied. An et al. [52] fabricated a covalently bonded (–C–N– and –CO–NH–) 3D BN/rGO thermally conductive network structure by surface functionalization and ice-template methods (Figure 7.7e). The specific preparation process includes five steps: (i) prepare amino-functionalized BN by ball milling urea and BN; (ii) add carboxyl activator (EDC) and catalyst (NHS) to prepare GO solution; (iii) mix and react the prepared BN/GO solution; (iv) prepare BN/rGO 3D network by ice-template method; and (v) mix with NR and vulcanize to obtain BN/rGO/NR composites. The covalent bond connection greatly reduces the ITR of filler/filler and filler/matrix. The out-of-plane thermal conductivity of the prepared composites is 1.28 W/m K when the volume fraction of filler is 4.9 vol%. Different from the establishment of covalent bonds between fillers, a further study has recently been

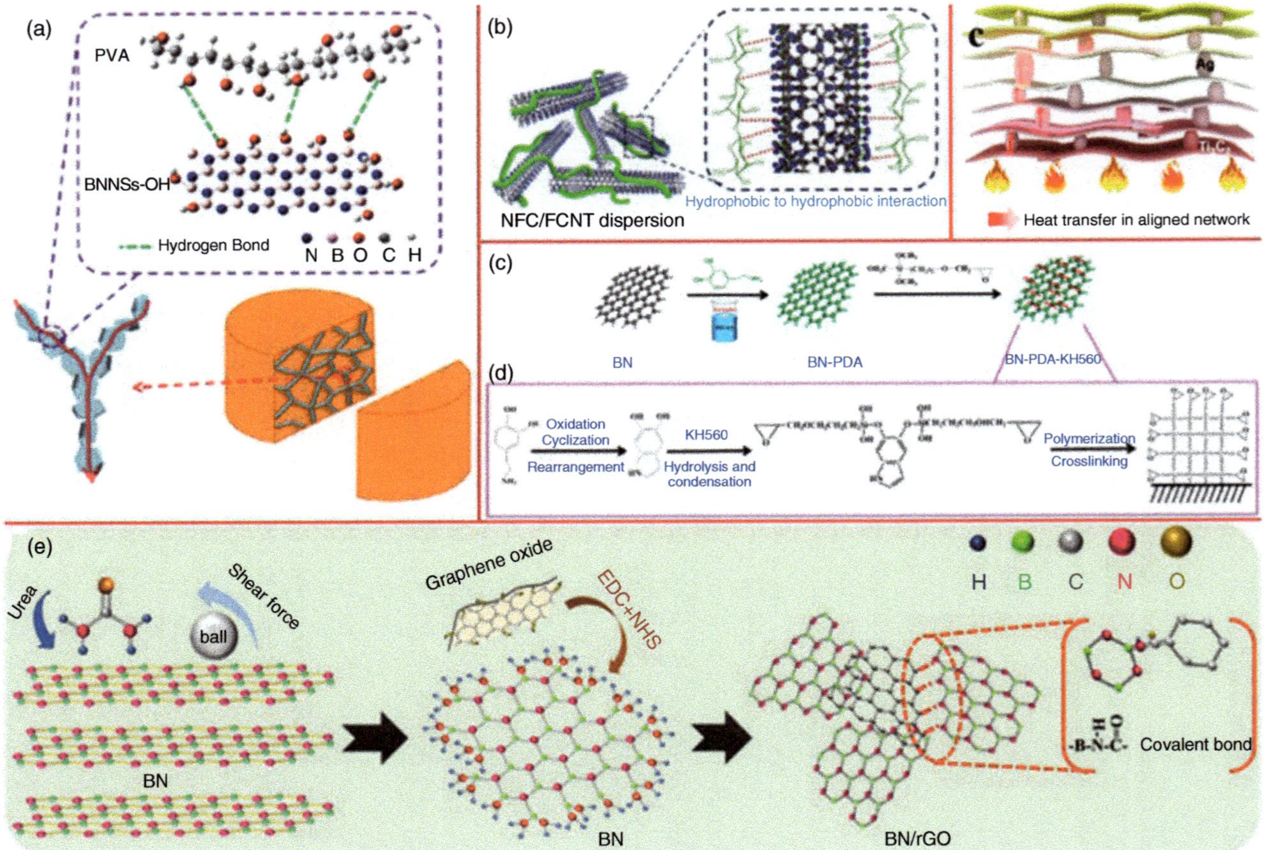

Figure 7.7 (a) Schematic of the hydrogen bond between PVA and hydroxyl functionalized BN NSs, (b) the hydrophobic-to-hydrophobic interaction between nanofibrillated cellulose (NFC) and fluorinated CNT (FCNT), (c) thermal transfer model in welded MXene/Ag aerogel skeleton, (d) schematic illustration and mechanism of preparation route of BN/poly (dopamine)/γ-(2,3-epoxy propoxy) propyltrimethoxysilane (BN-PDA-KH560) platelets, (e) preparation process of the BN/rGO hybrid with covalent linkage. Source: Reprinted from Refs. [49–53]; © 2023 American Chemical Society.

conducted on establishing covalent bonds between fillers and polymer matrix by An's group [55]. BN/NR composites were prepared by ice-template method, where the BN was functionalized by KH-590. Owing to the covalent bond between BN and NR enhancing phonon transmission, the thermal conductivity of the composite is 0.79 W/m K when the filler content is 25 wt%.

7.2.4.3 Welding and Thermally Conductive Pathways

Welding is a technique that uses heat, pressure, or a combination of both to achieve atomic bonding of materials. In recent years, some researchers have applied this technology to the fabrication of TIMs considering the interatomic bonding between fillers can greatly enhance the heat transfer capability of the fillers. Ji et al. [53] prepared 3D MXene/Ag aerogels arranged along the plane as thermally conductive network of epoxy composites. The prepared aerogel is heated at 200 °C to weld MXene and Ag, which effectively enhances the thermal transport of electrons and phonons (Figure 7.7c). Therefore, at the addition of 15.1 vol% Ag, the out-of-plane thermal conductivity of MXene/Ag/epoxy composites is 2.65 W/m K, which is 1225% higher than that of pure epoxy resin. Yao et al. [56] constructed an interconnected and vertically aligned BN-SiC hybrid skeleton by the combination of ice-template assembly and welding technology, and then prepared the 3D BN-SiC/PDMS composites. The welding of SiC NWs transforms the loose connected BN sponge into a robust 3D skeleton via thin interfacial borosilicate glass phases, which enhances the transfer of phonons between the adjacent BN plates and reduces the inter-skeleton phonon scattering. The vertically oriented BN-SiC skeleton contributes to an out-of-plane thermal conductivity (3.87 W/m K) when the volume fraction is 8.35 vol%.

7.3 Metal-Based Thermal Interface Materials

Traditional TIM is a polymer matrix composite containing thermal conductive filler. The polymer matrix gives the TIM flexibility to fill the air gap at the interface. However, their thermal properties are usually limited by low heat transport in polymers [8]. In contrast, metal TIM has some of the advantages of the highest bulk thermal conductivity of TIM materials. These TIMs can be in the form of solders, liquids or phase change metals, compressible materials (plastic deformation to object surface characteristics), and mixed metals (including phase change wetting). Many metal TIMs use metals with high isotropic thermal conductivity and low yield and flow strength. The low yield strength and flow strength make TIM conform to the surface roughness and irregularity of the object, so it has low thermal interface resistance. In addition, these TIMs will recover from low-temperature deformation. Hot grease may be "pumped out," which occurs during the use of electronic products, because the mismatch between thermal cycle and CTE will actually pump hot grease out of its application space, sometimes leading to thermal failure. However, solid metal TIMs are not affected by pump-out failure modes [57]. Solder TIM has high bulk thermal conductivity and will have low interface thermal resistance after the intermetallic compound is formed between the welding objects. Due to its high thermal conductivity, low melting point, and low shear strength, indium-based materials are ideal candidates for TIM applications. However, solders with high In content are expensive and have low compressive creep strength, which may lead to structural instability after heat sink attachment. I.·Dutta and colleagues proposed a completely different production method of microelectronic solder TIM based on liquid phase sintering (LPS), which not only solved the above problems but also paved the way for the next generation of packaging to develop TIM solder with higher K than In. LPS Sn-In solder was processed and characterized, and its microstructure was composed of high melting point phase (HMP) Sn and a small amount of intergranular low melting point phase In. The flow stress close to pure In and the conductivity/thermal conductivity of half of pure In were obtained. LPS solder joints between Cu substrates are produced by a single-step process combining LPS and bonding (Figure 7.8) [58].

Although In–Cu composite solder can achieve high thermal conductivity and flexibility theoretically, its mechanical and thermal properties are reduced due to the formation of a large number of Cu–In intermetallic compounds (IMC). There are also the following problems: (i) The distribution of Cu particles in the In matrix is uneven; (ii) The minimum thickness of In–Cu composite cannot be arbitrarily small. In addition, LPS in a short time is an important condition to reduce the formation of IMC, which usually leaves a large number of residual pores, resulting in poor contact between In and Cu and high interface thermal resistance. Deepak Sharma and colleagues addressed the above challenges by proposing to implement a two-step processing route consisting of LPS and cumulative welding [59]. Gong Wei and his colleagues demonstrated a high-performance TIM (Figure 7.9a), which is composed of super flexible and thermal conductive heterogeneous copper–tin nanowire arrays, which we call "super scaffolding" to simulate the role of traditional solder in bonding various surfaces, but its thermal and mechanical properties are superior to any existing TIM. The measured compressive elastic modulus and shear modulus of copper nanowires are in the range of 200 MPa to 1.5 GPa and 2 to 15 MPa, respectively,

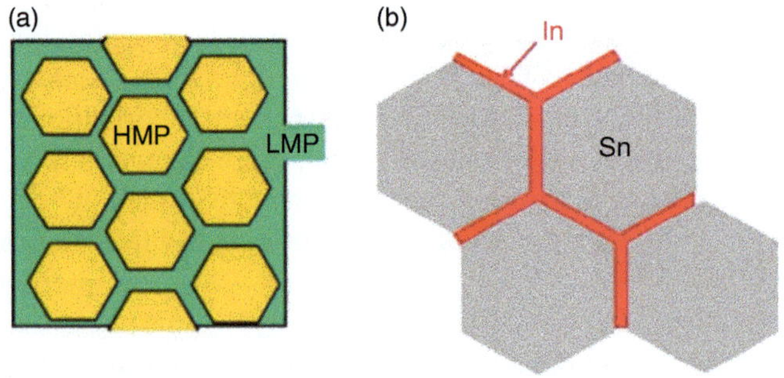

Figure 7.8 (a) Schematic of idealized LPS solder microstructure showing HMP and LMP distribution; (b) schematic of a Sn-In LPS solder showing a low LMP volume fraction (0.2–0.3) forming a continuous network around HMP grains. Source: Reprinted from Ref. [58]; © 2023 Springer Nature Switzerland AG.

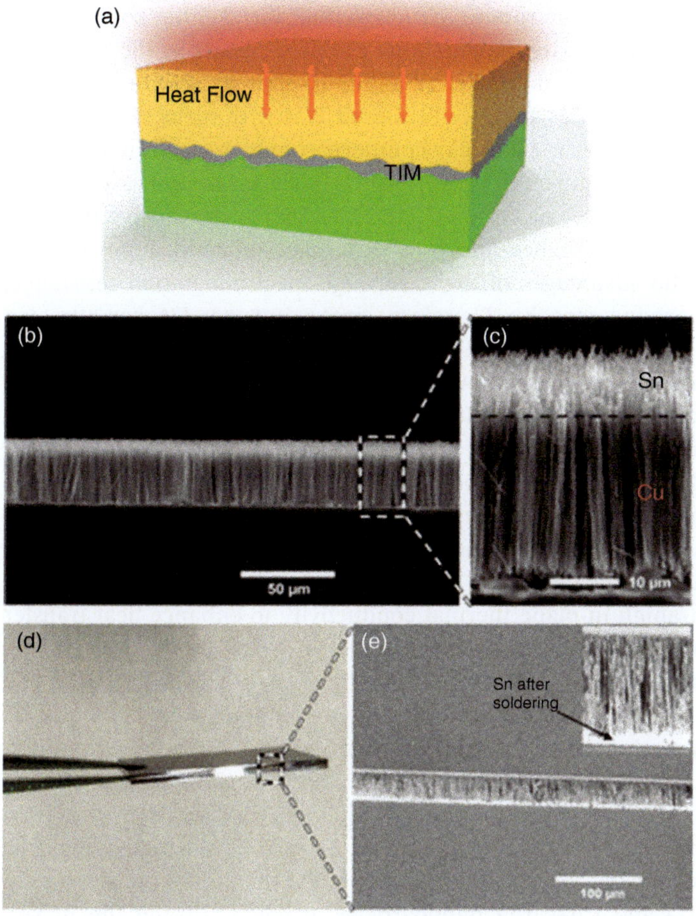

Figure 7.9 Vertically aligned copper–tin nanowire arrays as supersolder. (a) Schematic of thermal interface material. (b, c) SEM images of heterogeneous copper–tin nanowire arrays. In (c), darker portion, copper nanowires; brighter portion, tin nanowires. (d) Optical image of two silicon substrates bonded by the supersolder. (e) SEM image of the cross-section of the supersolder. On top of the nanowires is a thin layer of copper from which the nanowires are grown, whereas a tin layer is formed at the bottom of the nanowires after soldering. Inset: zoomed in image of the continuous tin layer formed by tin nanowires after soldering. Source: Reprinted from Ref. [60]; © 2023 American Chemical Society.

which are 2–3 orders of magnitude smaller than those of bulk copper. The super rack also has an extremely low thermal resistance of ~0.5 mm^2 K/W, which is about twice smaller than the most advanced TIM [61]. More importantly, the thermal power cycle test conducted in a wide temperature range shows that the super rack can work stably for more than 1200 times, thus proving its long-term reliability [60].

Theoretically, sintered metal nanoparticles have become ideal TIM candidate materials for high-power electronic packaging (≤300 °C) and good high-temperature thermal stability (>300 °C) due to their high thermal conductivity and low sintering temperature. The high surface of nanoscale metal particles allows them to be sintered at a temperature much lower than the melting point of bulk metals [62], which helps alleviate the reliability problem caused by the mismatch of thermal expansion coefficient (CTE). In addition, compared with traditional solder alloy and polymer binder TIM, well-sintered TIM with metal nanoparticles has a high melting point close to its blocky counterpart, which makes it have good thermal stability at high temperatures [63, 64]. Nanosilver paste has become a new DA in the field of power electronics due to its high performance [65], but its low electromigration resistance and high cost limit its application in some cases. Copper has the advantages of low cost, high electromigration resistance, and high current-carrying capacity. Therefore, sintered copper DA has attracted much attention in recent years [66]. However, the oxidation tendency of copper nanoparticles makes sintered copper DA have more voids and reduces its thermal, electrical, and mechanical properties. It is difficult to completely remove copper oxide. Therefore, in order to improve the performance of sintered copper, it is necessary to develop other methods besides removing copper oxide. Graphene has excellent thermal, electrical, and mechanical properties and is a promising candidate material. Its unique thermal conductivity promotes its application prospects in thermal management [67]. Its high aspect ratio makes it easy to form a network with ultrahigh in-plane thermal conductivity to enhance heat transfer between copper particles. In addition, its negative CTE is helpful for sintering DA to reduce CTE mismatch with chips and related reliability problems [68, 69].

Deng Shaojia and his colleagues proposed a method to prepare nanoscale Cu–Cu_xO particles on the surface of plasma-treated graphene to reduce the interface thermal resistance between graphene and copper matrix, thereby improving the thermal conductivity of sintered copper, which can be used as a chip TIM for packaging high-temperature power electronic devices. The oxygen-containing functional groups of plasma-treated graphene promoted the formation of C–O–Cu bonds between PTG and Cu–Cu_xO nanoparticles at 450 °C during the in situ preparation of Cu–Cu_xO nanoparticles on PTG prior to TIM sintering (Figure 7.10). It solves the problem that it is difficult to form C–O–Cu bond between graphene and copper at sintering temperature below 300 °C. As a result, the thermal conductivity of sintered composite TIM was increased by 123.5% [68].

Liquid metal has high thermal conductivity and good wettability to object surface. In addition, welding or surface metallization is not required, and very thin bonding wires can be obtained. However, liquid metals must be controlled to prevent diffusion and are difficult to handle and apply. Fortunately, once applied, its surface tension usually keeps it in the desired position. The performance of liquid metal is obviously better than that of hot grease (Figure 7.11).

There are many liquid metal TIM options. Most liquid metal TIMs are gallium-based metals with the addition of indium, tin, and zinc. By changing these metal components, a melting point of 8 to 30 °C can be obtained [57, 71].

Gallium-based liquid metals are promising candidates for high-performance TIMs [72]. Their metallic properties ensure inherent high thermal conductivity and good heat transfer performance. The fluidity of liquid metals allows them to fill the

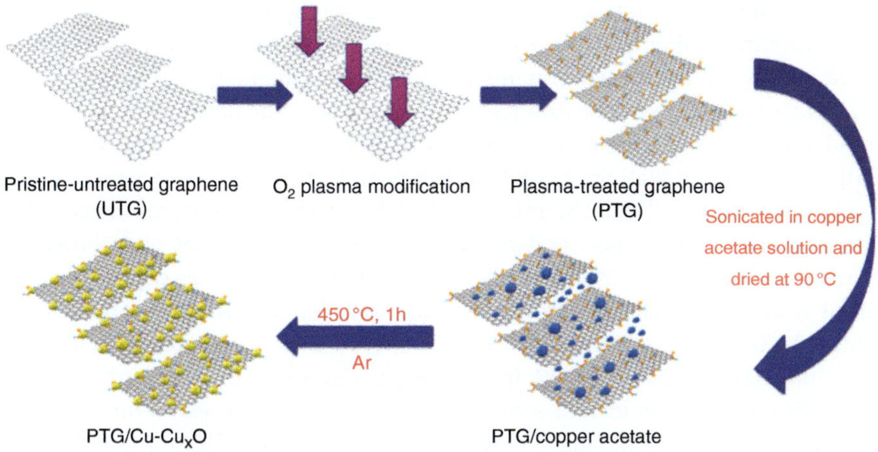

Figure 7.10 Schematic process procedure of preparing Cu–Cu_x–O nanoparticles on PTG. Source: Reprinted from Ref. [68]; © Copyright 2023 IOP Publishing.

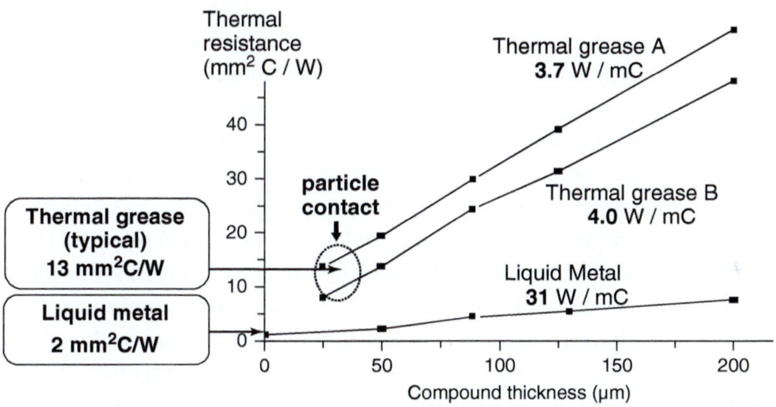

Figure 7.11 The thermal resistance of liquid metals is close to a factor of ten less than that of thermal greases [70]. Source: Reprinted from Ref. [70]; © Copyright 2023 IOP Publishing.

air gap between the electronic chip and the radiator. However, this liquidity may cause leakage problems during operation [73]. The relatively high surface tension of liquid metal also makes it difficult to apply it evenly on the surface of electronic devices [74]. Adding solid filler to liquid metal matrix through mechanical mixing is an effective method to further improve the thermal conductivity of liquid metal composites and solve the leakage problem. During the mixing process in air, gallium-based liquid metal can be easily oxidized to generate a thin (approximately several nanometers) layer of amorphous gallium oxide (Ga_2O_3), which helps encapsulate solid fillers into the liquid metal matrix [71, 75]. The formation of Ga_2O_3 also decreases the liquid metal surface tension, leading to better adhesion properties on various substrates. This enhanced adhesion is particularly useful for reducing contact thermal resistance in TIM applications [76]. By changing the amount of solid filler, the mixture can change from liquid to paste, thus reducing the possibility of liquid metal leakage. The method of adding solid particles to liquid metals has been used in many thermal management applications [77]. Liu and his co-workers used the iron particles coated with the silver shell to mix into liquid metal and successfully extended the stability of liquid metal composite with a thermal conductivity of 48.95 ± 0.79 W/m K. However, the coating process is relatively complex [78]. The mixture of high thermal conductivity metal fillers (such as silver and copper) and liquid metal matrix also helps to improve the thermal conductivity of liquid metal composites. For instance, copper particles mixed with eutectic gallium indium (EGaIn; 75% Ga; 25% In) in NaOH solution achieved a thermal conductivity of 50 W/m K [79]. Copper nanoparticles incorporated into GaInSn (66% Ga; 22% In; 16% Sn) via an oxide-free ultrasonic-assisted internalization process in HCl solution reached a high κ of 64.8 W/m K [80]. In these efforts involving mixing Ga-based liquid metal with fillers of high thermal conductivity, gallium, however, can react with many metals such as copper [81, 82], silver [83], aluminum [84], and iron [78], to form IMCs due to its reactive nature. For example, in the eutectic gallium indium–copper (EGaIn-Cu) composite system, the reaction between copper and gallium produces $CuGa_2$ IMCs. The formation of such IMCs consumes gallium in the EGaIn matrix and results in a compositional shift and melting point increase since $CuGa_2$ IMC has a high melting point [85]. The above reaction leads to a gradual solidification of EGaIn-Cu composites. If the EGaIn surface oxide is chemically removed by NaOH (aq), copper and gallium will come into direct contact and $CuGa_2$ IMCs start to form within several minutes after mixing [86]. Due to the rapid alloying process, the as-synthesized EGaIn-Cu composites solidify within days and are unable to be used as an effective TIM [87].

Wang Han and co-workers present an interfacial engineering approach to enhance both the thermal conductivity and stability of EGaIn-copper particle (EGaIn-CuP) composites, in which 3-chloropropyltriethoxysilane (CPTES) was used to chemically modify the surface of CuPs. CPTES molecules not only serve as effective thermal linkers at CuP/Ga_2O_3 interfaces to enhance the bulk thermal conductivity of the composite but also promote the stability of composites by suppressing the formation of $CuGa_2$ IMCs. The thermal conductivity of EGaIn-CuP composites modified with the chlorine-terminated silane (EGaIn-CuP@Cl) reaches as high as 65.9 W/m K. Most of the studies on liquid metal-based composites enhance material performance through direct mixing of various fillers into liquid metals with no interfacial modification [79, 80, 88, 89], which may further advance their applications in various fields including thermal management systems [90], flexible electronics [91, 92], consumer electronics [80], and biomedical systems [87, 93].

Gallium alloys react very quickly with most high thermal conductivity metal materials (such as copper, aluminum, silver, and tungsten). Therefore, with the drying of liquid metals, the composites will be transformed into intermetallic compounds in a very short time. As for inorganic nonmetal fillers, such as diamond, CNTs, graphene, or boron nitride, the current bottleneck is not their intrinsic heat conductivity but their poor wettability with liquid metal. The thermal conductivity of gallium/chromium-coated diamond composite reaches up to 112.5 W/m K which is approximately four times higher than that of pure gallium. In addition, the chromium diffusion barrier coating on the surface of diamond particles and metal substrate can greatly avoid excessive consumption of liquid metal and maintain the long-term reliable use of gallium-based TIM [94]. GaIP-containing diamond filler has excellent thermal conductivity and heat transfer performance, superior to commercial liquid metal-based thermal paste [89].

Although the previous studies have reported the improved thermo-physical properties of the LM matrix-fillers composites, solder-level thermal conductivity (>60 W/m K) with maintained fluidity has not been reported yet. The cooling performance results of the actual TIM as a test vehicle have not been provided, which is critical for practical applications. In addition, due to the remarkable aggregation and oxidation of nanofillers, little research has been done on the internalization of nanoparticles (<100 nm) in Ga-based LM. Seokkan Ki and co-workers developed the oxide-free ultrasonication-assisted particle internalization method to incorporate Cu nanofillers into GaInSn LM, preventing the issues mentioned above. Due to its high thermal conductivity and wettability, at the junction temperature of about 150 °C, its thermal resistance is about 46% lower than that of commercial heat transfer grease, and its thermal stability is high, indicating that the developed TIM has a strong application potential [80].

7.4 Carbon-Based Thermal Interface Materials

Carbon-based TIMs, ranging from one-dimensional CNTs, two-dimensional Gr films, to three-dimensional Gr foams (GF), Gr aerogels (GA), and vertical Gr (VG) (Figure 7.12), are the most promising candidates for the heat dissipation of electronic devices. The carbon-based materials with intrinsically high thermal conductivity, such as carbon black [96, 97], graphite [98], graphite nanoplatelets [97, 99], diamond [94, 100], and CNTs, are widely employed as TIMs for heat dissipation of electronic devices.

The conductivity of individual CNTs is about 3000 W/m K for multi-walled CNTs (MWCNTs) at room temperature [101] and about 2400 W/m K for single-walled CNTs (SWCNTs) [102]. In addition to the high thermal conductivity, low transverse elastic modulus of CNTs allows them to adapt to and conform to considerable amounts of external loads without any permanent atomic alterations [103, 104]. Thus, they can absorb stresses induced by the mismatched coefficient of thermal expansion (CTE) in a device [105], resulting in lower interfacial thermal resistance. The thermal resistance of the CNT arrays is as low as 7 mm^2 K/W, reduced significantly compared with the state-of-art commercial TIM. Furthermore, CNTs possess excellent properties, such as good chemical stability and low CTE [106].

Aligned CNTs or CNT arrays exhibit enhanced thermal conductivity compared to the CNTs of random distribution. Vertically aligned CNTs (VACNTs) are ideal TIMs, since they offer both high thermal conductivity and good mechanical compliance [105]. It is reported that CNT arrays enhance the contact thermal conductivity and reduce the thermal interface resistance [107]. The thermal interface resistance

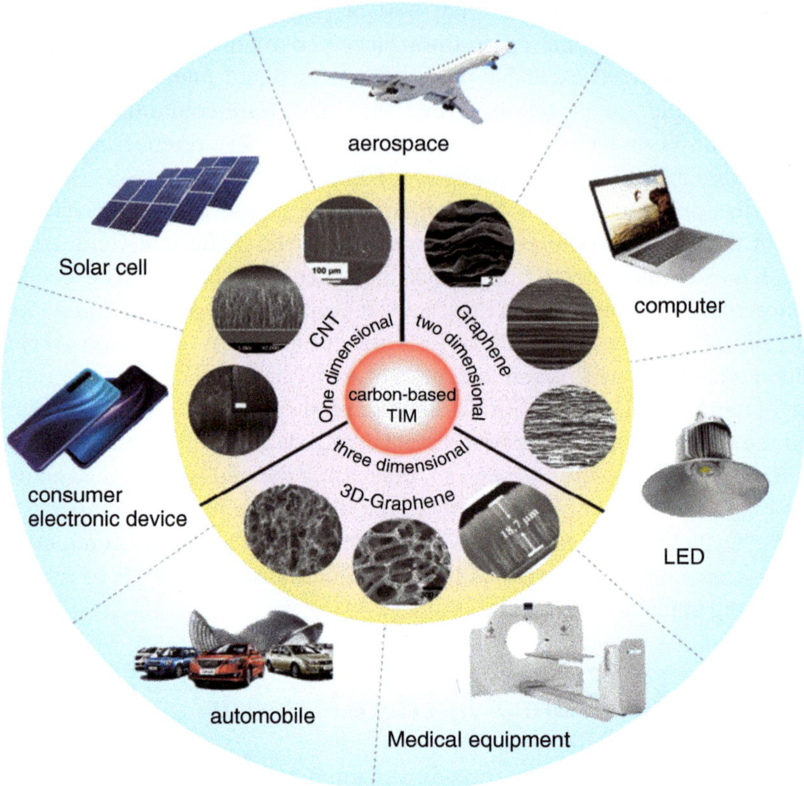

Figure 7.12 Carbon-based TIMs and their applications. Source: Reprinted from Ref. [95]; © 2023 Elsevier B.V.

will be reduced by making better contact at the interface or densified the CNTs. The vertically aligned CNTs that are metallic bonded with a substrate provide the thermal performance of a solder joint while maintaining a thermal resistance as low as 2 mm^2 K/W and good compliance. CNTs grown in arrays with relatively high density can also further improve the thermal conductivity or reduce the contact thermal resistance [108]. Furthermore, it is noted that the VACNT-TIMs show good stability when subjected to thermal cycling and high-temperature baking, thus assuring long-term reliability in application [61, 109].

Currently, CNT-TIMs are as good as typical commercially available thermal greases [110]. However, in practical applications, it is hard to guarantee 100% contact using CNTs. Hence, the thermal contact resistance between the CNTs and the substrate is rather high. The bulk thermal conductivity of CNTs is affected by aspects including but not limited to the quality of the CNTs [111], the volume density [112], and the degree of tangling. In addition, the harsh fabricating conditions prevent the CNTs to be grown on the semiconductor directly. Therefore, a CNT transfer process is inevitable, which may increase thermal resistance [95].

The thermal contact between the heat sink and the heat source plays a critical role in electronic systems [113, 114]. It is therefore highly desirable to design and fabricate TIMs to facilitate efficient heat transport. This can be achieved by reducing the thermal contact resistance caused by the air pockets between two contacting surfaces [115]. A highly conformable thermal fluid such as a mineral and/or paste is commonly applied at the interface to enhance the thermal contact [116]. However, thermal paste such as silicone filled with metallic particles is highly conductive but less conformable [117]. These conditions limit the performance of the system by leaving behind the air pockets at the interface. Therefore, an ideal TIM should provide minimum interface resistance and exhibit high thermal conductivity for efficient heat dissipation. CNTs and carbon nanofibers (CNFs) with their lightweight and high thermo-mechanical properties can be perfect candidates for TIMs.

CNTs and CNFs are new forms of carbon with outstanding electrical and thermal properties, high specific tensile strength and modulus [118]. These nanostructures are produced by many techniques including chemical vapor deposition (CVD) process [119]. Wang et al. [120] fabricated CNT-reinforced buckypapers by dispersing CNTs into water with the aid of epoxy resin. Multiple layers of these buckypapers with high CNT loading (up to 39 wt%) were used to produce nanocomposites. The storage moduli of these nanocomposites were 429% higher compared to neat resin modulus. However, the major limitation of this material lies in its rigidness. Zaeri et al. [121] developed a mechanical model to determine the elastic properties of buckypaper composites. Various parameters were investigated including Young's modulus, shear modulus, and porosity of the buckypaper composite. The Young's modulus of buckypaper composite was 132.2 MPa compared to the predicted value of 148.6 MPa with the porosity of 67.5%. Hong and Tai [122] synthesized SWCNTs and MWCNTs to form freestanding buckypapers. The thermal conductivities at room temperature of SWCNT and MWCNT buckypapers were found to be 2.24 W/m K, and 0.36 W/m K, respectively. Xu and Fisher [107] synthesized CNT arrays directly on silicon wafers using plasma-enhanced CVD to enhance thermal contact conductance of the system. Results show that dry CNT arrays produce a minimum thermal interface resistance of 19.8 mm^2 K/W, while the combination of a CNT array and a phase change material produces a minimum resistance of 5.2 mm^2 K/W. Cross et al. [123] grew CNTs on Si substrate using CVD process followed by metallization using Ti/Au. The coated CNTs were bonded to metallized substrates at 220 °C. Measurements were carried out by photoacoustic technique which showed that at no applied pressure, the thermal resistance for bonded CNT films of 30 μm in length was 1.7 mm^2 K/W, whereas that of CNTs up to 130 μm in length was 10 mm^2 K/W. It was hence suggested by the authors [123] that the interface resistance at the two bonded interfaces in the CNT samples dominates the overall resistance of the structure. Cola et al. [124] developed a contact resistance model to describe heat transfer across CNT array interfaces. The model revealed that the thermal resistances at single-CNT contacts dominate the thermal transport across CNT array interfaces. It was suggested that the total thermal resistance of CNT array interfaces can be reduced by optimizing the array conformability and density such that the true contact area established in the interface is maximized. Wang et al. [125] fabricated a TIM by synthesizing aligned CNTs on both sides of a thin copper foil. The laser flash and hot disk methods were used to measure the thermal conductivity of the CNT-TIMs.

Results showed that a thicker copper foil substrate or CNT layer led to a lower overall thermal resistance of 8.78 mm^2 K/W. It was suggested that an enhancement in thermal conductivity of more than 290% could be obtained under an applied contact pressure of 0.01 MPa, as compared with two copper plates in direct contact. Shaikh et al. [126] fabricated a thin low density aligned CNT as TIM to explore the effect of thermal contact resistance in aluminum and graphite pieces. Compared to the direct contact, the thermal conductivity of CNT-TIM with aluminum surfaces was enhanced by 385% with an 80% decrease in thermal resistance, while, with graphite surfaces, the thermal conductivity was increased by 362% with a 78% decrease in thermal resistance. All these data show the superiority of carbon-based TIMs over conventional materials. High fabrication cost is, however, a significant limitation of CNT-based materials.

Conversely, CNFs are relatively cost effective and have competitive thermal properties [127]. Mahanta et al. [128] measured the thermal conductivity of vapor-grown carbon nanofiber mats at different heat treatment temperatures and increasing volume fractions. The authors found that the in-plane thermal conductivity of the mats varied from 12 to 157 W/m K for volume fractions of 0.067 and 0.462, respectively. Thermal conductivities of individual CNFs at temperatures of 1100 and 3000 °C were calculated to be close to 1400 and 1600 W/m K, respectively. Patton et al. [129] measured thermal conductivities of CNF composites in epoxy of up to 0.8 W/m K for 40 vol%. This increment over the neat resin value of 0.26 W/m K was attributed to the complexity of thermal energy transfer within fibers. Lafdi and Matzek [130] demonstrated a significant increase in thermal conductivity of CNF composite from 0.2 W/m K for epoxy resin to 2.8 W/m K for a 20 wt%. Zhou et al. [131] infused various percentages of CNFs into epoxy matrix through sonic cavitations at room temperature. It was seen that the storage modulus steadily increased with an increasing weight percent of the fiber. The addition of 3 wt% of CNF yielded a 65% increase in the storage modulus. These properties can, however, be improved by tuning CNFs as a function of heat treatment temperatures. Tibbetts et al. [132] found that when CNFs are heat treated at temperatures above 1500 °C, a significant rearrangement of the core morphology is occurred [132]. When the temperature is up to 1873 K, the crystallite increases resulting in more conductive fiber [133]. Endo et al. [134] reported that there was a sharp decrease in the interlayer spacing from 0.342 to 0.337 nm after heat treatment to temperatures above 2100 °C. Heat treatment at graphitizing temperatures (>2400 °C) is an effective technique for purifying and annealing the graphitic structure [135]. To achieve high thermal conductivity, CNFs at various heat treatment temperatures (PR24-PS, PR24-LHT, and PR24-HHT) were used [132]. These CNFs have special structures with large diameters (less than 100 nm) and some crystal imperfections. The performance of the CNFs will be related to their degree of graphitizability. Muhammad Omar Memon and co-workers focused on the fabrication and characterization of CNF buckypapers. The analysis is performed by varying the ratio of CNF and binder and heat treatment temperatures of CNFs. These altering parameters should lead to a buckypaper with specific density (i.e., thickness) and thermal conductivity [136].

Some inorganic materials have a high intrinsic thermal conductivity, for example, AlN (130–260 W/m K) and Cu (400 W/m K), but their rigidity (compressive modulus: around 100 GPa) leads to high contact thermal resistance for TIM applications [137]. In contrast, organic materials such as resins and silicone rubbers can easily fill the gap at low applied pressure (<650 kPa), but the TIMs made up of purely organic materials show poor performance due to their low intrinsic thermal conductivity (0.1–0.3 $W m^{-1} K^{-1}$) [138]. Therefore, conventional TIMs combine the advantages of the above two by using a polymer matrix reinforced with highly thermally conductive fillers, such as AlN, Al_2O_3, and BN, to fabricate composites with a thermal conductivity of 1–5 W/m K at a filler content of 50–70 wt% [139]. However, conventional TIMs face a great challenge to meet the future demand for cooling high-power and high-frequency devices with significantly increased power consumption [140]. Recently, low-dimensional materials with ultrahigh thermal conductivity, for instance, hexagonal-BN nanosheets (up to 600 W/m K), carbon nanotubes (CNTs, higher than 3000 W/m K), and graphene (3500–5300 W/m K), have been employed extensively to prepare composites with polymers as matrix by melt or solution blending [141]. As a result, while the microstructure of the fillers in the polymer matrix is randomly distributed, the thermal conductivity of the composite TIMs can approach approximately 10 W/m K (filler content: 20–50 wt%) [142]. Nevertheless, these low-dimensional fillers, in which phonons can efficiently propagate only along a certain direction in the material lattice, have highly anisotropic thermal conductivity [143]. Therefore, in order to further improve the through-plane thermal conductivity, the microstructural modulation toward vertical alignment of low-dimensional materials, instead of random orientation, is highly desirable to break through the performance bottleneck of current TIMs.

Some techniques for the formation of a vertically aligned architecture of CNTs or graphene have been proposed [144]. For the case of CNTs, a popular approach is to grow vertically aligned CNT arrays on the substrate by catalytic chemical vapor deposition [145]. With low packing density (0.02–0.07 $g cm^{-3}$), the contact thermal resistance of CNT array-based TIMs could be very low owing to their good compressibility [146]. Correspondingly, the through-plane thermal conductivity is limited (1.21–4.87 W/m K) because of the small contact area between vertically aligned CNTs and the mating surfaces, leading to a reduced heat flux through the basal plane of CNT array-based TIMs [147]. In addition, the graphene monolith with a vertically aligned arrangement was fabricated by slicing paper consisting of layer-by-layer stacked graphene, followed by making graphene strips stand and then lateral compaction to form a densely packed structure with a density of 1.3–1.6 $g cm^{-3}$ [144, 148]. In contrast to CNT arrays, the vertically aligned graphene monolith shows ultrahigh through-plane thermal conductivity (615 W/m K) but is still not suitable for use as TIMs, due to its high contact thermal resistance originating from the surface rigidity (compressive modulus: up to 0.5 GPa) [149].

Wen Dai and co-workers developed a feasible machining-based process to prepare a graphene monolith with an aligned microstructure using conventional graphene paper as starting material (Figure 7.13). By applying a lateral mechanical force to the graphene paper, the resulting monolith is mainly constituted by vertically aligned

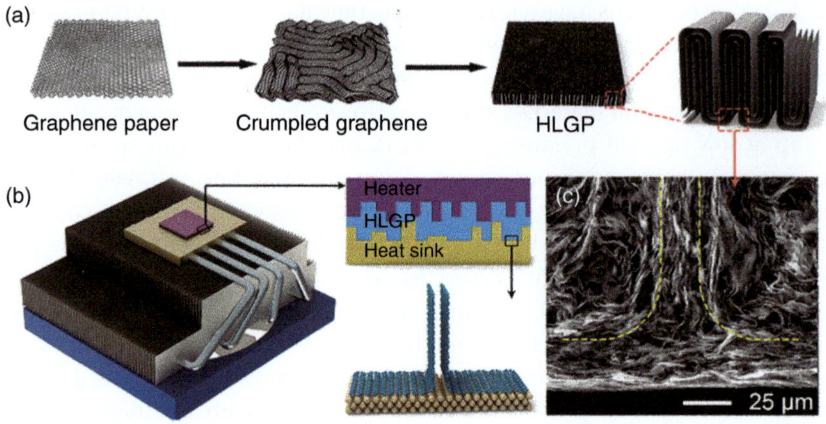

Figure 7.13 (a) Schematic illustration of the structural change of the graphene based on the proposed method. (b) Configuration for TIM performance measurement. (c) Detailed microstructure close to the monolith surface. Source: Reprinted from Ref. [150]; © 2023 American Chemical Society.

graphene and a thin cap of horizontal graphene layers at both the top and bottom sides. The resultant graphene monolith exhibits both an ultrahigh through-plane thermal conductivity of 143 W/m K (exceeding that of many metals) and a low compressive modulus of 0.87 MPa, which is much smaller than that of silicone-based composites (2–10 MPa).

Graphene is a single layer of densely packed sp2-hybridized carbon atoms that exhibits an ultrahigh intrinsic thermal conductivity of 3500–5300 W/m K and large specific surface area of ~2630 m^2 g^{-1} [151]. Due to the above characteristics, graphene/polymer composites with enhanced thermal conductivity [152], graphene coatings for efficient radiative cooling [153], and graphene papers as heat spreaders [154] have been extensively studied in recent years for development of the next generation of thermal management applications. Among them, graphene papers composed of layer-by-layer stacked graphene sheets have been commercially used as high-performance heat spreaders in portable devices because of their excellent in-plane thermal conductivity (κ: 1000–2000 W/m K) [155, 156] and ease of mass production [157]. Lian et al. fabricated a graphene paper with the size of 18 in. × 4.5 in. and in-plane thermal conductivity of 1434 W/m K by integration of electro-spray deposition and continuous roll-to-roll process [158]. However, in contrast to the superior in-plane thermal conductivity, conventional graphene paper exhibits a much-lower through-plane conductivity (κ: 0.1–3.4 W/m K) [157, 159] due to the high barrier for phonon transmission across a van der Waals interface between graphene layers [157]. Recently, some efforts have been done to solve this issue by improvement of the through-plane thermal conductivity of graphene papers. Jiang et al. constructed a graphene hybrid paper by growth of carbon nanoring between graphene sheets through annealing with the assistance of metal catalysts [157]. Although the through-plane thermal conductivity of the paper was improved by a factor of 3 after hybridization, the obtained value of

5.8 W/m K is just close to that of commercial TIMs [160]. Drzal et al. applied a cold compaction process with 100 psi to graphene paper and found that the through-plane thermal diffusivity decreased by 60% [159]. Wen Dai and co-workers fabricated a graphene/SiC nanorod hybrid paper with hierarchical structure by vacuum filtration of a mixture of SiO_2 nanoparticles-decorated graphene sheets, followed by high-frequency heat treatment to grow SiC nanorods between graphene layers (Figure 7.14). The through-plane thermal diffusivity of as-prepared graphene hybrid paper (GHP) can be improved by a factor of ~2.2 compared to pristine paper. Interestingly, under compression of 75 psi, the through-plane thermal conductivity of GHP can be further increased up to 17.6 W/m K, which significantly outperforms conventional TIMs [160].

The controllability of thermal/electrical conductivity under compressive deformation renders elastic materials potentially applicable to flexible piezoresistive devices such as sensors, actuators, and detectors [162, 163]. The performance of these devices mainly depends on the combined properties such as the electrical/thermal sensitivity and responsivity and cyclic elasticity of the materials [164]. Particularly, stress-enhanced conductivity (heat conduction or electron transport) at different compressive strains reflects the sensitivity or responsivity. Structural stability (little compression-induced change) under repeated alternating pressure ensures long-term cyclic elasticity [164, 165]. Moreover, high modulus and thermal stability determine their applicability in harsh environments such as under a strong external force and/or at high temperatures [166].

Achieving a high modulus and good elasticity simultaneously is greatly challenging for elastic polymers due to the strength-elasticity (compressibility/resilience) trade-off [167]. In thermoplastic elastomers, the free movement of flexible chains

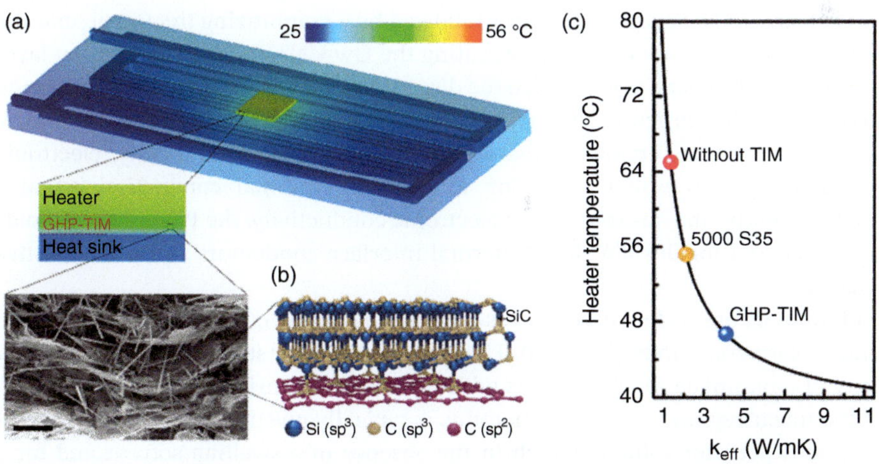

Figure 7.14 (a) Heat dissipation capability of GHP-TIM. (b) The calculated thermal conductance at the graphene–SiC nanorod interface bonded by covalent and van der Waals (vdW) interactions. (c) The calculated heater temperature as a function of effective thermal conductivity. Source: Reprinted from Ref. [161]; © 2023 American Chemical Society.

results in low mechanical load, poor heat resistance, and easy aging [168]. In contrast, in high-modulus polymers [169], like PI, the interchain interaction (entanglement and crosslinking) limits the segmental movement under stress, thus decreasing the compressibility and resilience. An effective strategy for rendering high-modulus polymer elasticity is fabricating a 3D elastic porous template. Among the potential templates, CNT sponge is regarded as one of the ideal 3D architectures to assemble polymers into elastic nanocomposites [170]. Such nanocomposites combine high thermal/electrical conductivity, good cyclic elasticity, and high thermal stability. The conductivity and elasticity are mainly determined by the microstructure of the CNT sponge.

The core problem of a 3D CNT network is that weak bonding or contact at the junction results in a high interfacial thermal/electrical resistance and inter-tube slippage under compression/resilience cycles [167]. Graphite welding at the junction is used to improve the contact between adjacent nanotubes [167, 171]. The welding or crosslinking effect mainly depends on the conjugated structure and interfacial compatibility with polymers. According to previous studies, a low graphitization degree affects the phonon/electron transfer, and the weak interaction leads to random deposition of graphite structure [171, 172]. Furthermore, a 3D CNT sponge usually tends to shrink largely when coated with polymers during solvent removal because the frameworks are not robust enough to resist the solvent evaporation capillary pressure [40, 173]. This effect leads to a large decrease in the sponge's porosity. Structure uniformity and interfacial contact are crucial for stress-controllable thermal/electrical conductivity of elastic CNT/polymer networks during compression and resilience cycles.

Fei Zhang and co-workers prepared a 3D elastic graphene-crosslinked CNT sponge/PI (Gw-CNT/PI) nanocomposite with stress-controllable thermal and electrical conductivities. The 3D interconnected Gw-CNT hybrid network architecture was constructed by coating a PI layer followed by a graphitizing treatment, and the PI nanocomposites are obtained by coating the Gw-CNT framework layer-by-layer with PI. The graphene layers bridge the discontinuous CNTs in the sponge, forming a continuous 3D structure that acts as a conductive network and elastic template for the composites, endowing them with high thermal conductivity, electronic conductivity, and resilience. Owing to the high thermal conduction, suitable compressibility, and stress-sensitive electronic conductivity, the PI composite holds great potential for flexible elastic thermal interface conductors and piezoresistive sensors.

Elastomers are widely used because of their large reversible deformability up to strains of several hundred percent. Their elastic behavior stems from entropic elasticity of crosslinked flexible chains with a low glass transition temperature [174]. In the linear regime, the elastic modulus is typically given by the number of elastic chains per unit volume, which in the absence of a swelling solvent and for a homogeneously crosslinked system is inversely proportional to the average molecular weight between crosslinks Mx [175]. Yet in uniaxial extension the majority of unfilled elastomers soften slightly relative to the linear behavior and then break in a brittle fashion well before the chains reach their maximum extensibility [176]. As a result, and with the notable exception of natural rubber, it is quite difficult to prepare an unfilled, well-crosslinked elastomer, displaying a sufficiently high strain at break to characterize the distinctive strain stiffening in uniaxial elongation due to the finite extensibility of the chains.

This intrinsic limitation of conventionally crosslinked elastomers is due to the very localized nature of the breakage of bonds near the interfacial plane. In the absence of viscoelastic dissipation, the threshold fracture energy Γ0 of the elastomers is quantitatively explained by the sequential fracture of highly stretched elastic strands in the plane of the interface as the crack propagates [177, 178]. To avoid this damage localization process leading to early fracture, it is possible to modify the network architecture to create additional energy dissipation mechanisms. Many strategies have been tried to increase the toughness of unfilled elastomers by network design [177].

One of the key advantages of these interesting new materials is their very pronounced strain stiffening in uniaxial extension which now occurs before the material breaks. In some specific applications such as resistance to cavitation [179], or dielectric elastomers [180], one expects this feature to be very useful to fine-tune the properties. Furthermore, the characterization of the strain stiffening of the material may be directly related to the progressive stiffening of the prestretched polymer chains as they approach their limiting extensibility. Etienne Ducrot and co-workers have explored how, by changing the monomer composition and the degree of crosslinking used in the first network, it is possible to control the mechanical properties of the final material and to characterize the strain stiffening of an otherwise brittle elastic polymer network.

Significant efforts have been expended in designing graphene-based composites for next-generation TIMs, including graphene-organic-polymer-resin composites (such as epoxies, urethanes, and silicones) [181] and graphene-inorganic-thermal-conductive-fillers composites (such as boron nitride and zinc oxide) [40, 182]. However, the random alignment of graphene nanoflakes, loose contact between graphene and mating surfaces, as well as additives with inferior heat conductivity tend to restrain the excellent anisotropic thermal conductance of graphene.

In principle, vertical graphene (VG) has a higher heat transfer capability than that of the randomly arranged graphene-based composites. A number of approaches have been developed for the construction of VG structures, including top-down methods represented by physical–mechanical machining [40, 150, 182, 183] and bottom-up methods represented by plasma-enhanced chemical vapor deposition (PECVD) [184–187]. Using mechanical machining, vertically aligned graphene foams [183] and vertical graphene monoliths [150] can be obtained by directional freezing-drying and shrinking-compressing, respectively. Both of these vertical graphene structures exhibit excellent thermal conductivity. Apart from mechanical methods, VG has been grown directly on various substrates using PECVD, which can be classified into numerous subclasses, including microwave PECVD [185, 188], inductively coupled PECVD [184, 189, 190], capacitively coupled PECVD [186, 187], and direct current PECVD [191]. Small petal-like and forest-like graphene, constructed by the random assembly of nanosheets, are typically observed in inductively

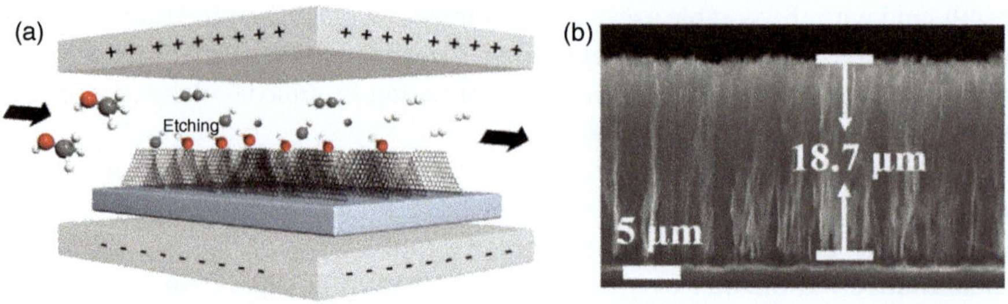

Figure 7.15 Experimental setup and typical morphology of VG arrays. (a) Schematic illustrating the experimental setup of an electric-field-assisted methanol-based PECVD system. (b) Cross-sectional SEM image of VG arrays after 6 hours of growth. The height of the VG arrays can reach 18.7 μm. Source: Reprinted from Ref. [95]; © 2023 Elsevier B.V.

coupled PECVD and microwave PECVD [111, 184, 189, 190]. However, enhancing the growth rate and accurately controlling the orientation of VG simultaneously still remains challenging. Shichen Xu and Co-workers developed an alcohol-based electric-field-assisted (AEF)-PECVD method, which can rapidly grow highly crystalline graphene sheets that are fully perpendicular to the substrates (Figure 7.15). TIMs constructed with these VG arrays exhibit a high through-plane thermal conductivity of 53.5 W/m K and a low contact thermal resistance of 11.8 K mm^2/W.

7.5 Molecular Simulation Study of Interfacial Thermal Transfer

Molecular dynamics (MD), a computer simulation technique for analyzing the physical motions of atoms and molecules, has been extensively used in the scientific literature to study heat transfer. For MD simulation formalization, however, previous knowledge of heat transfer is not required; it is only based on Newton's second law, records the positions, velocities, and forces of atoms and one can obtain the detailed heat flow information by extracting the evolutionary trajectory [192]. In addition, MD simulation has inherent anharmonicity based on the assumed potential and so includes inelastic effects due to numerous phonon interactions, making it a formidable tool for investigating thermal boundary conductance in a variety of material systems. Quantum statistics cannot be captured by conventional MD since all vibrational modes inside a material are excited [193]. Therefore, MD is the best suited for analyzing vibrational transport in a domain where most phonon modes involved in energy transfer are thermally generated. The maturity of MD software like LAMMPS [194] and the advancement of computing hardware have made related simulations routine tasks. It is a powerful tool for investigating a variety of atomic-level effects on heat transport, including temperature [195, 196], strain [197], defects [198], and functional surfaces [199]. However, with MD simulation, the limitation on the size of the computational domain and the short time scales for simulations is an additional significant obstacle. Moreover, since electron dynamics cannot be directly included in MD, and electrical conductors and electron–phonon coupling events in materials cannot be explored using a pure MD method.

Three MD approaches exist for calculating heat transport at the interface of different materials. These are equilibrium MD (EMD) simulations, the heat pulse technique, and nonequilibrium MD (NEMD) simulations. EMD is based on fluctuation–dissipation and linear response theory. By averaging over time, macroscopic thermodynamics traits may be represented as microscopic ones. This approach uses the Green–Kubo formula to compute the thermal interface conductance, G [200]. The EMD assumes that the system is in equilibrium. The approach, however, is incapable of revealing the status of nonequilibrium phonon dispersion at the interface [201]. The second technique is the heat pulse approach, which is computed using the lumped heat-capacity model [202]. This approach assumes that the thermal resistance between fillers is significantly lower than the thermal resistance between polymer and fillers, or that the thermal conductivity of fillers, such as carbon tubes and octane, is much greater than that of polymer materials. First, the initial configuration of the system is equilibrated at an average temperature of $T + \frac{1}{2}\Delta T$, followed by the application of two independent thermostats to the polymer and filler at temperatures of T and $T + \Delta T$, respectively. After reaching a steady state, the thermostats are removed to ensure that the temperature difference ΔT between polymer and filler decays exponentially as

$$\Delta T(t) = \Delta T(0)e^{\frac{t}{\tau}}$$

The relaxation time τ is an important component of this method, and it can be obtained by exponentially fitting the curve.

There are three computational algorithms in the NEMD method, including (i) the local thermostat algorithm, which applies a temperature gradient to two regions and tracks the heat flux during the simulations. (ii) The momentum-swapping algorithm is proposed by Müller-Plathe [203]: it involves swapping the momentum or kinetic energy of atoms between two regions. (iii) velocity-rescaling algorithm is proposed by Ikeshioji, Hafskjold [204], Jund-Jullien [205], and Winsberger [206]: adding or removing constant energy to two regions. In general, the last two algorithms require a certain amount of trial-and-error work to establish an ideal and stable temperature distribution. If the heat flux $\vec{J}$ is too small, the temperature drop ΔT will be quite noisy, implying the possibility of a significant error. If the heat flux $\vec{J}$ is too large, the temperature drops ΔT near the heat sink or heat source will be nonlinear. To avoid these troubles, the most convenient approach, called the thermostat algorithm, is to impose a temperature on two regions heat source and heat sink. During the simulation, energy will be exchanged from one side of the interface to the other, and the temperature distribution will reach a steady state. Then, heat flux $\vec{J}$ and temperature jump ΔT will be generated. The interfacial thermal conductance G can be evaluated using the following equation:

$$G = \frac{\vec{J}}{\Delta T}$$

EMD and heat pulse technologies have several limitations and lack flexibility in simulating the system. Therefore, these two approaches are always used to describe thermal transport processes, but they are rarely employed to directly determine thermal conductance. The NEMD approach is based on disturbance of the system. It is easy and swiftly converges. As a consequence, it is often used in the majority of research to measure thermal resistance and thermal conductance. Theoretically, the thermal conductance values computed by both approaches are comparable. Kiarash Gordiz and Asegun Henry [207] investigated the interface conductance of two distinct Lennard–Jones solids.

In MD simulations for heat transfer across the solid–solid interface, the semiconductor materials of silicon and germanium have become prototypical systems for researching interfacial heat transfer, partly owing to the trustworthy interatomic potentials of such materials that can reliably duplicate their vibrational and structural physics [208]. Compared to the semiconductor, there are few MD simulation investigations of the interface conductance of metal–semiconductor interfaces because of the intrinsic difficulties involved with adding contributions of electrons and phonons to heat transport in MD modeling [209]. However, Cruz et al., Wang et al., and Yang et al. have successfully calculated the interfacial thermal conductance of Au/Si, Cu/Si, and Al/Si, respectively, using MD simulations [210]. The thermal transport of 2D materials at the solid–solid interface is also studied, such as silicene/graphene [202], boron nitride/graphene [211], MoS2/graphene [212], graphene/silicon [213], and graphene/copper [214].

In terms of the interface between solid and soft materials, soft materials (such as liquids and polymers) are mostly disordered. NEMD simulations have been used to examine the heat transmission at the solid–liquid interfaces of a variety of material combinations [215], often using water as the liquid component [216]. Currently, the thermal interface transfer between soft polymers and other materials, such as 2D materials [217, 218] and metals [219], is the primary area of research. The benefit of simulation is that it permits control over every material property and offers techniques for isolating particular parts in complicated experiments. However, these benefits do not overshadow the significance of experiments, and even advanced MD modeling cannot replace the experiment technique, which serves as the standard for theoretical prediction.

7.6 Conclusion

TIMs play an important and increasing role in the behavior of high-power electronic circuits. Device miniaturization and densification are driving the unceasing demand for ever-improving TIM performance. Cost of production, ease of application, safety around exposed circuit elements, and lifespan reliability all contribute to the wide-scale adoption of polymer-based TIMs in industry. A great deal of research has been performed in the area of composite filler thermal interface materials in recent years, with a strong focus on nanomaterial fillers. Researchers have made strides in producing composite materials, including nanocomposites, where the thermal conductivity of the material is significantly increased over the matrix alone. However, much of this research has not demonstrated the thermal resistance of said composites, which must of course include the interfacial contact resistance, thus not demonstrating the efficacy of the materials as potential TIMs. In general, loading a matrix with nano-materials increases its Young's modulus which can increase the interfacial contact resistance, and this can more than offset any gains in the bulk thermal conductivity. When a high-performance TIM is examined a low bond line thickness is preferred, and this accentuates the importance of low interfacial contact resistance as it becomes a higher percentage of the overall thermal resistance as the bond line thickness decreases. The future direction of filler-type spreadable TIM research must then carefully consider the mechanical behavior, ensuring a thin bond line at allowable assembly pressures, while ensuring as low as possible interfacial contact resistance and as high as feasible bulk thermal conductivity. This is a highly nonlinear mutliphysics problem involving mechanics, tribology, and heat transfer, and it would seem that modern simulation tools could be used to predict TIM behavior and subsequently optimize their performance. The generation of an accurate model would be a significant step forward in producing the next generation of filler-type TIMs and accelerate the advancement of the technology.

The future research in thermal interface materials with ever decreasing size of our electronic devices and components will need to take multi-disciplinary approach. Physics-based model and mechanistic understanding of the problem that the engineers are facing in thermal management of a device/component are essential. In terms of developing new TIMs, the boundary conditions in materials development are required to be set by the reliability engineers and ultimately by the materials/chemical manufacturer for volume production. Understanding of the reliability issues that decipher the degradation trends of the TIMs is yet to be well understood. Use of eco-friendly materials and processes for the fabrication of next-generation TIMs and understanding of recyclability issues are paramount.

References

1. Liu, Z. et al. (2019). Graphene foam-embedded epoxy composites with significant thermal conductivity enhancement. *Nanoscale* 11 (38): 17600–17606.
2. Wang, F. et al. (2016). Silver nanoparticle-deposited boron nitride nanosheets as fillers for polymeric composites with high thermal conductivity. *Scientific Reports* 6 (1): 1–9.
3. Zhou, J. et al. (2012). Surface modification for PDMS-based microfluidic devices. *Electrophoresis* 33 (1): 89–104.
4. Subramanian, J. et al. (2005). Room temperature soldering of microelectronic components for enhanced thermal performance. In: *EuroSimE 2005. Proceedings of the 6th International Conference on Thermal, Mechanial and Multi-Physics Simulation and Experiments in Micro-Electronics and Micro-Systems, 2005*. IEEE.
5. Lambert, M. and Fletcher, L. (2002). Thermal contact conductance of non-flat, rough, metallic coated metals. *Journal of Heat Transfer* 124 (3): 405–412.

6 Mahajan, R., Chiu, C.-p., and Chrysler, G. (2006). Cooling a microprocessor chip. *Proceedings of the IEEE* 94 (8): 1476–1486.

7 Watwe, A.A. and Prasher, R.S. (2001). Spreadsheet tool for quick-turn 3D numerical modeling of package thermal performance with non-uniform die heating. In: *ASME International Mechanical Engineering Congress and Exposition*. American Society of Mechanical Engineers.

8 Mahajan, R., Chiu, C., and Prasher, R. (2004). Thermal interface materials: a brief review of design characteristics and materials. *Electronics Cooling* 10 (1): 10.

9 Lei, L. et al. (2016). Preparation and properties of amino-functional reduced graphene oxide/waterborne polyurethane hybrid emulsions. *Progress in Organic Coatings* 97: 19–27.

10 Yao, Y. et al. (2015). The effect of interfacial state on the thermal conductivity of functionalized Al_2O_3 filled glass fibers reinforced polymer composites. *Composites Part A: Applied Science and Manufacturing* 69: 49–55.

11 Kong, J.-H. et al. (2014). Simple and rapid micropatterning of conductive carbon composites and its application to elastic strain sensors. *Carbon* 77: 199–207.

12 Hu, M., Keblinski, P., and Schelling, P.K. (2009). Kapitza conductance of silicon-amorphous polyethylene interfaces by molecular dynamics simulations. *Physical Review B* 79 (10): 104305.

13 Martin-Gallego, M. et al. (2013). Comparison of filler percolation and mechanical properties in graphene and carbon nanotubes filled epoxy nanocomposites. *European Polymer Journal* 49 (6): 1347–1353.

14 Guo, H. et al. (2019). Improved interfacial properties for largely enhanced thermal conductivity of poly(vinylidene fluoride)-based nanocomposites via functionalized multi-wall carbon nanotubes. *Applied Surface Science* 487: 379–388.

15 Ji, C. et al. (2019). Thermal conductivity enhancement of CNT/MoS2/graphene-epoxy nanocomposites based on structural synergistic effects and interpenetrating network. *Composites Part B: Engineering* 163: 363–370.

16 Liu, J., Alhashme, M., and Yang, R. (2012). Thermal transport across carbon nanotubes connected by molecular linkers. *Carbon* 50 (3): 1063–1070.

17 You, J. et al. (2019). Plasma-assisted mechanochemistry to produce polyamide/boron nitride nanocomposites with high thermal conductivities and mechanical properties. *Composites Part B: Engineering* 164: 710–719.

18 Hu, J. et al. (2017). Polymer composite with improved thermal conductivity by constructing a hierarchically ordered three-dimensional interconnected network of BN. *ACS Applied Materials & Interfaces* 9 (15): 13544–13553.

19 Cheng, L. and Feng, J. (2020). Flexible and fire-resistant all-inorganic composite film with high in-plane thermal conductivity. *Chemical Engineering Journal* 398.

20 Hu, J. et al. (2018). Polymer composite with enhanced thermal conductivity and mechanical strength through orientation manipulating of BN. *Composites Science and Technology* 160: 127–137.

21 Kwon, O.H. et al. (2018). Anisotropy-driven high thermal conductivity in stretchable poly(vinyl alcohol)/hexagonal boron nitride nanohybrid films. *ACS Applied Materials & Interfaces* 10 (40): 34625–34633.

22 Yao, Y. et al. (2018). Construction of 3D skeleton for polymer composites achieving a high thermal conductivity. *Small* 14 (13): 1704044.

23 Kang, D.-G. et al. (2018). Anisotropic thermal interface materials: directional heat transfer in uniaxially oriented liquid crystal networks. *ACS Applied Materials & Interfaces* 10 (41): 35557–35562.

24 Xu, S. et al. (2020). Electric-field-assisted growth of vertical graphene arrays and the application in thermal interface materials. *Advanced Functional Materials* 30 (34).

25 Vu, M.C. et al. (2020). High thermal conductivity enhancement of polymer composites with vertically aligned silicon carbide sheet scaffolds. *ACS Applied Materials & Interfaces* 12 (20): 23388–23398.

26 Putz, K.W. et al. (2010). High-nanofiller-content graphene oxide-polymer nanocomposites via vacuum-assisted self-assembly. *Advanced Functional Materials* 20 (19): 3322–3329.

27 Chen, Z. et al. (2015). Effect of interface evolution on thermal conductivity of vacuum hot pressed SiC/Al composites. *Advanced Engineering Materials* 17 (7): 1077–1085.

28 Sun, N. et al. (2018). Hot-pressing induced orientation of boron nitride in polycarbonate composites with enhanced thermal conductivity. *Composites Part A: Applied Science and Manufacturing* 110: 45–52.

29 Li, J. et al. (2020). Advanced flexible rGO-BN natural rubber films with high thermal conductivity for improved thermal management capability. *Carbon* 162: 46–55.

30 Cheng, L. and Feng, J. (2020). Flexible and fire-resistant all-inorganic composite film with high in-plane thermal conductivity. *Chemical Engineering Journal* 398: 125633.

31 Yuan, J. et al. (2019). Highly thermally conducting polymer-based films with magnetic field-assisted vertically aligned hexagonal boron nitride for flexible electronic encapsulation. *ACS Applied Materials & Interfaces* 11 (19): 17915–17924.

32 Han, J. et al. (2019). An anisotropically high thermal conductive boron nitride/epoxy composite based on nacre-mimetic 3D network. *Advanced Functional Materials* 29 (13).

33 Li, Y. et al. (2020). Boosting the heat dissipation performance of graphene/polyimide flexible carbon film via enhanced through-plane conductivity of 3D hybridized structure. *Small* 16 (8): 1903315.

34 Chang, T.-C., Kwan, Y.-K., and Fuh, Y.-K. (2020). A reduced percolation threshold of hybrid fillers of ball-milled exfoliated graphite nanoplatelets and AgNWs for enhanced thermal interface materials in high power electronics. *Composites Part B: Engineering* 191: 107954.

35 Chang, T.-C., Kwan, Y.-K., and Fuh, Y.-K. (2020). A reduced percolation threshold of hybrid fillers of ball-milled exfoliated graphite nanoplatelets and AgNWs for enhanced thermal interface materials in high power electronics. *Composites Part B: Engineering* 191.

36 Yao, Y. et al. (2018). Vertically aligned and interconnected sic nanowire networks leading to significantly enhanced thermal conductivity of polymer composites. *ACS Applied Materials & Interfaces* 10 (11): 9669–9678.

37 Yoon, B.-H. et al. (2008). Aligned porous alumina ceramics with high compressive strengths for bone tissue engineering. *Scripta Materialia* 58 (7): 537–540.

38 Fu, C. et al. (2019). Improving thermal conductivity through welding boron nitride nanosheets onto silver nanowires via silver nanoparticles. *Composites Science and Technology* 177: 118–126.

39 Lin, Z. et al. (2013). Magnetic alignment of hexagonal boron nitride platelets in polymer matrix: toward high performance anisotropic polymer composites for electronic encapsulation. *ACS Applied Materials & Interfaces* 5 (15): 7633–7640.

40 An, F. et al. (2018). Vertically aligned high-quality graphene foams for anisotropically conductive polymer composites with ultrahigh through-plane thermal conductivities. *ACS Applied Materials & Interfaces* 10 (20): 17383–17392.

41 Song, Q. et al. (2019). Enhanced through-plane thermal conductivity and high electrical insulation of flexible composite films with aligned boron nitride for thermal interface material. *Composites Part A: Applied Science and Manufacturing* 127: 105654.

42 Song, Q. et al. (2019). Enhanced through-plane thermal conductivity and high electrical insulation of flexible composite films with aligned boron nitride for thermal interface material. *Composites Part A: Applied Science and Manufacturing* 127.

43 Mao, D. et al. (2019). Spherical core-shell Al@Al_2O_3 filled epoxy resin composites as high-performance thermal interface materials. *Composites Part A: Applied Science And Manufacturing* 123: 260–269.

44 Huang, X. et al. (2012). Role of interface on the thermal conductivity of highly filled dielectric epoxy/aln composites. *Journal of Physical Chemistry C* 116 (25): 13629–13639.

45 Ma, J. et al. (2020). Through-plane assembly of carbon fibers into 3D skeleton achieving enhanced thermal conductivity of a thermal interface material. *Chemical Engineering Journal* 380.

46 Feng, Y. et al. (2018). Synergetic improvement in thermal conductivity and flame retardancy of epoxy/silver nanowires composites by incorporating "branch-like" flame-retardant functionalized graphene. *ACS Applied Materials & Interfaces* 10 (25): 21628–21641.

47 Tian, Z. et al. (2018). A thermal interface material based on foam-templated three-dimensional hierarchical porous boron nitride. *Journal of Materials Chemistry A* 6 (36): 17540–17547.

48 Wang, M. et al. (2019). Highly compressive boron nitride nanotube aerogels reinforced with reduced graphene oxide. *ACS Nano* 13 (7): 7402–7409.

49 Jiang, F. et al. (2018). Hydrogen bond-regulated boron nitride network structures for improved thermal conductive property of polyamide-imide composites. *ACS Applied Materials & Interfaces* 10 (19): 16812–16821.

50 Wang, X. and Wu, P. (2018). Fluorinated carbon nanotube/nanofibrillated cellulose composite film with enhanced toughness, superior thermal conductivity, and electrical insulation. *ACS Applied Materials & Interfaces* 10 (40): 34311–34321.

51 Yang, D. et al. (2019). Novel nitrile-butadiene rubber composites with enhanced thermal conductivity and high dielectric constant. *Composites Part A: Applied Science and Manufacturing* 124: 105447.

52 An, D. et al. (2019). A polymer-based thermal management material with enhanced thermal conductivity by introducing three-dimensional networks and covalent bond connections. *Carbon* 155: 258–267.

53 Ji, C. et al. (2020). Ice-templated mxene/ag-epoxy nanocomposites as high-performance thermal management materials. *ACS Applied Materials & Interfaces* 12 (21): 24298–24307.

54 Wang, Z.-G. et al. (2020). Highly thermal conductive, anisotropically heat-transferred, mechanically flexible composite film by assembly of boron nitride nanosheets for thermal management. *Composites Part B: Engineering* 180: 107569.

55 An, D. et al. (2020). Enhanced thermal conductivity of natural rubber based thermal interfacial materials by constructing covalent bonds and three-dimensional networks. *Composites Part A: Applied Science and Manufacturing* 135.

56 Yao, Y. et al. (2020). Achieving significant thermal conductivity enhancement via an ice-templated and sintered BN-SiC skeleton. *ACS Applied Materials & Interfaces* 12 (2): 2892–2902.

57 Jensen, T. and Lasky, R. (2020). The basics of metal thermal interface materials (TIMs). In: *2020 Pan Pacific Microelectronics Symposium (Pan Pacific)*. IEEE.

58 Dutta, I. et al. (2009). Liquid phase sintered solders with indium as minority phase for next generation thermal interface material applications. *Journal of Electronic Materials* 38 (12): 2735–2745.

59 Sharma, D. et al. (2015). Two-phase metallic thermal interface materials processed through liquid phase sintering followed by accumulative roll bonding. *IEEE Transactions on Components, Packaging and Manufacturing Technology* 6 (1): 58–66.

60 Gong, W. et al. (2018). Ultracompliant heterogeneous copper–tin nanowire arrays making a supersolder. *Nano Letters* 18 (6): 3586–3592.

61 Bar-Cohen, A., Matin, K., and Narumanchi, S. (2015). Nanothermal interface materials: technology review and recent results. *Journal of Electronic Packaging* 137 (4): 040803.

62 Peng, P. et al. (2015). Joining of silver nanomaterials at low temperatures: processes, properties, and applications. *ACS Applied Materials & Interfaces* 7 (23): 12597–12618.

63 Groza, J. and Zavaliangos, A. (2003). Nanostructured bulk solids by field activated sintering. *Reviews on Advanced Materials Science* 5 (1): 24–33.

64 Groza, J.R. (1999). Nanosintering. *Nanostructured Materials* 12 (5-8): 987–992.

65 Calata, J.N., Lei, T.G., and Lu, G.-Q. (2009). Sintered nanosilver paste for high-temperature power semiconductor device attachment. *International Journal of Materials and Product Technology* 34 (1–2): 95–110.

66 Zürcher, J. et al. (2015). Nanoparticle assembly and sintering towards all-copper flip chip interconnects. In: *2015 IEEE 65th Electronic Components and Technology Conference (ECTC)*. IEEE.

67 Zhang, Z. et al. (2020). Size-dependent phononic thermal transport in low-dimensional nanomaterials. *Physics Reports* 860: 1–26.

68 Deng, S. et al. (2021). Thermal interface material with graphene enhanced sintered copper for high temperature power electronics. *Nanotechnology* 32 (31): 315710.

69 Balandin, A. et al. (2008). Extremely high thermal conductivity of graphene: prospects for thermal management applications in silicon nanoelectronics. In: *2008 IEEE Silicon Nanoelectronics Workshop*. IEEE.

70 Martin, Y. and Kessel, T. (2007). High performance liquid metal thermal interface for large volume production. *IMAPS Thermal and Power Management, San Jose CA* 2: 4.

71 Dickey, M.D. et al. (2008). Eutectic gallium-indium (EGaIn): a liquid metal alloy for the formation of stable structures in microchannels at room temperature. *Advanced Functional Materials* 18 (7): 1097–1104.

72 Chiew, C., Morris, M., and Malakooti, M.H. (2021). Functional liquid metal nanoparticles: synthesis and applications. *Materials Advances*.

73 Huang, K. et al. (2020). An anti-leakage liquid metal thermal interface material. *RSC Advances* 10 (32): 18824–18829.

74 Ma, K.-Q. and Liu, J. (2007). Nano liquid-metal fluid as ultimate coolant. *Physics Letters A* 361 (3): 252–256.

75 Kazem, N., Hellebrekers, T., and Majidi, C. (2017). Soft multifunctional composites and emulsions with liquid metals. *Advanced Materials* 29 (27): 1605985.

76 Doudrick, K. et al. (2014). Different shades of oxide: from nanoscale wetting mechanisms to contact printing of gallium-based liquid metals. *Langmuir* 30 (23): 6867–6877.

77 Lin, Z. et al. (2018). High thermal conductivity liquid metal pad for heat dissipation in electronic devices. *Applied Physics A* 124 (5): 1–6.

78 Lu, Y. et al. (2021). Mussel-inspired multifunctional integrated liquid metal-based magnetic suspensions with rheological, magnetic, electrical, and thermal reinforcement. *ACS Applied Materials & Interfaces* 13 (4): 5256–5265.

79 Tang, J. et al. (2017). Gallium-based liquid metal amalgams: transitional-state metallic mixtures (TransM2ixes) with enhanced and tunable electrical, thermal, and mechanical properties. *ACS Applied Materials & Interfaces* 9 (41): 35977–35987.

80 Ki, S. et al. (2021). Gallium-based liquid metal alloy incorporating oxide-free copper nanoparticle clusters for high-performance thermal interface materials. *International Journal of Heat and Mass Transfer* 170: 121012.

81 Sarfo, D.K., Taylor, R.R., and O'Mullane, A.P. (2020). Investigating liquid metal galinstan as a high current carrier and its interaction with collector electrodes. *ACS Applied Electronic Materials* 2 (9): 2921–2928.

82 Parekh, D.P. et al. (2020). Liquid–solid mixtures of Ga metal infused with Cu microparticles and nanoparticles for microscale and nanoscale patterning of solid metals at room temperature. *ACS Applied Nano Materials* 3 (12): 12064–12070.

83 Lereu, A.L. et al. (2015). Optical properties and plasmonic response of silver-gallium nanostructures. *Journal of Applied Physics* 117 (6): 063110.

84 Trenikhin, M. et al. (2006). The penetration of indium-gallium melt components into aluminum. *Russian journal of physical chemistry* 80 (7): 1110–1114.

85 Tikhomirova, O. et al. (1972). An investigation of structural transformation during the solidification of copper-gallium alloys. *Soviet materials science: a transl. of Fiziko-khimicheskaya mekhanika materialov/Academy of Sciences of the Ukrainian SSR* 5 (4): 355–358.

86 Hong, S.-J. and Suryanarayana, C. (2004). Mechanism of low-temperature θ-Cu Ga 2 phase formation in Cu-Ga alloys by mechanical alloying. *Journal of Applied Physics* 96 (11): 6120–6126.

87 Wang, H. et al. (2021). Liquid metal composites with enhanced thermal conductivity and stability using molecular thermal linker. *Advanced Materials* 33 (43): 2103104.

88 Kong, W. et al. (2019). Oxide-mediated formation of chemically stable tungsten–liquid metal mixtures for enhanced thermal interfaces. *Advanced Materials* 31 (44): 1904309.

89 Wang, C. et al. (2021). A general approach to composites containing nonmetallic fillers and liquid gallium. *Science Advances* 7 (1): eabe3767.

90 de Castro, I.A. et al. (2017). A gallium-based magnetocaloric liquid metal ferrofluid. *Nano Letters* 17 (12): 7831–7838.

91 Guo, R. et al. (2018). Ni-GaIn amalgams enabled rapid and customizable fabrication of wearable and wireless healthcare electronics. *Advanced Engineering Materials* 20 (10).

92 Guo, R. et al. (2020). Cu–EGaIn enabled stretchable e-skin for interactive electronics and CT assistant localization. *Materials Horizons* 7 (7): 1845–1853.

93 Wang, X.L. et al. (2018). Soft and moldable Mg-doped liquid metal for conformable skin tumor photothermal therapy. *Advanced Healthcare Materials* 7 (14).

94 Wei, S. et al. (2019). Investigation on enhancing the thermal conductance of gallium-based thermal interface materials using chromium-coated diamond particles. *Journal of Materials Science: Materials in Electronics* 30 (7): 7194–7202.

95 Guo, X. et al. (2021). A review of carbon-based thermal interface materials: Mechanism, thermal measurements and thermal properties. *Materials & Design* 209: 109936.

96 Leong, C.-K., Aoyagi, Y., and Chung, D. (2006). Carbon black pastes as coatings for improving thermal gap-filling materials. *Carbon* 44 (3): 435–440.

97 Lin, C. and Chung, D. (2009). Graphite nanoplatelet pastes vs. carbon black pastes as thermal interface materials. *Carbon* 47 (1): 295–305.

98 Sharma, M. and Chung, D. (2015). Solder–graphite network composite sheets as high-performance thermal interface materials. *Journal of Electronic Materials* 44 (3): 929–947.

99 Raza, M. et al. (2012). Texture, transport and mechanical properties of graphite nanoplatelet/silicone composites produced by three roll mill. *Composites science and technology* 72 (3): 467–475.

100 Molina-Jordá, J.M. (2018). Multi-scale design of novel materials for emerging challenges in active thermal management: Open-pore magnesium-diamond composite foams with nano-engineered interfaces. *Composites Part A: Applied Science and Manufacturing* 105: 265–273.

101 Kim, P. et al. (2001). Thermal transport measurements of individual multi-walled nanotubes. *Physical Review Letters* 87 (21): 215502.

102 Li, Q. et al. (2009). Measuring the thermal conductivity of individual carbon nanotubes by the Raman shift method. *Nanotechnology* 20 (14): 145702.

103 Razeeb, K.M. et al. (2018). Present and future thermal interface materials for electronic devices. *International Materials Reviews* 63 (1): 1–21.

104 Peacock, M.A. et al. (2016). Characterization of transferred vertically aligned carbon nanotubes arrays as thermal interface materials. *International Journal of Heat and Mass Transfer* 97: 94–100.

105 Yao, Y. et al. (2014). High-quality vertically aligned carbon nanotubes for applications as thermal interface materials. *IEEE Transactions on Components, Packaging and Manufacturing Technology* 4 (2): 232–239.

106 Lee, Y., Shanmugan, S., and Mutharasu, D. (2014). Thermal resistance of CNTs-based thermal interface material for high power solid state device packages. *Applied Physics A* 114 (4): 1145–1152.

107 Xu, J. and Fisher, T.S. (2006). Enhancement of thermal interface materials with carbon nanotube arrays. *International Journal of Heat and Mass Transfer* 49 (9-10): 1658–1666.

108 Lin, W. et al. (2012). Parametric study of intrinsic thermal transport in vertically aligned multi-walled carbon nanotubes using a laser flash technique. *Carbon* 50 (4): 1591–1603.

109 Wasniewski, J.R. et al. (2012). Characterization of metallically bonded carbon nanotube-based thermal interface materials using a high accuracy 1D steady-state technique. *Journal of Electronic Packaging* 134 (2): 020901.

110 Ganguli, S. et al. (2013). Superior thermal interface via vertically aligned carbon nanotubes grown on graphite foils. *Journal of Materials Research* 28 (7): 933–939.

111 Liu, J. et al. (2013). Carbon nanotubes for electronics manufacturing and packaging: from growth to integration. *Advances in Manufacturing* 1 (1): 13–27.

112 Na, N. et al. (2015). Denser and taller carbon nanotube arrays on Cu foils useable as thermal interface materials. *Japanese Journal of Applied Physics* 54 (9): 095102.

113 Wolff, E. and Schneider, D. (1998). Prediction of thermal contact resistance between polished surfaces. *International Journal of Heat and Mass Transfer* 41 (22): 3469–3482.

114 Ouellette, T. and De Sorgo, M. (1985). Thermal performance of heat transfer interface materials. In: *Proceedings of the Power Electronics Design Conference, Power Sources Users Conference, Cerritos, CA*.

115 Gwinn, J.P. and Webb, R.L. (2003). Performance and testing of thermal interface materials. *Microelectronics Journal* 34 (3): 215–222.

116 Vogel, M. (1995). Thermal performance for a miniature heat sink cooled by microPCM slurry. In: *Proceedings of the International Intersociety of Electronic Packaging Conference, Advances in Electronic Packaging*. American Society of Mechanical Engineers.

117 Wilson, S. et al. (1996). Thermally conductive adhesives for high thermally stressed assembly. In: NEPCON WEST.

118 Heinemann, H. (1968). Catalysis Reviews. *Soil Science* 106 (4): 326.

119 Rutkofsky, M. et al. (2005). Using a carbon nanotube additive to make electrically conductive commercial polymer composites. *SAMPE Journal* 41 (2): 54–55.

120 Wang, Z. et al. (2004). Processing and property investigation of single-walled carbon nanotube (SWNT) buckypaper/epoxy resin matrix nanocomposites. *Composites Part A: Applied Science and Manufacturing* 35 (10): 1225–1232.

121 Zaeri, M. et al. (2010). Mechanical modelling of carbon nanomaterials from nanotubes to buckypaper. *Carbon* 48 (13): 3916–3930.

122 Hong, W.-T. and Tai, N.-H. (2008). Investigations on the thermal conductivity of composites reinforced with carbon nanotubes. *Diamond and Related Materials* 17 (7–10): 1577–1581.

123 Cross, R. et al. (2010). A metallization and bonding approach for high performance carbon nanotube thermal interface materials. *Nanotechnology* 21 (44): 445705.

124 Cola, B.A., Xu, J., and Fisher, T.S. (2009). Contact mechanics and thermal conductance of carbon nanotube array interfaces. *International Journal of Heat and Mass Transfer* 52 (15–16): 3490–3503.

125 Wang, H. et al. (2010). Reducing thermal contact resistance using a bilayer aligned CNT thermal interface material. *Chemical Engineering Science* 65 (3): 1101–1108.

126 Shaikh, S., Lafdi, K., and Silverman, E. (2007). The effect of a CNT interface on the thermal resistance of contacting surfaces. *Carbon* 45 (4): 695–703.

127 Rodriguez, N. (1993). A review of catalytically grown carbon nanofibers. *Journal of materials research* 8 (12): 3233–3250.

128 Mahanta, N.K. et al. (2010). Thermal conductivity of carbon nanofiber mats. *Carbon* 48 (15): 4457–4465.

129 Patton, R. et al. (1999). Vapor grown carbon fiber composites with epoxy and poly (phenylene sulfide) matrices. *Composites Part A: Applied Science and Manufacturing* 30 (9): 1081–1091.
130 Lafdi, K. and Matzek, M. (2003). Carbon nanofibers as a nano-reinforcement for polymeric nanocomposites. In: The 35th International SAMPE Technical Conference.
131 Zhou, Y. et al. (2006). Fabrication and evaluation of carbon nano fiber filled carbon/epoxy composite. *Materials Science and Engineering: A* 426 (1–2): 221–228.
132 Howe, J.Y. et al. (2006). Heat treating carbon nanofibers for optimal composite performance. *Journal of Materials Research* 21 (10): 2646–2652.
133 Zhu, H. et al. (2003). Hydrogen storage in heat-treated carbon nanofibers prepared by the vertical floating catalyst method. *Materials Chemistry and Physics* 78 (3): 670–675.
134 Endo, M. et al. (2003). Microstructural changes induced in "stacked cup" carbon nanofibers by heat treatment. *Carbon* 41 (10): 1941–1947.
135 Endo, M. et al. (2001). Comparison study of semi-crystalline and highly crystalline multiwalled carbon nanotubes. *Applied Physics Letters* 79 (10): 1531–1533.
136 Oberlin, A., Goma, J., and Rouzaud, J. (1984). Techniques for studying structures and textures (microtextures) of carbon materials. *Journal de Chimie Physique et de Physico-Chimie Biologique* 81 (11–1): 701–710.
137 Barako, M.T. et al. (2017). Dense vertically aligned copper nanowire composites as high performance thermal Interface materials. *ACS Applied Materials & Interfaces* 9: 42067.
138 Razeeb, K.M. et al. (2018). Present and future thermal Interface materials for electronic devices. *International Materials Reviews* 63: 1.
139 Shahil, K.M. and Balandin, A.A. (2012). Graphene-multilayer graphene nanocomposites as highly efficient thermal interface materials. *Nano Letters* 12: 861.
140 Loeblein, M. et al. (2017). High-density 3D-boron nitride and 3D-graphene for high-performance nano-thermal Interface material. *ACS Nano* 11: 2033.
141 Lindsay, L. and Broido, D. (2011). Enhanced thermal conductivity and isotope effect in single-layer hexagonal boron nitride. *Physical Review B: Condensed Matter and Materials Physics* 84: 155421.
142 Tang, B. et al. (2015). Application of graphene as filler to improve thermal transport property of epoxy resin for thermal Interface materials. *International Journal of Heat and Mass Transfer* 85: 420.
143 Cai, W. et al. (2010). Thermal transport in suspended and supported monolayer graphene grown by chemical vapor deposition. *Nano Letters* 10: 1645.
144 Yoon, Y. et al. (2014). Vertical alignments of graphene sheets spatially and densely piled for fast ion diffusion in compact supercapacitors. *ACS Nano* 8: 4580.
145 Semet, V. et al. (2002). Field electron emission from individual carbon nanotubes of a vertically aligned array. *Applied Physics Letters* 81: 343.

146 Futaba, D.N. et al. (2006). 84% Catalyst activity of water-assisted growth of single walled carbon nanotube forest characterization by a statistical and macroscopic approach. *The Journal of Physical Chemistry B* 110: 8035.

147 Huang, H. et al. (2005). Aligned carbon nanotube composite films for thermal management. *Advanced Materials* 17: 1652.

148 Liang, Q. et al. (2011). A three-dimensional vertically aligned functionalized multilayer graphene architecture: an approach for graphene-based thermal interfacial materials. *ACS Nano* 5: 2392.

149 Zhang, Y.F. et al. (2016). High-performance thermal interface materials consisting of vertically aligned graphene film and polymer. *Carbon* 109: 552.

150 Dai, W. et al. (2019). Metal-level thermally conductive yet soft graphene thermal interface materials. *ACS Nano* 13 (10): 11561–11571.

151 Zheng, C. et al. (2014). Synthesis of porous graphene/activated carbon composite with high packing density and large specific surface area for supercapacitor electrode material. *Journal of Power Sources* 258: 290.

152 Lian, G. et al. (2016). Vertically aligned and interconnected graphene networks for high thermal conductivity of epoxy composites with ultralow loading. *Chemistry of Materials* 28: 6096.

153 Hsiao, T.J. et al. (2013). Monolayer graphene dispersion and radiative cooling for high power LED. *Nanotechnology* 24: 395401.

154 Peng, L. et al. (2017). Ultrahigh thermal conductive yet superflexible graphene films. *Advanced Materials* 29: 1700589.

155 Shen, B., Zhai, W., and Zheng, W. (2014). Ultrathin flexible graphene film: an excellent thermal conducting material with efficient EMI shielding. *Advanced Functional Materials* 24: 4542.

156 Teng, C. et al. (2017). Ultrahigh conductive graphene paper based on ball milling exfoliated graphene. *Advanced Functional Materials* 27: 1700240.

157 Zhang, J. et al. (2015). 3D bridged carbon nanoring/graphene hybrid paper as a high performance lateral heat spreader. *Small* 11: 6197.

158 Xin, G. et al. (2014). Large area freestanding graphene paper for superior thermal management. *Advanced Materials* 26: 4521.

159 Xiang, J. and Drzal, L.T. (2011). Electron and phonon transport in au nanoparticle decorated graphene nanoplatelet nanostructured paper. *ACS Applied Materials & Interfaces* 3: 1325.

160 Hansson, J. et al. (2018). Novel nanostructured thermal interface materials: a review. *International Materials Reviews* 63: 22.

161 Dai, W. et al. (2019). A paper-like inorganic thermal interface material composed of hierarchically structured graphene/silicon carbide nanorods. *ACS Nano* 13 (2): 1547–1554.

162 Wu, X. et al. (2018). Reducing structural defects and oxygen-containing functional groups in GO-hybridized CNTs aerogels: simultaneously improve the electrical and mechanical properties to enhance pressure sensitivity. *ACS Applied Materials & Interfaces* 10 (45): 39009–39017.

163 Kim, K.H., Oh, Y., and Islam, M.F. (2012). Graphene coating makes carbon nanotube aerogels superelastic and resistant to fatigue. *Nature Nanotechnology* 7 (9): 562–566.

164 Qin, Y.Y. et al. (2015). Lightweight, superelastic, and mechanically flexible graphene/polyimide nanocomposite foam for strain sensor application. *ACS Nano* 9 (9): 8933–8941.

165 Qiu, L. et al. (2012). Biomimetic superelastic graphene-based cellular monoliths. *Nature Communications* 3 (1): 1–7.

166 Li, S. et al. (2016). Transparent and flexible films of horizontally aligned carbon nanotube/polyimide composites with highly anisotropic mechanical, thermal, and electrical properties. *Carbon* 109: 131–140.

167 Zhang, F. et al. (2019). Stress-sensitive thermally conductive elastic nanocomposite based on interconnected graphite-welded carbon nanotube sponges. *Carbon* 145: 378–388.

168 Kuang, Z. et al. (2015). Fabrication of highly oriented hexagonal boron nitride nanosheet/elastomer nanocomposites with high thermal conductivity. *Small* 11 (14): 1655–1659.

169 Kim, B. et al. (1999). Semicrystalline blends of polyethylene and isotactic polypropylene: Improving mechanical performance by enhancing the interfacial structure. *Journal of Polymer Science Part B: Polymer Physics* 38 (0): 108–121.

170 Gui, X. et al. (2010). Carbon nanotube sponges. *Advanced Materials* 22 (5): 617–621.

171 Cui, Y. and Zhang, M. (2013). Fabrication of cross-linked carbon nanotube foam using polymethylmethacrylate microspheres as templates. *Journal of Materials Chemistry A* 1 (44): 13984–13988.

172 Papkov, D. et al. (2013). Extraordinary improvement of the graphitic structure of continuous carbon nanofibers templated with double wall carbon nanotubes. *ACS Nano* 7 (1): 126–142.

173 Xu, X. et al. (2016). Naturally dried graphene aerogels with superelasticity and tunable poisson's ratio. *Advanced Materials* 28 (41): 9223.

174 Treloar, L.G. (1975). *The Physics of Rubber Elasticity*.

175 Rubinstein, M. and Colby, R.H. (2003). *Polymer Physics*, vol. 23. New York: Oxford University Press.

176 Gent, A.N. (2012). *Engineering with Rubber: How to Design Rubber Components*. Carl Hanser Verlag GmbH Co KG.

177 Ducrot, E. et al. (2014). Toughening elastomers with sacrificial bonds and watching them break. *Science* 344 (6180): 186–189.

178 Lake, G. and Thomas, A. (1967). The strength of highly elastic materials. *Proceedings of the Royal Society of London. Series A* 300 (1460): 1905–1934.

179 Cristiano, A. et al. (2010). An experimental investigation of fracture by cavitation of model elastomeric networks. *Journal of Polymer Science Part B: Polymer Physics* 48 (13): 1409–1422.

180 Zhao, X. and Suo, Z. (2010). Theory of dielectric elastomers capable of giant deformation of actuation. *Physical Review Letters* 104 (17): 178302.

181 Liang, Q. et al. (2011). A three-dimensional vertically aligned functionalized multilayer graphene architecture: an approach for graphene-based thermal interfacial materials. *ACS Nano* 5 (3): 2392–2401.
182 An, F. et al. (2018). Highly anisotropic graphene/boron nitride hybrid aerogels with long-range ordered architecture and moderate density for highly thermally conductive composites. *Carbon* 126: 119–127.
183 Li, X.-H. et al. (2018). Vertically aligned, ultralight and highly compressive all-graphitized graphene aerogels for highly thermally conductive polymer composites. *Carbon* 140: 624–633.
184 (a) Zhu, M. (2007). Synthesis of carbon nanosheets and carbon nanotubes by radio frequency plasma enhanced chemical vapor deposition. *Diamond and Related Materials* 16: 196. (b) Wang, J., Zhu, M., Outlaw, R.A. et al. (2004). Synthesis of carbon nanosheets by inductively coupled radio-frequency plasma enhanced chemical vapor deposition. *Carbon* 42: 2867–2872.
185 Shang, N.G. et al. (2008). Catalyst-free efficient growth, orientation and biosensing properties of multilayer graphene nanoflake films with sharp edge planes. *Advanced Functional Materials* 18 (21): 3506–3514.
186 Hiramatsu, M. et al. (2004). Fabrication of vertically aligned carbon nanowalls using capacitively coupled plasma-enhanced chemical vapor deposition assisted by hydrogen radical injection. *Applied Physics Letters* 84 (23): 4708–4710.
187 Shiji, K. et al. (2005). Vertical growth of carbon nanowalls using RF plasma-enhanced chemical vapor deposition. *Diamond and Related Materials* 14 (3–7): 831–834.
188 Wu, Y. et al. (2002). Carbon nanowalls grown by microwave plasma enhanced chemical vapor deposition. *Advanced Materials* 14 (1): 64–67.
189 French, B. et al. (2005). Structural characterization of carbon nanosheets via X-ray scattering. *Journal of Applied Physics* 97 (11): 114317.
190 Ma, Y. et al. (2019). Experimental investigation on vertically oriented graphene grown in a plasma-enhanced chemical vapor deposition process. *ACS Applied Materials & Interfaces* 11 (10): 10237–10243.
191 Ci, H.N. et al. (2018). 6-inch uniform vertically-oriented graphene on soda-lime glass for photothermal applications. *Nano Research* 11 (6): 3106–3115.
192 Song, L. et al. (2022). Interfacial thermal resistance in polymer composites: a molecular dynamic perspective. *Molecular Simulation* 1–24.
193 Allen, M.P. and Tildesley, D.J. (2017). *Computer Simulation of Liquids*. Oxford University Press.
194 Plimpton, S. (1995). Fast parallel algorithms for short-range molecular dynamics. *Journal of Computational Physics* 117 (1): 1–19.
195 Moeini, M. et al. (2020). Molecular dynamics simulations of the effect of temperature and strain rate on mechanical properties of graphene–epoxy nanocomposites. *Molecular Simulation* 46 (6): 476–486.
196 Jackson, N., Rubi, J.M., and Bresme, F. (2016). Non-equilibrium molecular dynamics simulations of the thermal transport properties of Lennard-Jones fluids using configurational temperatures. *Molecular Simulation* 42 (14): 1214–1222.

197 Chen, J., Walther, J.H., and Koumoutsakos, P. (2014). Strain engineering of Kapitza resistance in few-layer graphene. *Nano Letters* 14 (2): 819–825.

198 Li, M. et al. (2018). Effect of defects on thermal conductivity of graphene/epoxy nanocomposites. *Carbon* 130: 295–303.

199 Wu, X. et al. (2016). Thermal transport across surfactant layers on gold nanorods in aqueous solution. *ACS Applied Materials & Interfaces* 8 (16): 10581–10589.

200 Barrat, J.-L. and Chiaruttini, F. (2003). Kapitza resistance at the liquid—solid interface. *Molecular Physics* 101 (11): 1605–1610.

201 Feng, T. et al. (2019). Unexpected high inelastic phonon transport across solid-solid interface: modal nonequilibrium molecular dynamics simulations and Landauer analysis. *Physical Review B* 99 (4): 045301.

202 Liu, B. et al. (2014). Interfacial thermal conductance of a silicene/graphene bilayer heterostructure and the effect of hydrogenation. *ACS Applied Materials & Interfaces* 6 (20): 18180–18188.

203 Müller-Plathe, F. (1997). A simple nonequilibrium molecular dynamics method for calculating the thermal conductivity. *The Journal of Chemical Physics* 106 (14): 6082–6085.

204 Ikeshoji, T. and Hafskjold, B. (1994). Non-equilibrium molecular dynamics calculation of heat conduction in liquid and through liquid-gas interface. *Molecular Physics* 81 (2): 251–261.

205 Jund, P. and Jullien, R. (1999). Molecular-dynamics calculation of the thermal conductivity of vitreous silica. *Physical Review B* 59 (21): 13707.

206 Wirnsberger, P., Frenkel, D., and Dellago, C. (2015). An enhanced version of the heat exchange algorithm with excellent energy conservation properties. *The Journal of Chemical Physics* 143 (12): 124104.

207 Gordiz, K. and Henry, A. (2015). A formalism for calculating the modal contributions to thermal interface conductance. *New Journal of Physics* 17 (10): 103002.

208 Tersoff, J. (1988). Empirical interatomic potential for carbon, with applications to amorphous carbon. *Physical Review Letters* 61 (25): 2879.

209 Heino, P. (2007). Thermal conduction simulations in the nanoscale. *Journal of Computational and Theoretical Nanoscience* 4 (5): 896–927.

210 Abs da Cruz, C., Chantrenne, P., and Kleber, X. (2012). Molecular dynamics simulations and Kapitza conductance prediction of Si/Au systems using the new full 2NN MEAM Si/Au cross-potential. *Journal of Heat Transfer* 134 (6).

211 Zhang, J., Hong, Y., and Yue, Y. (2015). Thermal transport across graphene and single layer hexagonal boron nitride. *Journal of Applied Physics* 117 (13): 134307.

212 Liu, B. et al. (2015). Thermal transport in a graphene–MoS2 bilayer heterostructure: a molecular dynamics study. *RSC Advances* 5 (37): 29193–29200.

213 Zhang, J., Wang, Y., and Wang, X. (2013). Rough contact is not always bad for interfacial energy coupling. *Nanoscale* 5 (23): 11598–11603.

214 Hong, Y. et al. (2015). Tuning thermal contact conductance at graphene–copper interface via surface nanoengineering. *Nanoscale* 7 (14): 6286–6294.

215 Alosious, S. et al. (2021). Nanoconfinement effects on the Kapitza resistance at water–CNT interfaces. *Langmuir* 37 (7): 2355–2361.

216 Sarode, A. et al. (2017). Role of carbon nanotube on the interfacial thermal resistance: a molecular dynamics approach. In: *2017 16th IEEE Intersociety Conference on Thermal and Thermomechanical Phenomena in Electronic Systems (ITherm)*. IEEE.

217 Luo, T. and Lloyd, J.R. (2012). Enhancement of thermal energy transport across graphene/graphite and polymer interfaces: a molecular dynamics study. *Advanced Functional Materials* 22 (12): 2495–2502.

218 Verma, A., Kumar, R., and Parashar, A. (2019). Enhanced thermal transport across a bi-crystalline graphene–polymer interface: an atomistic approach. *Physical Chemistry Chemical Physics* 21 (11): 6229–6237.

219 Dinpajooh, M. and Nitzan, A. (2022). Heat conduction in polymer chains: effect of substrate on the thermal conductance. *The Journal of Chemical Physics* 156 (14): 144901.

8

Study on Simulation of Thermal Conductive Composite Filling Theory

Bin Wu, Peng Chen, and Jiasheng Qian

Anhui University, Key Laboratory of Environment-Friendly Polymeric Materials of Anhui Province, School of Chemistry & Chemical Engineering, 111 Jiulong Road, Hefei, Anhui 230601, China

8.1 Molecular Simulation Algorithms for Thermal Conductivity Calculating

8.1.1 MD (Green–Kubo) Method

Experimentally, the thermal conductivity is determined by measuring the stationary heat flux necessary to maintain a temperature gradient generated by two heaters in the sample. This is readily described as Fourier's law as the Eq. (8.1):

$$k = \frac{\delta Q_t}{A\delta T} \tag{8.1}$$

In Eq. (8.1), k means the thermal conductivity, A is the area of the heat flux passing through the system, $\frac{\delta Q_t}{\delta T}$ is the instantaneous heat flux at a stationary gradient temperature. So, the heat flux is the most important parameter for both experiment and simulation studies. Experimentally, it is hard to make absolute thermal insulation of the sample against heat loss to the surroundings. And so, various technologies such as transient methods and steady-state methods are developed to get an accurate testing result on thermal conductivity [1]. Of course, such problems are not existed in computer simulation studies, while it is necessary to make sure the studied issues are treated with proper simulation methods.

Molecular dynamics (MD) simulations are ideal theoretical method for addressing thermal conductivity issues of polymers, crystals, and/or composites [2]. Two most popular algorithms of MD simulation, that is, equilibrium molecular dynamics (EMD) and non-equilibrium molecular dynamics (NEMD) simulations, are widely used to describe heat transporting in those solid materials [3]. Although researchers have shown a good agreement between the EMD and NEMD methods in calculating thermal conductivity of crystals, both methods found their convenience for individual cases [4].

For EMD method, the fluctuation–dissipation theorem relates to the spontaneous energy fluctuations in a thermodynamical system and is described with the

Thermal Management Materials for Electronic Packaging: Preparation, Characterization, and Devices, First Edition. Edited by Xingyou Tian.

Green–Kubo theory, by which the correlation function of the heat current and the thermal conductivity is written as Eq. (8.2):

$$\kappa = \frac{V}{3k_B T^2} \int_0^t \langle \vec{J}(0) \cdot \vec{J}(t) \rangle \mathrm{d}t \tag{8.2}$$

where is the k_B Boltzmann's constant, T is the temperature, V is the system volume, $J(t)$ is the heat flux at time t, and the notation <...> indicates ensemble averaging. The expression for the heat flux is usually written as Eq. (8.3):

$$\vec{J} = \frac{1}{V} \left[\sum_i \vec{e}_i \vec{v}_i + \frac{1}{2} \sum_{i<j} (\vec{f}_{ij} \cdot (\vec{v}_i + \vec{v}_j)) \vec{x}_{ij} \right] \tag{8.3}$$

where V is the volume of the simulation cell, i and j are the indexes of the atoms, $\vec{e}_i$ is the total energy of the atom i, $\vec{v}_i$ and $\vec{v}_j$ are the atom velocities of atom i and j, respectively, $\vec{f}_{ij}$ is the force between atom i and j, and $\vec{x}_{ij}$ is the relative position of atoms i and j. For many-body potential cases, the expression for heat flux can be revised accordingly.

EMD method is thought to be reliable as it solves Green–Kubo equations in equilibrium state, and the finite-size effect in computer simulation may be relieved by using a periodic boundary condition [5]. Besides those advantages, EMD method is apt to study anisotropic effect of heat flux. However, interfacial effect of heat flux transporting in inhomogeneous system, that is, the generous case for composites, is covered by Green–Kubo method, because it gives out an average thermal conductivity over the system. Furthermore, the correlation function of Green–Kubo method may converge very slowly. Those make researchers turn to a NEMD method in studying systems composed of polymers and composites.

8.1.2 NEMD Method

In spite of successes of EMD method in determining structural, dynamical, and thermodynamical quantities of molecular systems, NEMD method exhibits specific capability in simulating inhomogeneous system by imposing energy disturbance into it. The energy disturbance from NEMD method can be used to check the interfacial problem and get a steady state of heat current flow.

Experimentally, the thermal conductivity is determined by measuring the stationary heat flux necessary to maintain a temperature gradient generated by two heaters in the sample. NEMD method models this phenomenon by synergistically rescaling particle velocities in predetermined parts of the system; those are heat source and heat sink regions, respectively.

A linear temperature gradient along with heat flux direction is developed during the simulation. This means the simulated system reach a steady non-equilibrium state. The heat flux and thermal conductivity can be readily calculated by directly using Fourier's law, as shown in Eq. (8.4), where the heat flux j necessary to maintain the temperature difference $2\Delta T$ is given by:

$$\vec{J} = \frac{\Delta E}{A \Delta t} = -\kappa \frac{2\Delta T}{(L/2)} \tag{8.4}$$

where Δt is the time step used in the MD simulation, A is the interface area of the sample perpendicular to the heat flux, and ΔE is the average energy per time step Δt which is added and subtracted, respectively, in the layers representing the heat source and sink. The energy ΔE is determined by rescaling the velocities of the particles comprising the layers.

NEMD method is strongly limited by finite-size effect, because of the grain-boundary condition used along heat transporting direction, as well as the distance between heat source and sink. Usually, NEMD simulation is carried out with varied system dimensions to estimate the finite-size effects and extrapolate simulating results into macroscopic states [5].

8.1.3 e-DPD Method

Dissipative particle dynamics (DPD) is a well-known mesoscopic method to model complex fluids. The primary DPD algorithms are isothermal and cannot calculate heat flux. Qiao and He developed an e-DPD method [6] according to the theory by Espanol [7]. The e-DPD schedule is similar to the standard DPD approach but involves energy and temperature variables [8]. In e-DPD method, the motion and heat transporting of e-DPD beads (coarse-grained model particle) are guided by Eqs. (8.5)–(8.7):

$$\frac{d\vec{r}_i}{dt} = \vec{v}_i \tag{8.5}$$

$$m\frac{d\vec{v}_i}{dt} = \sum_{j\neq i}\left(\vec{f}_{ij}^{C} + \vec{f}_{ij}^{D} + \vec{f}_{ij}^{R}\right) \tag{8.6}$$

$$C_v\frac{\mathrm{d}T_i}{\mathrm{d}t} = q_i = \sum_{j\neq i}\left(q_{ij}^{visc} + q_{ij}^{coll} + q_{ij}^{rand}\right) \tag{8.7}$$

In those equations, m, t, r_i, v_i, and f_{ij} represent the mass, time, position, velocity, and force, of e-DPD beads, respectively. Because of the coarse-grained nature, the thermal capacity at constant volume of an e-DPD bead is signed as C_v. q_i and T_i mean the heat flux and temperature of e-DPD bead, respectively. The force vectors $\vec{f}_{ij}^{C}$, $\vec{f}_{ij}^{D}$, and $\vec{f}_{ij}^{R}$ represent conservative, dissipative, and random forces, and those are the same with a primary DPD form:

$$\vec{f}_{ij}^{C} = a_{ij}\omega(r_{ij})\vec{e}_{ij} \tag{8.8}$$

$$\vec{f}_{ij}^{D} = -\gamma_{ij}\omega^2(r_{ij})(\vec{e}_{ij}.\vec{v}_{ij})\vec{e}_{ij} \tag{8.9}$$

$$\vec{f}_{ij}^{R} = \sigma_{ij}\omega(r_{ij})\xi_{ij}\Delta t^{\frac{1}{-2}}\vec{e}_{ij} \tag{8.10}$$

In Eq. (8.8), a_{ij} is repulsion between beads i and j. $r_{ij} = r_i - r_j$, $v_{ij} = v_i - v_j$, and $\vec{e}_{ij} = \vec{r}_{ij}/|\vec{r}_{ij}|$. The parameters r_{ij} and σ_{ij} in Eqs. (8.9) and (8.10) represent strength of dissipative and random forces, and those serve the fluctuation–dissipation correlation. To make sure the momentum conservation, the value of ξ_{ij} is a random number with a zero mean and a unit variance as $\xi_{ij} = \xi_{ji}$.

The most important in e-DPD method are three kinds of heat fluxes among the particle-particle including viscous heat flux q_{ij}^{visc}, collision heat flux, q_{ij}^{coll}, and random heat flux q_{ij}^{rand}; those are written as

$$q_{ij}^{visc} = \frac{1}{2}\left\{\omega^2(r_{ij})\left[\gamma_{ij}(\vec{e}_{ij}.\vec{v}_{ij})^2 - \sigma_{ij}^2\right] - \sigma_{ij}\omega(r_{ij})(\vec{e}_{ij}.\vec{v}_{ij})\Delta t^{-1/2}\xi_{ij}\right\} \tag{8.11}$$

$$q_{ij}^{coll} = k_{ij}\omega^2(r_{ij})\left(\frac{1}{T_i} - \frac{1}{T_j}\right) \tag{8.12}$$

$$q_{ij}^{rand} = a_{ij}\omega(r_{ij})\Delta t^{-1/2}\xi_{ij}^e \tag{8.13}$$

The variables k_{ij} and a_{ij} in Eqs. (8.12) and (8.13) represent the strength of collisional and random heat fluxes. The non-symmetrical random number ξ_{ij}^e has zero mean and unit variance with ξ_{ij}^e=-ξ_{ij}^eor each pair of interaction beads being the same as that in Eq. (8.10). The weighting function $\omega(r_{ij})$ is simply described as the linear weighting function:

$$\omega(r_{ij}) = \begin{cases} 1 - r_{ij}/r_c & r_{ij} < r_c \\ 0 & r_{ij} \geq r_c \end{cases} \tag{8.14}$$

The relation between the parameter γ_{ij} and σ_{ij} is determined by the fluctuation–dissipation theorem, that is,

$$\gamma_{ij} = \sigma_{ij}^2(T_i + T_j)/4k_B T_i T_j \tag{8.15}$$

$$\alpha_{ij} = \sqrt{2k_B k_{ij}} \tag{8.16}$$

where k_B is a Boltzmann constant and k_{ij} is usually given as

$$k_{ij} = C_v^2 k_0 (T_i + T_j)^2/4k_B \tag{8.17}$$

where k_0 is the mesoscopic heat friction that describes thermal conductance nature of coarse-grained e-DPD bead. The heat capacity of an e-DPD bead is termed as the dimensionless heat capacity $\overline{C_v} = C_v/k_B$. In this setting, the heat flux between e-DPD beads is proportional to their temperature difference and does not depend on the absolute temperature.

The thermal diffusivity can be calculated as

$$D = \frac{\rho\overline{C}_v k_0 \int r^2\omega^D d\vec{r}}{2d} \tag{8.18}$$

In this expression, D is the thermal diffusivity, d is the dimension of the model, and ρ is the density of e-DPD beads. The weighting function ω^D is also a linear function as

$$\omega^D = \omega^2(r) = \begin{cases} (1-r)^s & r < r_c \\ 0 & r \geq r_c \end{cases} \tag{8.19}$$

To make Eq. (8.19) agree with Eq. (8.14), the value of s is often set to be 2. The heat flux is calculated using Eq. (8.20) when the e-DPD system being reached a state temperature gradient.

$$Q = \sum_{i=1}^{N}\sum_{j>i} k_{ij} w^2(r_{ij})\left(\frac{1}{T_i} - \frac{1}{T_j}\right)\vec{r}_{ij} \tag{8.20}$$

Q is total heat flux of the e-DPD simulation system. Finally, the Fourier law $k_c = -Q/(A\Delta T)$can be used to calculate thermal conductivity of the system.

e-DPD method makes thermal conductivity possible in a mesoscopic system. However, the heat flux is dimensionless owning the coarse-grain nature of DPD method. The interfacial thermal conductance in composites and other inhomogeneous systems is covered in e-DPD simulation. Those need to be further addressed in simulating thermal transporting process by using e-DPD method.

8.2 Molecular Simulation Study on Polymers

Most polymers are thought to be intrinsically thermal insulating with thermal conductivity in the order of 0.1 W/m K, such as thermal conductivity of polyethylene (PE) bulk is about 0.2–0.5 W/m K. However, Henry and Chen found that an individual PE chain may have extremely high thermal conductivity or even be divergent [9]. The ultra-drawn PE nanofibers got thermal conductivity values about 104 W/m K that is far beyond those of many metal and ceramic materials [10, 11]. This means polymers may turn out to be high thermal conductive if the chains' configuration being well arranged.

With NEMD simulation, thermal conductivities of two interacting adjacent PE chains with different drawn ratios are studied as shown in Figure 8.1a. As shown in Figure 8.1b, thermal conductivity of polymer decreases with the drawn ratio decreasing. Especially, when drawn ratio is more than 0.95, thermal conductivity values

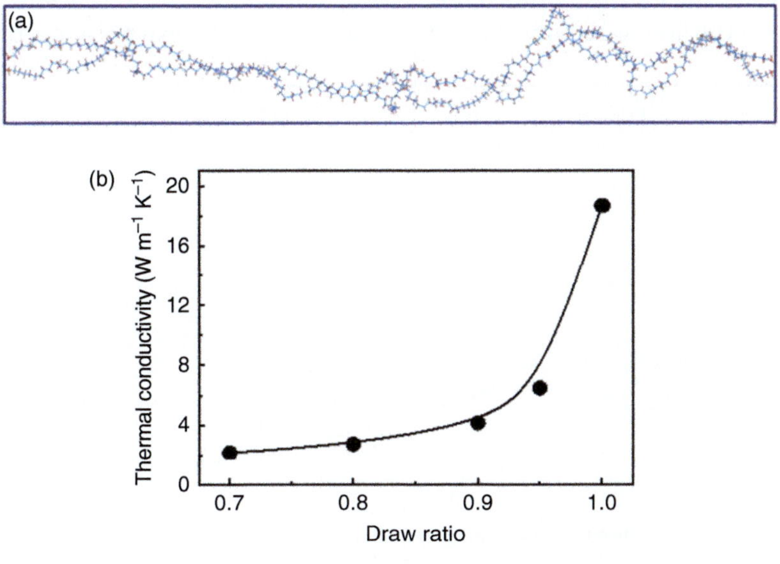

Figure 8.1 Thermal conductivity of drawn PE chains. (a) Representative morphologies of drawn PE chains. (b) Thermal conductivity of drawn PE chains with varied drawn ratio. Source: Reproduced with permission from Ref. [12]; © 2020 Springer Nature Switzerland AG. Part of Springer Nature.

of PE chains are very high. This means almost fully stretched PE chains will show excellent thermal conductivity, thermal conductivity of coiled chains is relatively low [12].

PE chains are easy to crystallize. Usually, crystalline cells in polymer are randomly arranged, which is known as polycrystalline state. It is interesting to see how the arrangement of PE crystalline cells affects thermal conductivity. In Figure 8.2a, PE bulk models with same crystalline degree are constructed, each crystalline bulk has two crystalline cells those being deliberately arranged into predetermined angels. Thermal conductivity of crystalline PE as in Figure 8.2a, the corresponding thermal conductivity values are shown in Figure 8.2b. It is obviously that crystalline polymer exhibits good thermal conductivity only in the case of polymer chains in both crystalline cells parallel to the thermal transporting direction, that is 0° in Figure 8.2a. This means crystallization contributes to polymer thermal conductivity only in cases of thermal transporting along the chain.

Be similar with those in composites, interfaces between crystal and amorphous zones affect thermal conductivity of the bulk. Two models with different interfaces between crystal-amorphous zones are shown in Figure 8.3. Figure 8.3a shows the model with continuous interface, in which polymer chains running through crystal and amorphous zones. Figure 8.3b shows the model with the fractured interface, in which polymer chains are segregated from each other into either zone, respectively. In both cases, temperature gradients obviously break down at interfaces, which are corresponding to the depression of thermal conductivity. However, the slops of temperature gradient lines in amorphous zones of these two models are different. Thermal conductivity values of the amorphous zone with intra-zones chains (continuous interface model) are over two times of that without intra-zones chains (fractured interface model). This means intra-zones chains can decrease thermal resistance of

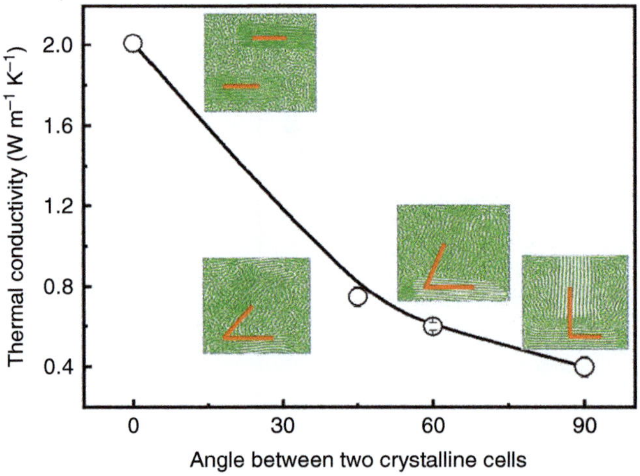

Figure 8.2 NEMD simulation study on thermal conductivity of polycrystalline PE bulk. (a) Four arrangements of crystalline cells in PE bulk. (b) Thermal conductivity of PE crystals. Source: Reproduced with permission from Ref. [12]; © 2020 Springer Nature Switzerland AG. Part of Springer Nature.

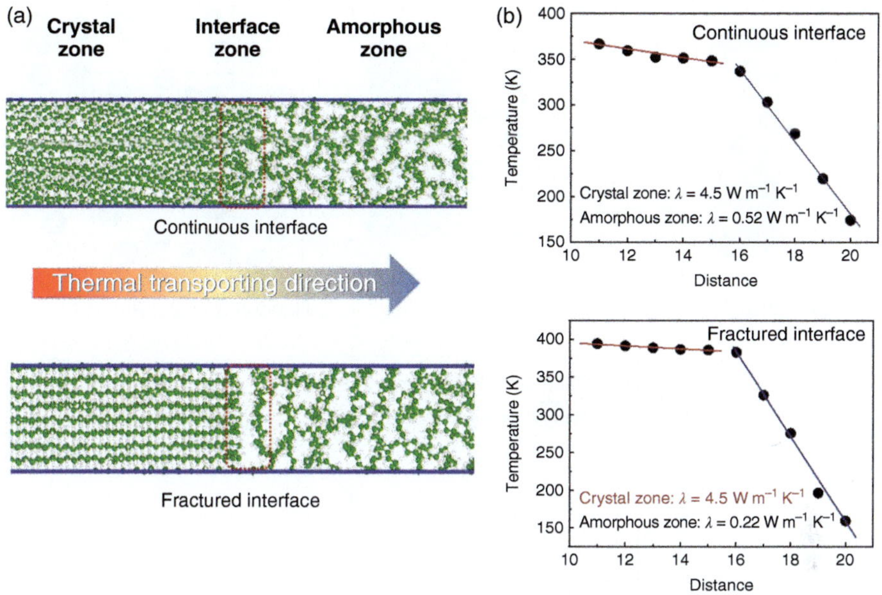

Figure 8.3 Effect of inter-zone chain on thermal conductivity of crystalline PE. The PE bulk is composed of 16 PE chains each of 200 segments. Only part of the model is shown here for clarity. (a) Continuous interfaces in crystalline polymer. (b) Fractured interfaces in crystalline polymer. Source: Reproduced with permission from Ref. [12]; © 2020 Springer Nature Switzerland AG. Part of Springer Nature.

the zones interface, which is similar with the effect of surface modification onto inorganic particle in a polymer matrix those are found to effectively improve thermal conductivity of polymer composites.

For polymer composites, filling particles with high thermal conductivity are used to improve that of the system. There have been a lot of filling theory on this issue. Using e-DPD methods, the distribution patterns of filling particles are deliberately designed to check thermal transporting efficient across polymer matrix [13]. In Figure 8.4, three distribution states of filling particles represent a cylindrical heat transporting pathway by filling particles, a lamellar pathway, as well as a randomly distributed state. e-DPD simulation shows that with increasing of filling content thermal conductivity of all composites increase, while cylindrical and lamellar distribution states exhibit excellent heat transporting efficiency as show in Figure 8.4b. This is because both cylindrical and lamellar pathways directly connect hot and cold zones of the system, which are more efficient than randomly dispersed filler particles in transporting heat flow. In experiment, researchers observed that adding thermal conductive fillers into polymer matrix have limited effect on composites' thermal conductivity even till the filling content is very large, which is similar with modeled composites with randomly dispersed fillers. So, building an effective heat transporting pathway across polymer matrix will intensively increase composites' thermal conductivity even at low filling content.

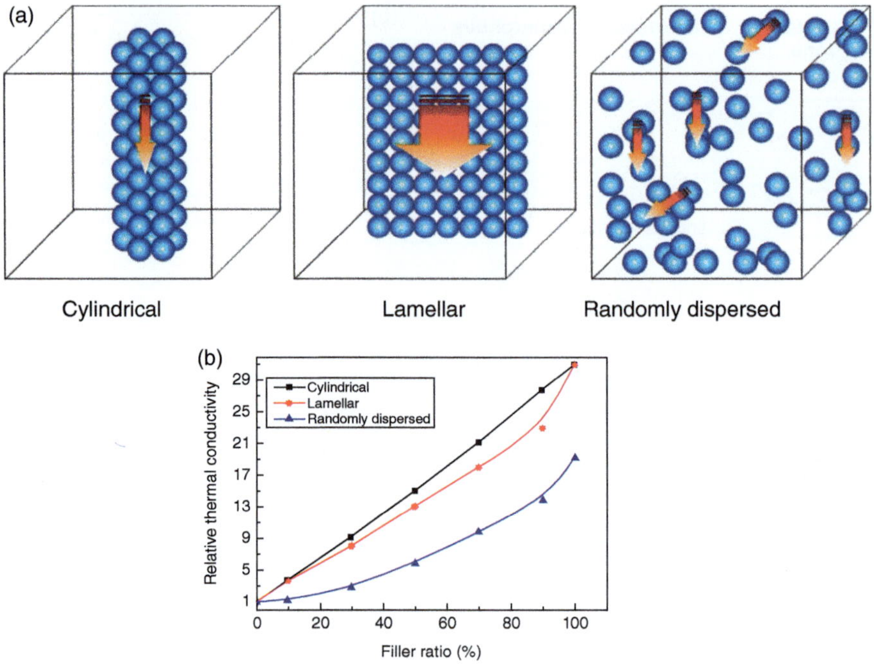

Figure 8.4 e-DPD study on thermal conductivity of polymer composites with varied distribution of filling particles. (a) Distribution state of filling particles in polymer matrix; (b) Variation of composites' thermal conductivity along with filling contents. Source: Reproduced with permission from Ref. [13]; © 2017 Springer Nature Switzerland AG. Part of Springer Nature.

Although thermal conductivity of polymer composites increases with increasing of filling content, mechanical strength and elasticity will be damaged at high filling content. Researchers pay much attention to improving thermal conductivity of polymer composites at low filling content. Theoretically, filling particles in polymer composites are at low filling content and evenly dispersed state, and it is impossible for filling particles to form an ideal heat conduction pathway. However, if these filler particles are arranged along temperature gradient they may form "pseudo" heat conduction pathways. In Figure 8.5, three arrangements of filling particles are designed by e-DPD simulation, those are: Arrange A (particles being arranged into a line along temperature gradient), Arrange B (particles being arranged into a plane parallel with temperature gradient), and Arrange C (particles being evenly distributed in the system). The thermal conductivities of corresponding composites Arrange A, B, and C show that polymer composites with thermal conductive fillers being arranged into a line (Arrange A) or a plane (Arrange B) along temperature gradient exhibit higher thermal conductivity values than that of Arrange C. Although thermal conductive filler particles are thermal-isotropic and filler particles are dispersed from each other, particles in arrange A and B form "pseudo" pathways for heat transporting those show excellent heat transporting efficiency. This means arranging filling particles into such "pseudo" heat transporting pathways in polymer matrix can fabricate high thermal conductive composites at low filling content.

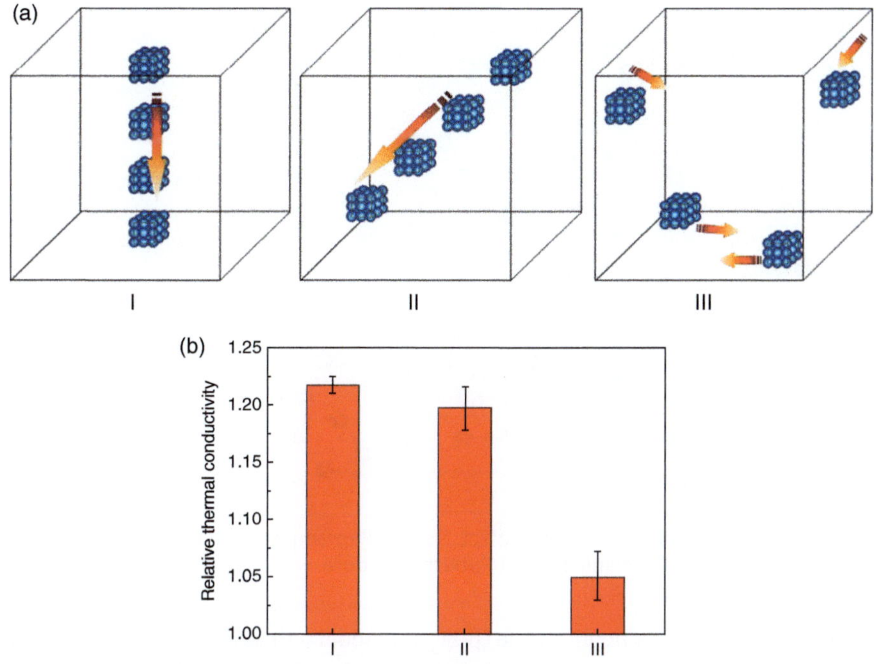

Figure 8.5 e-DPD study on "pseudo" heat transporting pathways formed at low filling content of polymer composites. (a) Three kinds of arrangement of filling particles in polymer matrix. (b) Relative thermal conductivity of polymer composites. Source: Reproduced with permission from Ref. [13]; © 2017 Springer Nature Switzerland AG. Part of Springer Nature.

8.3 Molecular Simulation Study on TC of Si_3N_4 Ceramic

Owing to the excellent mechanical properties and thermal conductivity, silicon nitride ceramics have received a lot of attention [14]. Naoto Hirosaki and collaborators calculated ideal thermal conductivity of single crystal α- and β-Si_3N_4 those two crystal structures are generally found in experiment. They used Green–Kubo formulation to calculate the thermal conductivity of the single crystals as a function of temperature. The thermal conductivity of the crystals is estimated with a high degree of accuracy, even though they employed a relatively small simulation super cell size. At a room temperature of 300 K, thermal conductivity of a-Si_3N_4 and β-Si_3N_4 is found approximately 105 and 225 in units of W/(m K), respectively [15]. It is clear from comparisons with existing experimental data that β-Si_3N_4 single crystal exhibits the higher thermal conductivity.

Experimentally fabricated silicon nitride ceramics are often polymorphous or polycrystalline, which means the bulk ceramics are mixture of crystals and amorphous silicon nitride constituents. Furthermore, oxygen atoms are easily involved in the crystalline sites and replace nitrogen atoms on their sites. To mimic thermal conductivity of silicon nitride ceramic with defects, oxygen replacement and nitrogen vacancy (loss of nitrogen atoms in crystal sites) should be considered.

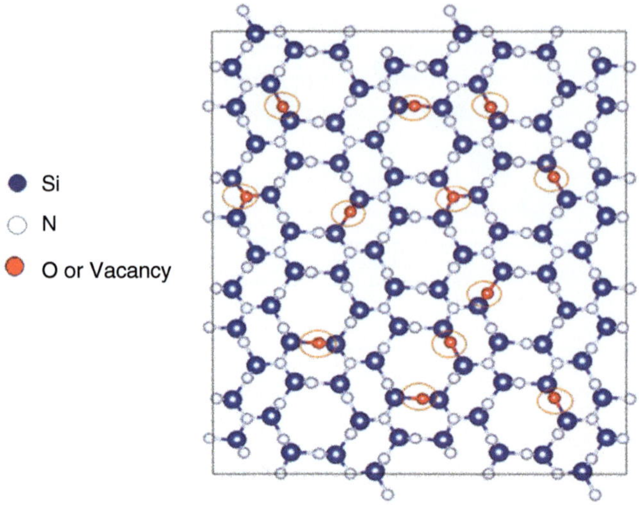

Figure 8.6 Silicon nitride model with defect. Source: Reproduced with permission from Ref. [16]; © 2022 China Academic Journal Publishing House.

As shown in Figure 8.6, silicon nitride (β-Si_3N_4) model with some N atoms replaced by O atoms or just vacancy is constructed to study the effect of oxygen replacement and nitrogen vacancy on to thermal conductivity of silicon nitride ceramic [16].

NEMD simulation study reveals that, both oxygen replacement and nitrogen vacancy will strongly decrease thermal conductivity of β-Si_3N_4 ceramic. As shown in Figure 8.7, with only about 2.5% nitrogen atoms being replaced by oxygen atoms, thermal conductivity of the ceramic decreases to almost two-thirds of its

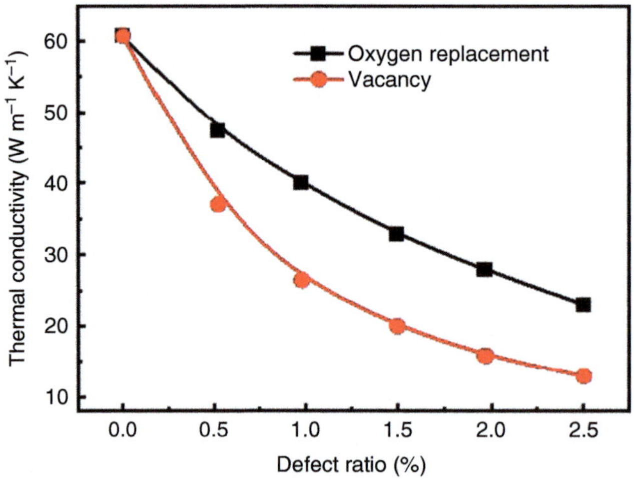

Figure 8.7 Effect of defect ratio on thermal conductivity of silicon nitride. Source: Reproduced with permission from Ref. [16]; © 2022 China Academic Journal Publishing House.

original values. Missing nitrogen atoms show more depressing effect on thermal conductivity of the ceramic than that of oxygen replacement.

Besides the proportion of defects, the distribution of defects in the ceramic also plays an important effect on thermal conductivity. Figure 8.8 shows three kinds of defect distribution those are P-aggregation (defects arranged along the heat transporting direction), disperse (defects are evenly arranged), and V-aggregation (defects arranged vertically to the heat transporting direction).

It is clear, both oxygen replacement and nitrogen vacancy defects are strongly dependent on the position and distribution states affecting thermal conductivity of the ceramics. P-aggregation state of defects arranged along the heat transporting direction shows the least disturbance on heat transporting pathway, while V-aggregation state of defects nearly cuts down the heat transporting pathway. This phenomenon may be used to fabricate transform devices those changing from heat transporting to heat insulting and vice versa.

The simulation studies on silicon nitride ceramics show that thermal conductivity of ceramics is affected by crystal formation, defects as well as distribution of defects. As defects are hard to avoid in ceramic fabrication process, making a good control of the distribution of defects is a good choice to improve thermal conductivity and develop new function for ceramic materials.

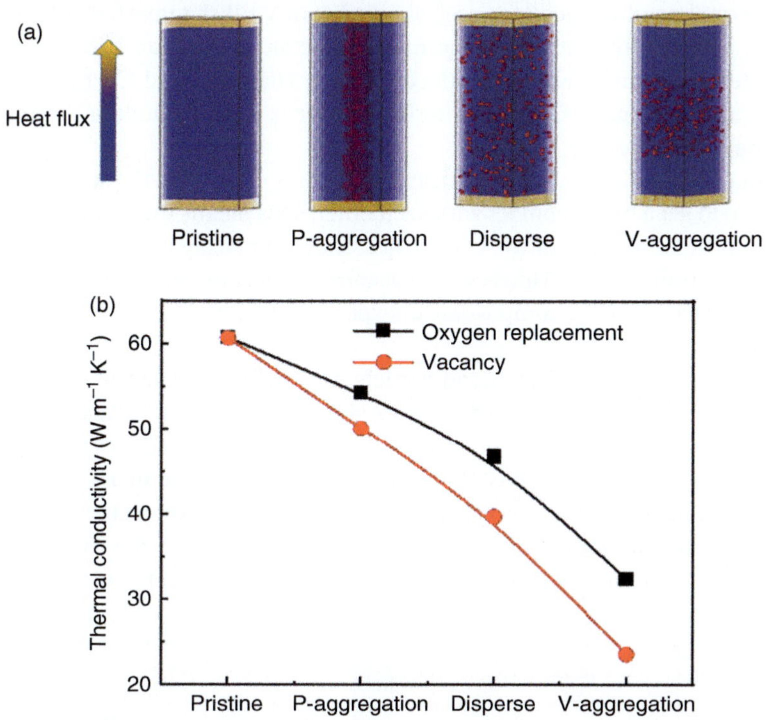

Figure 8.8 NEMD simulation on β-Si_3N_4 with defects of oxygen replacement or vacancy on nitrogen sites. (a) Schematic diagram of the model. (b) Influence of defect distribution on thermal conductivity of silicon nitride. Source: Reproduced with permission from Ref. [16]; © 2022 China Academic Journal Publishing House.

8.4 Molecular Simulation Study on TC of Diamond/Copper Composites

Diamond/copper (diamond/Cu) composites, due to their high thermal conductivity and tailorable coefficient of thermal expansion, have garnered considerable interest. However, the low wettability between Diamond and Cu makes the interfacial bonding of the composite hard to be formed, resulting in poor thermal conductivity. To solve the problem mentioned above, surface metallization of diamond surface or alloying of Cu matrix is thought to be effective method to improve the bonding state between Diamond and Cu interface and the thermal conductivity of Diamond/Cu composites by forming chemical bonds at the diamond/Cu interface [17]. Besides experiments exploration of the interfacial conductance of diamond/Cu composites, simulation study on diamond/Cu composites can examine more detailed approach to improve interfacial heat transporting efficiency of composites.

TiC was experimentally found to effectively improve interfacial thermal conductance of diamond/Cu composites [12]. Figure 8.9 shows models of diamond/Cu composites and that with TiC interfacial layer. NEMD simulation shows that the size of interfacial TiC constituent affects the interfacial thermal conductance (ITC) [18]. In Figure 8.10, with adding of TiC interfacial layer into diamond/Cu composites, interfacial thermal conductance increases almost linearly till the interfacial layers reaching at 6, and then the interfacial thermal conductance decreases slowly. This means a moderate modification of interface between Diamond and Cu matrix is appropriate for enhancing ITC, while interfacial layer is thick to build a thermal obstacle in the composites.

Experimentally, to combine metal particles tightly, it is often to make surfaces of particles coarse to get a better contact with each other. This means the defects being introduced into the interfaces of the composites; those are easily thought to build obstacles for heat transporting. However, as vacancy defects are introduced into the interface of the Diamond/Cu composites, unexpected phenomena are found in a NEMD study.

In Figure 8.11, three Cu/TiC/diamond models with defects located in different regions are deliberately designed. The defects are simplified as crystal vacancy; those are located in interface layer by TiC or surface of Diamond/Cu matrix, respectively. It is reasonable to find in Figure 8.12 that vacancy defects increase in TiC and Cu regions, and ITC values decrease correspondingly. However, in cases vacancy defects reside on the surface of diamond matrix, ITC values increase with the increasing of vacancy defects first and then those decrease rapidly. This means introducing

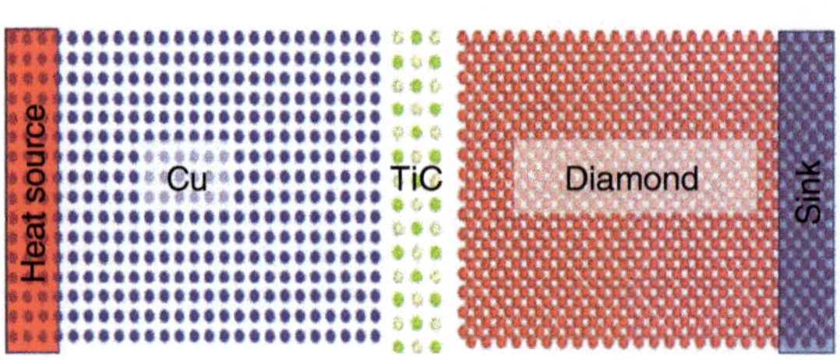

Figure 8.9 Schematic diagram of models: (a) the diamond/Cu system and (b) the diamond/TiC/Cu system. Source: Reproduced with permission from Ref. [18]; © 2023 China Academic Journal Publishing House.

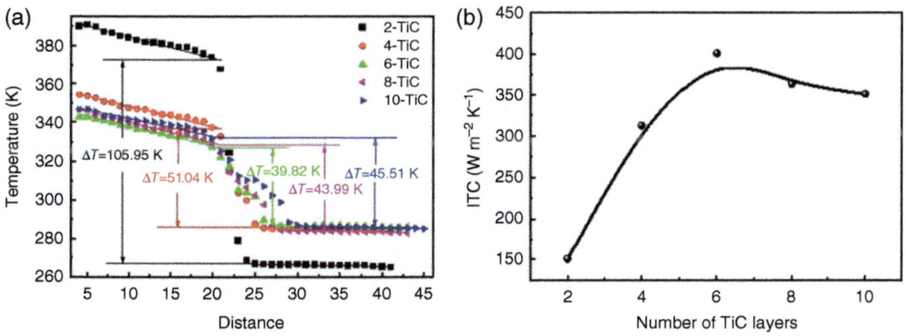

Figure 8.10 Diamond/TiC/Cu composites with varied TiC layer thickness. (a) Temperature gradient diagram; (b) The interfacial thermal conductance of the composites. Source: Reproduced with permission from Ref. [18]; © 2023 China Academic Journal Publishing House.

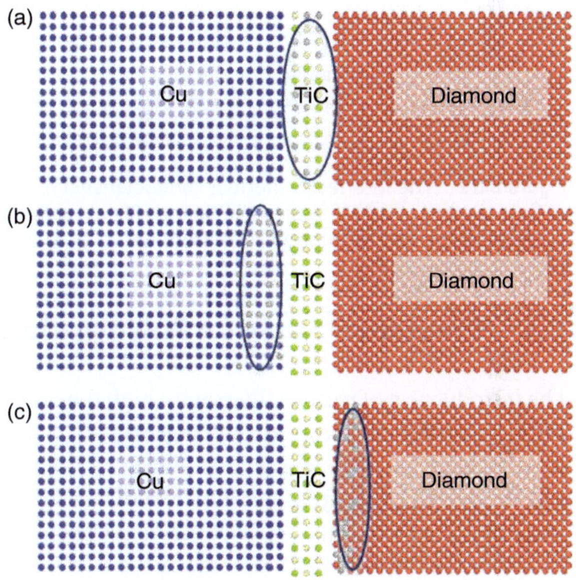

Figure 8.11 Schematic diagram of the model of the Diamond/TiC/Cu systems with interfacial defects located in (a) TiC region, (b) Cu interfacial region, and (c) diamond interfacial region. Source: Reproduced with permission from Ref. [18]; © 2023 China Academic Journal Publishing House.

moderate vacancy defects into surface region of diamond may improve ITC of the composites. The mechanism for this phenomenon is still unclear, and more studies including experimental efforts should be considered on this issue.

The simulation demonstrated that introducing the interfacial materials such as TiC compound at the diamond/Cu interface will enhance interfacial thermal conductance. The thickness of interfacial layer and defects in interfacial region play an important effect on ITC values. Of course, simulation predictions are not always consistent with experimental result, especially for quantity cases such as thickness of interfacial layer. The importance of simulation is that it gives out theoretical guides to improve thermal conductivity of the composites.

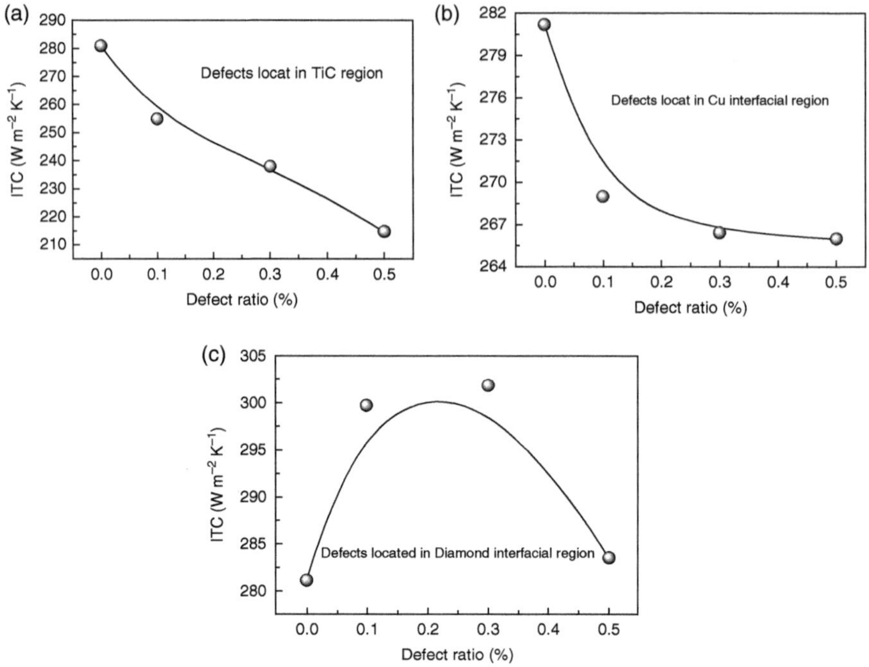

Figure 8.12 The interfacial thermal conductance (ITC) of diamond/TiC/Cu systems with defects located in (a) TiC region, (b) Cu interfacial region, and (c) diamond interfacial region. Source: Reproduced with permission from Ref. [18]; © 2023 China Academic Journal Publishing House.

8.5 Simulation Study on Polymer-Based Composites

The introduction of various fillers with high thermal conductivity (such as graphene, carbon nanotubes, boron nitride, and so on) in the polymer endows the composite with high heat transfer ability. However, the arrangement of fillers in polymer and the interface thermal resistance between fillers, between fillers and polymer, are important factors that restrict the heat transfer performance of composites. The control of the filler's orientation along the heat transfer direction and the construction of the interface structure to reduce the interface thermal resistance are effective ways to solve the above issues.

8.5.1 Simulation Analysis in Heat Transfer Pathways Construction

External force driving is the main way to promote orderly arrangement of fillers, such as ice template method to adjust fillers arrangement by controlling the growth direction of ice crystals, magnetic orientation method to control magnetic fillers by external magnetic field, and layer-by-layer self-assembly method to realize fillers stacking by using pressure difference.

Zeng et al. constructed an orientation pathway composed of boron nitride (BN) flaks in silicone rubber by layer-by-layer self-assembly method to show high out-

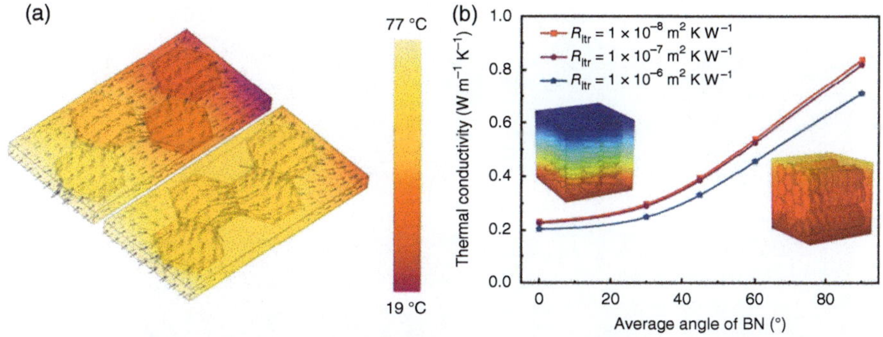

Figure 8.13 (a) The finite element simulation is used to simulate the heat transfer behavior when the BN flakes in SiR are discrete and interconnected. (b) The relationship between the average angle of the BN flakes and the thermal conductivity obtained by the finite element simulation. Source: Adapted with permission from Ref. [19]; © 2022 Elsevier.

plane thermal conductivity [19]. As shown in Figure 8.13, transient finite element simulation verifies that the BN alignment is more beneficial to heat transfer. Furthermore, the influence of different alignment angles on the heat transfer performance of BN in the orientation channel at a constant temperature field of 380 K was analyzed.

Using NH_4HCO_3 as a template, Wang et al. construct a heat transfer pathway of BN to form a three-dimensional network structure by sacrificing the template method [20]. The transient-state finite element simulation showed that this structure can form an effective heat transfer channels and exhibit a higher surface temperature than pure epoxy resin (EP) and composites with randomly distributed boron nitride (RBN/EP) (Figure 8.14). For EP and RBN/EP, as a result of temperature distribution, the heat transfers uniformly and slowly from the bottom to the top, while in EP-based composites with three-dimensional BN structure (3D-BN/EP), the heat transfers fast and comparatively from one side to another. Combined with the heat flux arrows, heat flows equally from each region for EP and RBN/EP. But for 3D-BN/EP, the heat arrows are mainly concentrated in the region of BN, indicating that most heat is quickly transferred through network structure.

Li et al. constructed BN into a 3D interconnected structure network by aerogel template method. The honeycomb 3D structure built by finite element simulation clearly exhibited that the heat can move rapidly along the heat transfer pathways formed by boron nitride, which effectively improves the heat dissipation performance of the composite [21]. It can be seen from Figure 8.15 that the temperature at the top of PDMS-based composites with 3D BN network (3D-BN/PDMS) was significantly higher than that of PDMS-based composites with random BN (RBN/PDMS), suggesting an improved heat-transfer capacity endowed by the 3D network. More meaningful, the intensity of the heat flow in 3D-BN/PDMS was significantly higher, and a connected path was formed along the heat-transfer direction.

Yang et al. attached aluminum (Al) sheet to the skeleton of PU sponge and constructed a 3D heat transfer pathways depending on the structure [22]. When a stable heat flow of 10 pW mm^2 was applied to the structure formed by the Al sheet formed by finite element simulation, it revealed faster thermal conductivity and

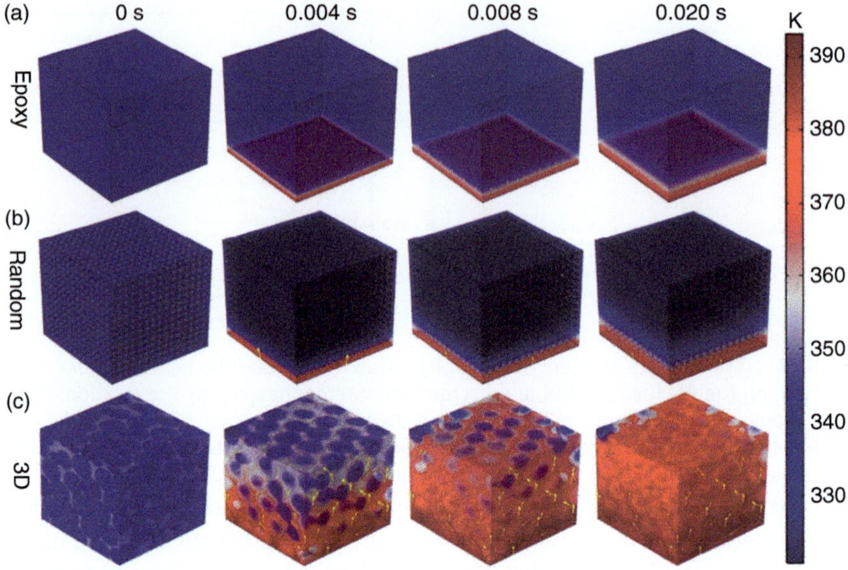

Figure 8.14 Heat transfer process simulated by finite element simulation. Temperature distribution and heat flow arrows of (a) pure epoxy resin, (b) randomly distributed BN/epoxy composite with BN volume ratio of 20%, (c) 3D-BN/epoxy composite with BN volume ratio of 20%. Source: Adapted with permission from Ref. [20]; © 2020 Elsevier.

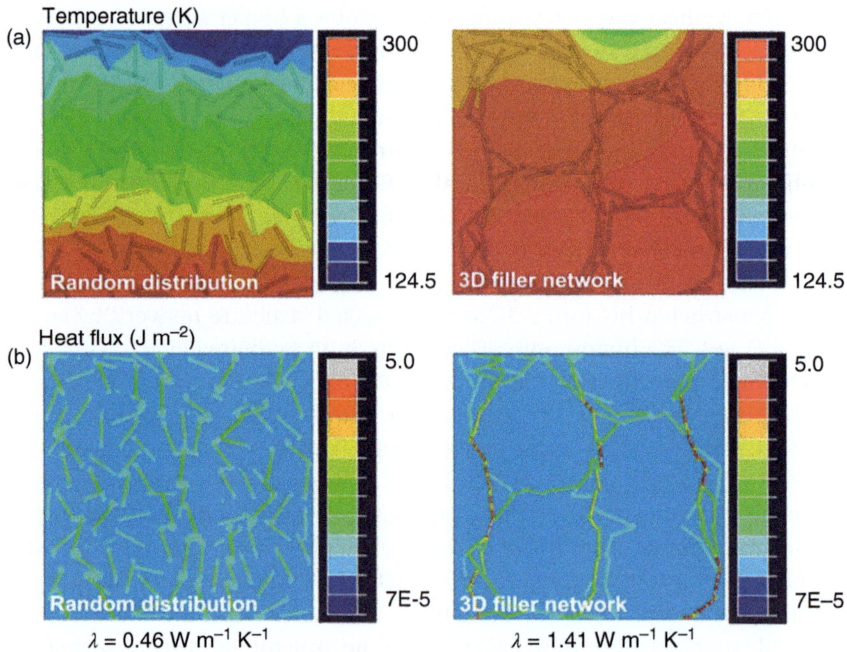

Figure 8.15 Simulated (a) temperature distribution and (b) heat flux distribution as well as the thermal conductivity values of the random BN-PDMS and 3D BN-PDMS composites based on a heat source acquired by finite element analysis. Source: Reproduced with permission from Ref. [21]; © 2007 American Chemical Society.

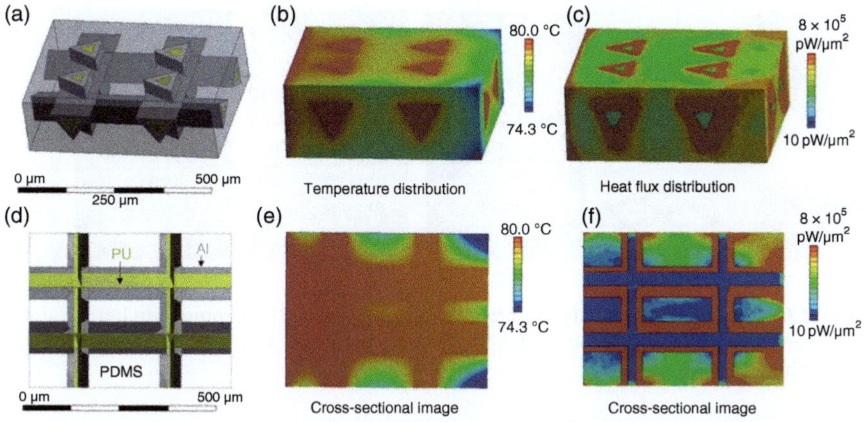

Figure 8.16 (a) Finite element models, (b) simulated temperature distribution, and (c) heat flux distribution results of PDMS/PU@Al_{30} composites; (d–f) the corresponding cross-sectional images of (a–c). Source: Reproduced with permission from Ref. [22]; © 2021 Royal Society of Chemistry.

more uniform temperature than the composite with random distribution of Al sheets (Figure 8.16). The simulation demonstrated that the heat transferred mainly along the Al network in the PDMS-based composites (PU@Al/PDMS). The tightly connected Al flake network is mainly responsible for the effective diffusion of heat flux and causes effective thermal conductivity in PU@Al/PDMS. The results visually highlight the significance of morphology engineering for achieving high heat transfer efficiency and are helpful to understand the heat conduction process for the composites with BN networks.

In the carbon nanotubes (CNTs) sponge structure formed by thermochemical vapor deposition, Feng et al. used the graphene formed by pyrolysis to fill the gaps between the CNTs, thus constructing the all-graphene multi-dimensional heat transfer path [23]. Moreover, finite element simulation was used to further analyze the compressed all-graphene network, which could obtain higher thermal conductivity because it could provide a denser thermal conductivity pathway (Figure 8.17). The all-graphene skeleton showed a higher temperature than the matrix at the same distance due to efficient phonon transport. The results indicated that the compression remarkably facilitates phonon transport throughout the composites, which was important for stress controllability of thermal conductivity under different strains.

Wu et al. introduced silver nanoparticles (AgNP) into the BN framework structure constructed by using ice templates to fill the gaps between BN sheets and obtain a long-range continuous thermal conductivity pathway [24]. The structure not only had an effective heat transfer channel, but also AgNP optimized the long-range network structure. The transient-state finite element simulation was used to verify the superiority of this structure in heat transport (Figure 8.18). Comparing with PDMS matrix and PDMS-based composite with randomly BN structure, heat selectively flowed along the BN sheets rather than the PDMS due to the formation of 3D filler

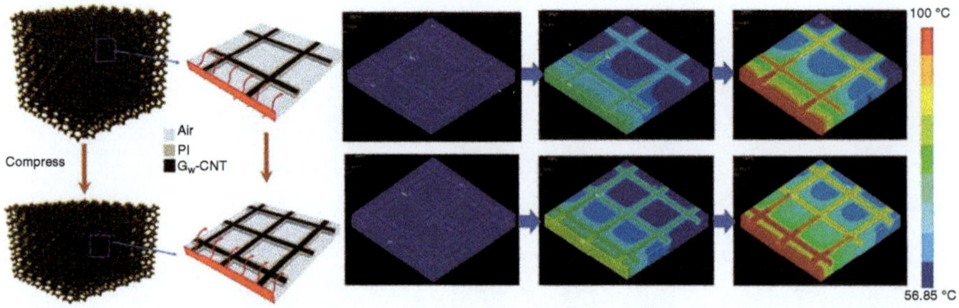

Figure 8.17 Finite element analysis of heat transfer in Gw-CNT/PI composite before and after compression. Source: Reproduced with permission from Ref. [23]; © 2019 Wiley-VCH Verlag GmbH & Co. KGaA.

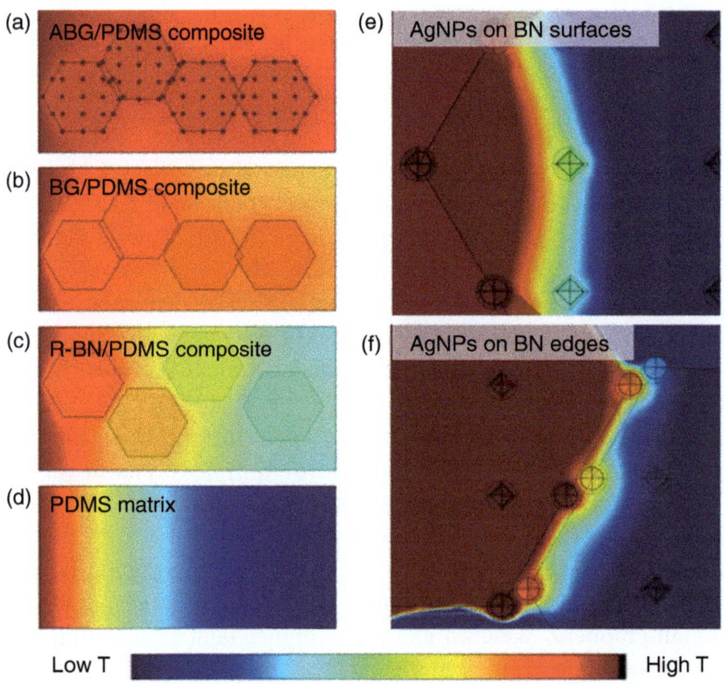

Figure 8.18 Finite element analyses. Temperature profile of the (a) ABG/PDMS composite, (b) BG/PDMS composite, (c) R-BN/PDMS composite, and (d) PDMS matrix obtained during heating. Fine temperature profile displaying the effect of distribution states of AgNPs on the heat transfer process in the ABG/PDMS composite: (e) AgNPs on BN planes and (f) AgNPs on BN edges. Source: Reproduced with permission from Ref. [24]; © 2021 American Chemical Society.

structures in the PDMS-based composite with orient BN sheet (BG/PDMS). In addition, the heat transmission between BN and BN interfaces was optimized, due to the AgNPs on BN edges connected adjoining BN flakes together to improve the heat flow along BN frameworks.

Qian et al. constructed carbon fiber cloth (CFC) into a coherent thermal conductivity network by using CNTs grown *in situ* on the surface of carbon fiber [25]. The topological network structure constructed by finite element simulation not only realizes the heat transfer between carbon fibers but also achieves the purpose of multi-direction heat transfer of composite materials (Figure 8.19). The surface temperatures at different sites of EP-based composites with carbon fiber cloth containing carbon tube (CFC@CNT/EP) and pure carbon fiber cloth (CFC/EP) were compared by simulation. It is further verified that the connected orientation structure is more conducive to the heat transfer inside the composite, and the heat can be evenly distributed inside.

8.5.2 Simulation Analysis of Low Thermal Resistance Interface Structure Construction

The covalent bonding or non-covalent bonding formed between fillers and polymer matrix constitutes an effective heat transfer "bridge" to optimize the heat transfer pathways in the composite and achieve the purpose of improving the thermal conductivity. On the basis of experimental characterization, simulation more profound analysis and verification of the above methods to enhance the thermal conductivity of composite materials from a theoretical point of view.

8.5.2.1 Covalent Bonding Construct Interface Structure

Qian et al. connected the graphene layer by growing a boron nitride nanotube (BNNT) on the surface of the graphene [26]. The interface structure formed by the covalent bonding between the BNNT and the graphene effectively reduced the interfacial thermal resistance. Meanwhile, molecular dynamics simulation verified that the covalent bonding exhibited a higher interfacial thermal conductivity than the van der Waals (VdW) force, indicating that the graphene layer had a better heat transfer effect due to the connection of the BNNT (Figure 8.20). Moreover, it can be seen from the phonon density of states (PDOS) obtained from the simulation that the covalent bonding can cause high-efficiency resonance of graphene and BNNT in the low-frequency stage below 30 THz, which can lead to good interfacial heat transfer effect.

Qian et al. have also realized the connection between BN nanosheets (BNNS) by *in situ* growing CNTs between the oriented BNNS constructed by the ice template, which not only built a through heat transfer path but also formed covalent bonding to alleviate the thermal resistance of the interface [27]. Compared with the VdW interaction, the molecular dynamics simulation showed that covalent bonding has advantages in regulating the thermal resistance of the interface. The interface thermal conductivity was as high as 1.97 GW m^2/K, which is more than 50 times that of the interaction of VdW interaction (Figure 8.21). More importantly, the simulation further clarified that the low-frequency region between 20 and 30 THz was an effective region where the BNNS and CNTs were prone to resonance to promote interfacial heat transfer.

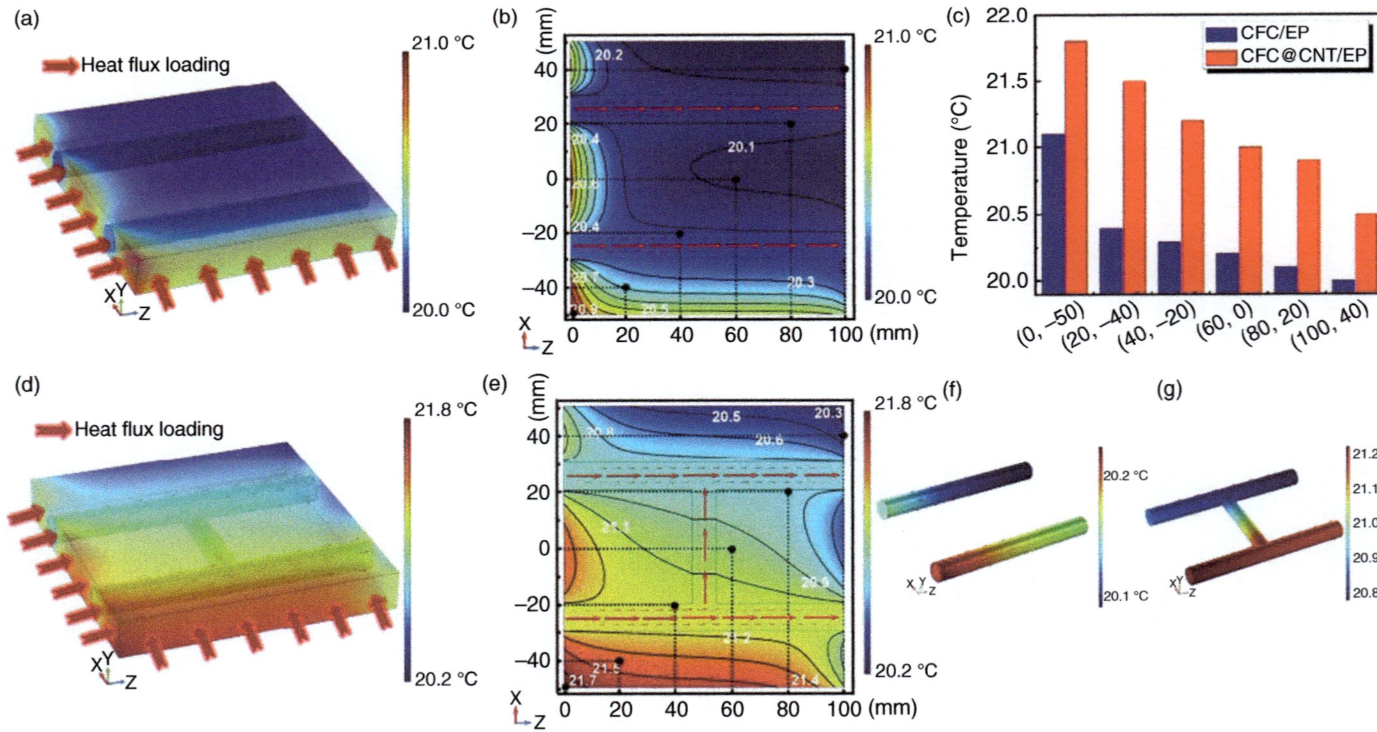

Figure 8.19 Finite element analysis and simulated results of heat transfer in (a) CFC/EP, (d) CFC@CNT/EP, (f) CFC, and (g) CFC@CNT. Thermal diffusion isotherms of (b) CFC/EP and (e) CFC@CNT/EP in simulation. (c) The temperature at the same coordinate point in the CFC/EP and CFC@CNT/EP. Source: Reproduced with permission from Ref. [25]; © 2022 American Chemical Society.

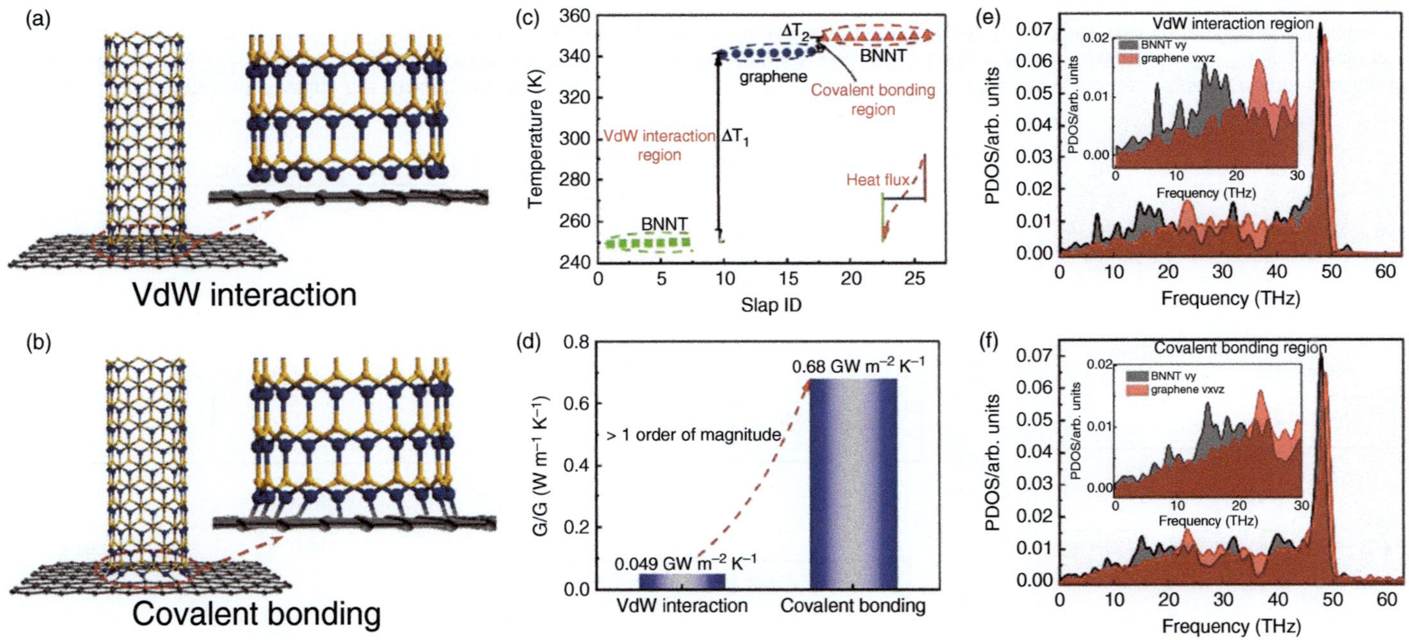

Figure 8.20 The graphene and BNNT interface bonded by VdW interaction (a) and covalent bonding (b), (c) Steady-state temperature profile of system obtained using the NEMD approach at 300 K, (d) the interfacial thermal conductance between graphene and BNNT connected by covalent C–N bonding and VdW interaction, the PDOS of VdW interaction (e) and covalent bonding (f) along the axial direction of BNNT (BNNT vy) and along the horizontal direction of graphene (graphene vxvz). Source: Reproduced with permission from Ref. [26]; © 2020 Royal Society of Chemistry.

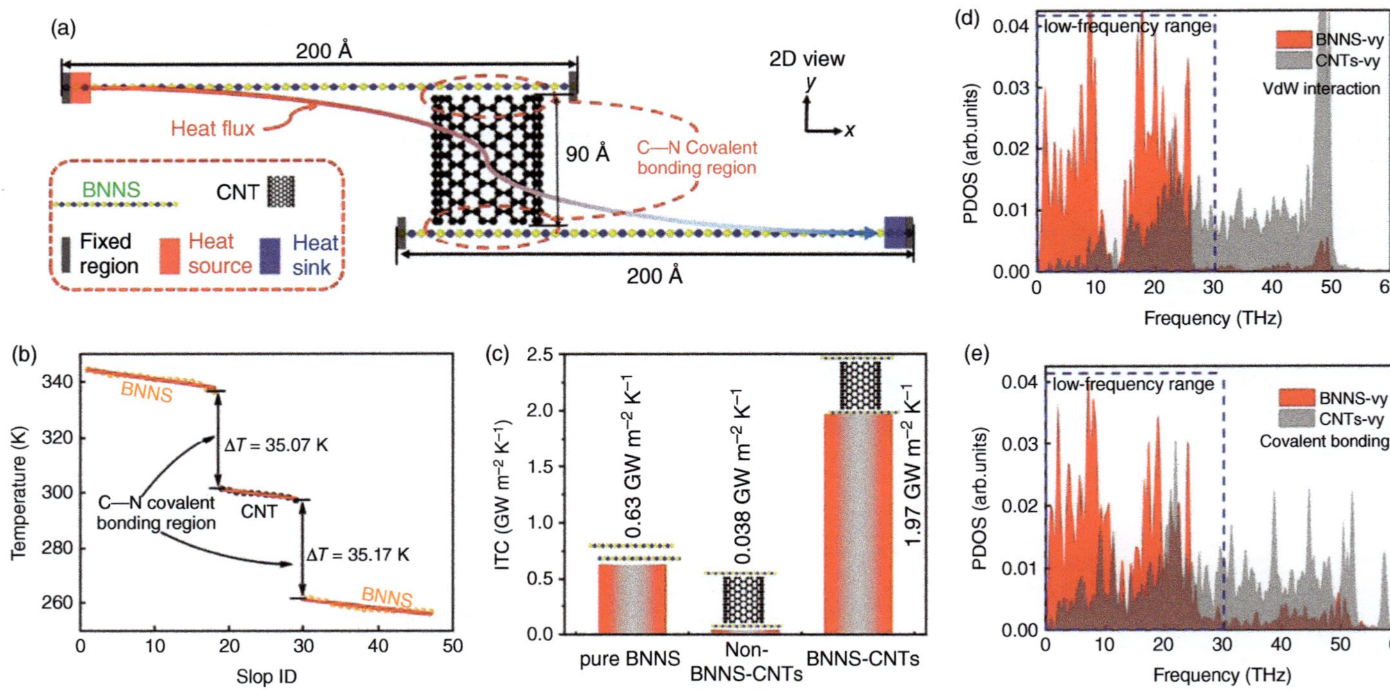

Figure 8.21 (a) Schematic illustrating the structure of the BNNS connected by CNTs. (b) Steady-state temperature profile of system obtained using the NEMD approach. (c) The interfacial thermal conductivity of BNNS layer, non-BNNS-CNTs, and BNNS-CNTs. PDOS of VdW interaction (d) and covalent bonding (e) along the *y*-axis (vy) in the BNNS-CNTs. Source: Reproduced with permission from Ref. [27]; © 2021 Wiley-VCH Verlag GmbH & Co. KGaA.

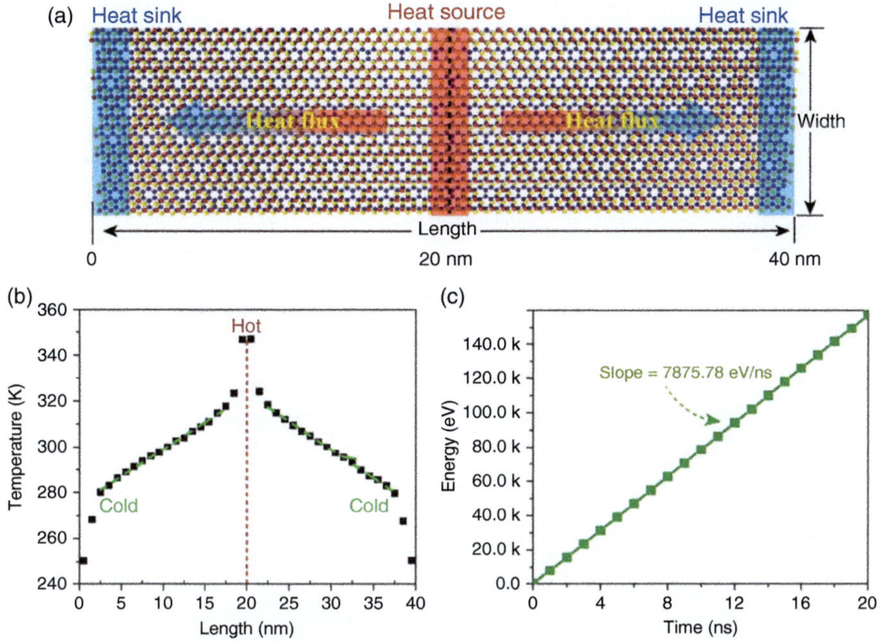

Figure 8.22 (a) The model diagram of calculation of rGO@CN thermal conductivity, (b) the temperature gradient with the system divided into 20 segments, (c) heat flow rate. Source: Adapted with permission from Ref. [28]; © 2021 Elsevier.

Tian et al. grew carbon nitride nanosheets (CN) between the horizontal layers of graphene nanosheets to connect graphene through covalent bonds to form a continuous structure to improve thermal conductivity [28]. The non-equilibrium molecular dynamics simulation further demonstrated the role of covalent bonding in improving the thermal conductivity of graphene nanosheets through the thermal conductivity of the integrated structure formed by graphene and CN (Figure 8.22).

Gu et al. used the imide reaction to connect the polyimide (PI) matrix to the surface of ammoniated graphene (NH_2-rGO), which effectively reduced the interfacial thermal resistance between the graphene and the polymer substrate and verified the efficient heat transfer effect in the integrated structure of graphene and polyimide composites by finite element simulation (FES) (Figure 8.23) [29]. The above simulation results proved that the NH_2-rGO/PI thermally conductive composite films present more excellent heat dissipation and efficient thermal management capabilities.

Lin et al. constructed an effective interfacial heat transfer structure by forming C–Si covalent bonding between graphene and silicon carbide (SiC) by *in situ* calcination method and calculated through molecular dynamics simulation that covalent bonding could provide an interfacial thermal conductivity 10 times greater than VdW interaction (Figure 8.24) [30]. This can be ascribed to the fact that the SiC nanorods covalently bonded between graphene sheets lower the phonon diffusion barrier across the basal planes of graphene, thus compensating

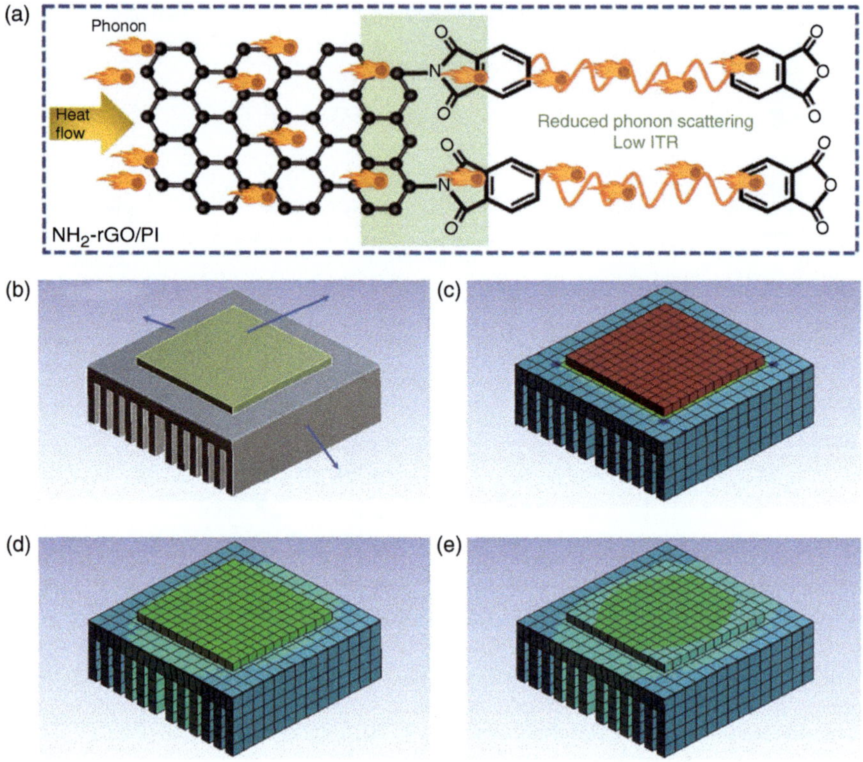

Figure 8.23 (a) Schematic diagrams of smooth heat flow, reduced phonon scattering, and low ITR at the interfaces in NH_2-rGO/PI thermally conductive composite films. (b) FES model. (c) FES results of pure PI film, (d) 15 wt% rGO/PI, and (e) 15 wt% NH_2-rGO/PI thermally conductive composite films. Source: Adapted with permission from Ref. [29]; © 2021 American Association for the Advancement of Science.

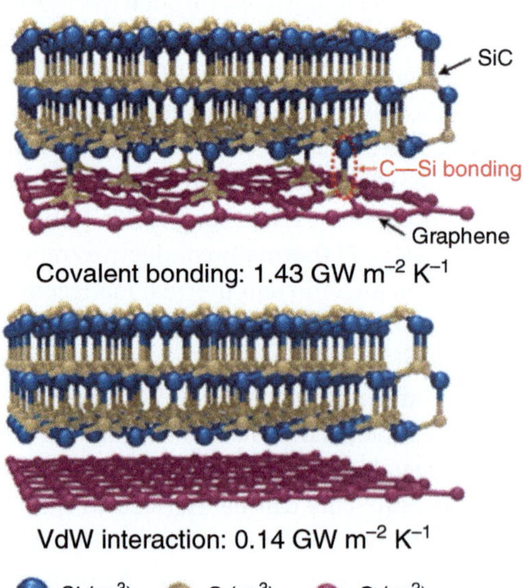

Figure 8.24 The calculated thermal conductance at the graphene–SiC nanorod interface bonded by covalent and VdW interactions. Source: Reproduced with permission from Ref. [30]; © 2019 American Chemical Society.

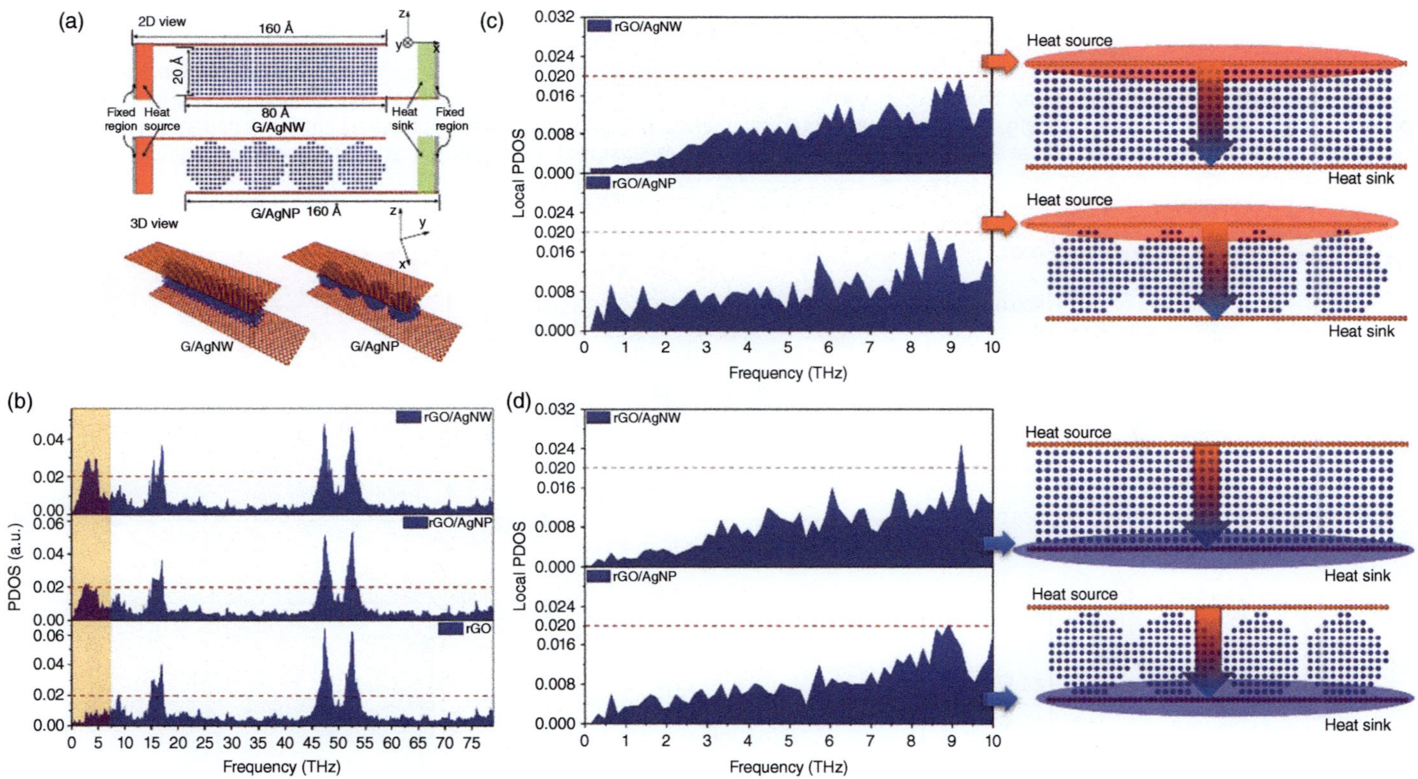

Figure 8.25 (a) Simulation modeling for comparing the effects of AgNW and AgNP on thermal conductivity between graphene layers. (b) Overall PDOS of G/AgNW, G/AgNP, and graphene to illustrate the LFP vibrations of Ag. (c) Local PDOS of interfacial C atoms of G/AgNW, and G/AgNP on the heat source side. (d) Local PDOS of interfacial C atoms of G/AgNW and G/AgNP on the heat sink side. Source: Reproduced with permission from Ref. [32]; © 2020 American Chemical Society.

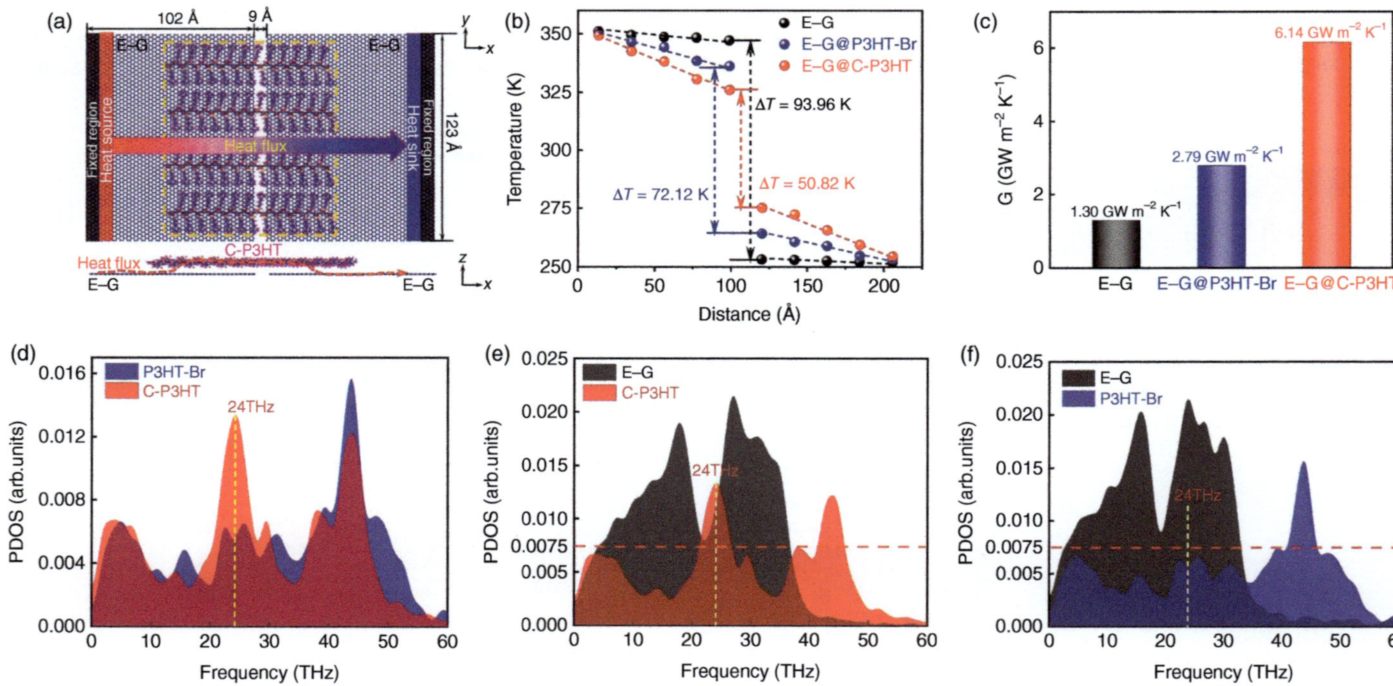

Figure 8.26 (a) Simulation modeling for exhibiting thermal conductivity of E–G@C-P3HT. (b) Steady-state temperature profile of E–G, E–G@ P3HT-Br, and E–G@C-P3HT obtained using the NEMD approach at 300 K. (c) Interfacial heat conductance of E–G, E–G@ P3HT-Br, and E–G@C-P3HT obtained by simulation. (d–f) PDOS form between P3HT-Br and C-P3HT, E–G and C-P3HT, and E–G and P3HT-Br. Source: Reproduced with permission from Ref. [33]; © 2022 Royal Society of Chemistry.

for the decrease of through-plane heat conduction capability caused by the rearrangement of graphene sheets.

Although the covalent bonding can alleviate the interface thermal resistance between the fillers by constructing an effective interface structure, the generation of covalent bonds may cause certain damage to the structure of the filler itself, resulting in a decrease in the thermal conductivity of the filler [31]. Therefore, the use of non-covalent bonding to construct the interface structure can not only preserve the filler structure itself but also exhibiteded an interface heat transfer effect comparable to that of covalent bonding.

8.5.2.2 Non-covalent Construct Bonding Interface Structure

Qian et al. have made use of the *in situ* silver nanowires (AgNW) to form non-covalent bond interactions between graphene sheets to provide an effective way for heat transfer along surfaces [32]. After using non-equilibrium molecular dynamics simulation, AgNW can provide more heat transfer channels, because they have more action sites than silver nanoparticles (AgNP). As a result, the composite structure containing AgNW exhibited higher in-plane thermal conductivity of containing AgNP, as depicted in Figure 8.25. At the same time, the simulation results also revealed that the AgNW can show better resonance effect with graphene in the low-frequency region, which can better promote the heat transfer at the interface.

Qian et al. used charged polythiophene (C-P3HT) to modify the surface of graphene (E–G) and established a dual-channel interface structure between graphene layers through the cation–π and π–π interactions generated between (C-P3HT) and E–G to improve interfacial heat transfer [33]. The non-equilibrium molecular dynamics simulation not only verifies the effect of charged structure on the heat transfer performance of C-P3HT with intrinsic thermal conductivity but also further illustrates the importance of dual-channel structure to improve the thermal conductivity between C–E–G nanosheets (Figure 8.26). The simulation further showed that the composite has a higher PDOS value at the 14 THz position, which indicates that the charged structure is more likely to produce resonance heat transfer with E–G.

References

1 Hans, M.R., Richard, A.P., and Arno, L. (2000). Absolute steady-state thermal conductivity measurements by use of a transient hot-wire system. *Journal of Research of the National Institute of Standards and Technology* 105: 221.

2 Erik, L. (2015) Molecular Dynamics Simulations. chapter 4, 63.

3 Aleksandr, V., Tommy, L., and Cornelia, B. (2022). Prediction of thermal conductivities of rubbers by MD simulations-new insights. *Polymers* 14: 2046.

4 Patrick, K.S., Simon, R.P., and Pawel, K. (2002). Comparison of atomic-level simulation methods for computing thermal conductivity. *Physical Review B* 65: 144306.

5 Patrice, C. and Jean-Louis, B. (2004). Finite size effects in determination of thermal conductivities: comparing molecular dynamics results with simple models. *Journal of Heat Transfer* 126: 577.

6 Qiao, R. and He, P. (2007). Simulation of heat conduction in nanocomposite using energy-conserving dissipative particle dynamics. *Molecular Simulation* 33: 677.

7 Espanol, P. (1997). Dissipative particle dynamics with energy conservation. *Europhys Letter* 40: 631.

8 Avalos, J.B. and Mackie, A.D. (1997). Dissipative particle dynamics with energy conservation. *Europhys Letter* 40: 141.

9 Asegun, H. and Chen, G. (2008). Thermal conductivity of polyethylene chains using molecular dynamics simulations. *Proceedings of 3rd Energy Nanotechnology International Conference* 1: 224.

10 Sheng, S., Asegun, H., Jonathan, T. et al. (2010). Polyethylene nanofibres with very high thermal conductivities. *Nature Nanotechnology* 5: 251.

11 Asegun, H. and Gang, C. (2008). High thermal conductivity of single polyethylene chains using molecular dynamics simulations. *Physical Review Letters* 101: 235502.

12 Zhang, X., Wang, Y., Xia, R. et al. (2020). Effect of chain configuration on thermal conductivity of polyethylene A molecular dynamic simulation study. *Chinese Journal of Polymer Science* 38: 1418.

13 Yang, Y., Shu, J., Chen, P. et al. (2017). Effect of microstructure on thermal conductivity of polymer composites. *Macromolecular Research* 25: 344.

14 Naoto, H., Yusuke, O., Fumio, M., and Yoshio, A. (1999). Effect of seeding on the thermal conductivity of self-reinforced silicon nitride. *Journal of the European Ceramic Society* 19: 2183.

15 Naoto, H., Shigenobu, O., Cenk, K. et al. (2002). Molecular dynamics calculation of the ideal thermal conductivity of single-crystal a- and b-Si_3N_4. *Physical Review B* 65: 134110.

16 Chen, D.Y., Chen, P., Qian, J.S. et al. (2022). Simulation study on effect of defects on thermal conductivity of silicon nitride. *Bulletin of the Chinese Ceramic Society* 41: 1797.

17 Zhang, D., Yuan, M., Tan, Z. et al. (2018). Progress in interface modification and nanoscale study of diamond/Cu composites. *Acta Metallurgica Sinica* 54: 11.

18 Zhao, T.Y., Chen, P., Qian, J.S. et al. (2023). Simulation study on interfacial thermal conductance of titanium carbide coated copper/diamond composites. *Hot Working Technology* https://doi.org/10.14158/j.cnki.1001-3814. 20220229.

19 Hu, Q., Bai, X., Zhang, C. et al. (2022). Oriented BN/silicone rubber composite thermal interface materials with high out-of-plane thermal conductivity and flexibility. *Composites Part A: Applied Science and Manufacturing* 152: 106681.

20 Xu, X., Hu, R., Chen, M. et al. (2020). 3D boron nitride foam filled epoxy composites with significantly enhanced thermal conductivity by a facial and scalable approach. *Chemical Engineering Journal* 397: 125447.

21 Li, J., Li, F., Zhao, X. et al. (2020). Jelly-inspired construction of the three-dimensional interconnected BN network for lightweight, thermally conductive, and electrically insulating rubber composites. *ACS Applied Electronic Materials* 1661–1669.

22 Wei, B., Chen, X., and Yang, S. (2021). Construction of a 3D aluminum flake framework with a sponge template to prepare thermally conductive polymer composites. *Journal of Materials Chemistry A* 9: 10979–10991.

23 Zhang, F., Feng, Y., Qin, M. et al. (2019). Stress controllability in thermal and electrical conductivity of 3D elastic graphene-crosslinked carbon nanotube sponge/polyimide nanocomposite. *Advanced Functional Materials* 29: 1901383.

24 Liu, C., Wu, W., Wang, Y. et al. (2021). Silver nanoparticle-enhanced three-dimensional boron nitride/reduced graphene oxide skeletons for improving thermal conductivity of polymer composites. *ACS Applied Polymer Materials* 3: 3334–3343.

25 Wu, B., Qian, G., Yan, Y. et al. (2022). Design of interconnected carbon fiber thermal management composites with effective EMI shielding activity. *ACS Applied Materials & Interfaces* 14: 49082–49093.

26 Li, X., Li, Y., Alam, M.M. et al. (2020). Enhanced through-plane thermal conductivity in polymer nanocomposites by constructing graphene-supported BN nanotubes. *Journal of Materials Chemistry C* 8: 9569–9575.

27 Li, X., Wu, B., Li, Y. et al. (2021). Construction of oriented interconnected BNNS skeleton by self-growing CNTs leading high thermal conductivity. *Advanced Materials Interfaces* 8: 2001910.

28 Wang, Y., Zhang, X., Ding, X. et al. (2021). Enhanced thermal conductivity of carbon nitride-doped graphene/polyimide composite film via a “deciduous-like” strategy. *Composites Science and Technology* 205: 108693.

29 Ruan, K., Guo, Y., Lu, C. et al. (2021). Significant reduction of interfacial thermal resistance and phonon scattering in graphene/polyimide thermally conductive composite films for thermal management. *Research (Washington DC)* 2021: 8438614.

30 Dai, W., Lv, L., Lu, J. et al. (2019). A paper-like inorganic thermal interface material composed of hierarchically structured graphene/silicon carbide nanorods. *ACS Nano* 13: 1547–1554.

31 Mehra, N., Mu, L., Ji, T. et al. (2018). Thermal transport in polymeric materials and across composite interfaces. *Applied Materials Today* 12: 92–130.

32 Li, Y., Li, X., Alam, M.M. et al. (2020). Incorporating Ag nanowires into graphene nanosheets for enhanced thermal conductivity: implications for thermal management. *ACS Applied Nano Materials* 3: 6061–6070.

33 Wu, B., Li, Y., Chen, W. et al. (2022). A dual non-covalent bonding constructed continuous interfacial structure for reducing interfacial thermal resistance. *Journal of Materials Chemistry A* 10: 13858–13867.

9

Market and Future Prospects of High Thermal Conductivity Composite Materials

Chen Hongda and Zhang Xu

Institute of Semiconductor, Chinese Academy of Sciences, Haidian District, Beijing 100083, China

9.1 Basic Concept of Composite Materials

With the advent of the 5G era, electronic integration technology has developed rapidly. Electronic equipment has gradually developed from large electronic devices to miniature electronic components. Highly integrated modern electronic products have put forward higher requirements for the heat dissipation performance of materials. During the use of electronic equipment, the gradually accumulated heat is not only detrimental to the maintenance of the original power of the equipment but also seriously affects the service life and reduces the work efficiency. Thermal conductive materials can solve the problem of heat accumulation and give full play to the effect of effective heat dissipation and temperature improvement, so it has a great impact on electronic components.

9.1.1 The History of Composite Materials

Humans have used composites for thousands of years. In 3400 BC, the first man-made composites were engineered by the Mesopotamians in Iraq. The ancient society glued wood strips on top of each other at different angles to create plywood. Following this, in around 2181 BC, the Egyptians started to make death masks out of linen or papyrus soaked in plaster. Later on, both of these societies started to reinforce their materials with straw to strengthen mud bricks, pottery, and boats [1].

In 1200 AD, the Mongols began to engineer composite bows which were incredibly effective at the time. These were made out of wood, bamboo, bone, cattle tendons, horn, and silk bonded with pine resin.

Following the industrial revolution, synthetic resins started to take a solid form by using polymerization. In the 1900s, this newfound knowledge about chemicals led to the creation of various plastics such as polyester, phenolic, and vinyl. Synthetics then started to be developed. Bakelite was created by the chemist Leo Baekeland. The fact that it did not conduct electricity and was heat resistance meant it could be widely used across many industries.

Thermal Management Materials for Electronic Packaging: Preparation, Characterization, and Devices, First Edition. Edited by Xingyou Tian.

The 1930s was an incredibly important time for the advancement of composites. Glass fiber was introduced by Owens Corning who also started the first fiber-reinforced polymer (FRP) industry. The resins engineered during this era are still used to this day and, in 1936, unsaturated polyester resins were patented. Two years later, higher-performance resin systems became accessible.

The first carbon fiber was patented in 1961 and then became commercially available. Then, in the mid-1990s, composites were starting to become increasingly common for manufacturing processes and construction due to their relatively cheap cost compared to materials that had been used previously.

The composites on a Boeing 787 Dreamliner in the mid-2000s substantiated their use for high-strength applications.

9.1.2 The Introduction of Composite Materials

Composite material is the combination of two materials with different physical and chemical properties. These constituent materials have notably dissimilar chemical or physical properties and are merged to create a material with properties unlike the individual elements.

Since composite materials can be manufactured into almost any shape, they allow great design flexibility and offer reduced parts count for articles. The opportunity to select the constituents, tailor them to obtain the required properties, and then through design make the optimum use of the properties is a situation that makes composites very attractive to many industries.

The raw materials of composite materials include matrix materials and reinforcing materials. Among them, matrix materials mainly come from metal materials, ceramic materials, and polymer materials, which often appear in the form of continuous phases in composite materials; reinforcing materials mainly include various high-performance fibers, whiskers, and particles.

9.1.3 The Application of Composite Materials

The aerospace and military markets are the two areas that have accounted for the largest effort in the development and advancement of composite technology. The need for stronger, stiffer, and lighter structures was an opportunity for composite materials to demonstrate their superiority over more commonly used materials. Durability and low maintenance are additional assets. These increase the service life and reduce the cost of maintaining systems. The development of new and the improvement of existing fabrication processes have brought about a reduction in manufacturing costs. There have been reductions in the number of parts required to construct some components by using molding and composite materials. The unique features of composites have enabled designers to formulate advanced systems that could be made only of composite materials. New military aircraft almost exclusively utilize advanced composites for structure. Rocket motor cases, nozzles, and nose cones are missile applications. Radar domes, rotor blades, propellers, and many secondary structure components such as fairings, doors, and access panels are also fabricated from advanced composites. Numerous pressure vessels, armaments, and items of space hardware are made of selected composite materials.

Nowadays, many composites are made for functions other than simply improved strength or other mechanical properties [2]. Many composites are tailored to be good conductors or insulators of heat or to have certain magnetic properties; properties that are very specific and specialized but also very important and useful. These composites are used in a huge range of electrical devices, including transistors, solar cells, sensors, detectors, diodes, and lasers as well as to make anti-corrosive and anti-static surface coatings.

Composites made from metal oxides can also have specific electrical properties and are used to manufacture silicon chips that can be smaller and packed more densely into a computer. This improves the computer's memory capacity and speed. Oxide composites are also used to create high-temperature superconducting properties that are now used in electrical cables. The preparation procedure for EG/PEI composites is shown in Figure 9.1.

Carbon fiber composites are light and much stronger than glass fibers but are also more expensive. Of the two, graphite fibers are cheaper and easier to produce than carbon nanotubes. They are used in aircraft structures and in high-performance sporting equipment like golf clubs, tennis rackets and rowing boats, and are increasingly being used instead of metals to repair or replace damaged bones [3].

Even stronger (and more costly) than carbon fibers are threads of boron. Nanotubes of boron nitride have the additional advantage of being much more resistant to heat than carbon fibers. They also possess piezoelectric qualities, which means they can generate electricity when physical pressure is applied to them, such as twisting or stretching.

Polymers can also be used as the reinforcement material in composites. For example, Kevlar, originally developed to replace steel in radial tires but best known for its use in bullet-proof vests and helmets, is a polymer fiber that is immensely

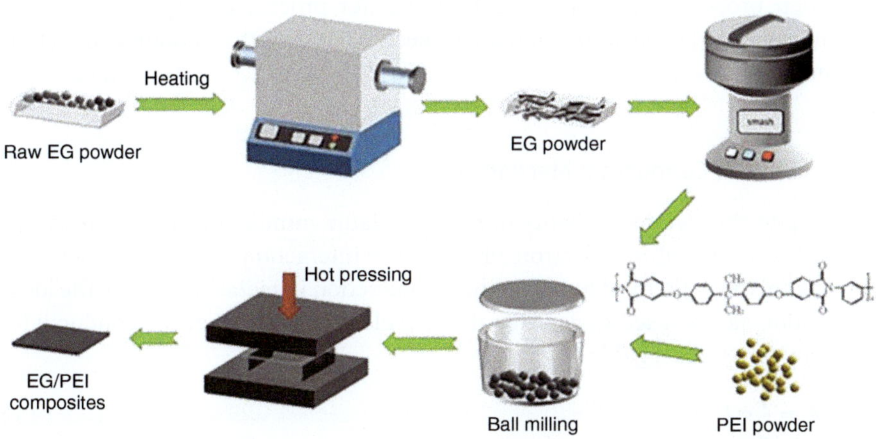

Figure 9.1 Preparation procedure for EG/PEI composites. Source: Reproduced with permission from Junyu et al. [2]; © 2022 Chemical Journal of Chinese Universities.

strong and adds toughness to a composite. It is used as the reinforcement in composite products that require lightweight and reliable construction (e.g., structural body parts of an aircraft). Even stronger than Kevlar is a substance made from a combination of graphene and carbon nanotubes.

9.2 Thermal Conductivity Mechanism and Thermal Conductivity Model

With the rapid development of the electronic information industry, electronic devices require miniaturization and high-density integration, and there are higher and higher requirements for the performance of substrates and packaging materials, such as high thermal conductivity, low thermal expansion coefficient, and lower dielectric constant. After research and verification, polymer composite materials can effectively meet the above conditions [4]. Polymers and polymer-based composite materials are the most common materials in the electronics industry at present, and their main performance advantages are shown in corrosion resistance, durability, lightweight, and lower production costs. However, due to its poor thermal conductivity, polymers and their composite materials are often not widely used in some specific applications. Therefore, seeking high thermal conductivity and high-performance composite materials has become a development trend in the electronics industry. This paper will give an overview of the composite materials used in the electronic industry from three aspects: thermal conductivity model, thermal conductivity mechanism, and methods to improve thermal conductivity.

For solid materials, the thermal conduction mechanism of solid materials can be divided into photon thermal conduction, electronic thermal conduction, and phonon thermal conduction according to the different thermal conduction carriers. The photon thermal conduction mechanism is mainly through the radiation conduction process of photons. The heat transfer process is very weak, so this paper focuses on the electron conduction mechanism and the phonon conduction mechanism [5].

9.2.1 Electron Conduction Mechanism

According to the electronic theory of metals, a large number of free electrons can be regarded as a metal-free electron air mass. The interaction and collision between electrons are the main mechanisms of heat conduction in metal crystals, so the ideal gas equation can be used to describe the free electrons. The thermal conductivity expression for an ideal gas is

$$K = \frac{1}{3}cvl \tag{9.1}$$

where c is the heat capacity of electrons per unit volume, l is the mean-free path of electrons, and v is the average speed of electron motion.

9.2.2 Phonon Heat Conduction Mechanism

In non-metallic materials, there are no free electrons, and electrons are bound and cannot become heat conduction carriers. The heat conduction process at this time is mainly transmitted by lattice waves of lattice vibration. According to quantum theory, the energy of lattice vibration is quantized, and the quantum excited by lattice wave is usually called phonon. Phonons are not real quasi-particles but the excitation unit of the collective motion state of lattice atoms. The interaction between lattice vibration waves and matter is understood as the collision between phonons and matter, and the process of lattice vibration waves being scattered by the propagation of lattice vibration waves. It can be understood as the collision between phonons and phonons, and between phonons and grain boundaries, lattice defects, etc.

9.2.3 Thermal Conduction Mechanism

In thermodynamic systems, when the temperature in a single object or system varies, heat will flow from the part with higher temperature to the part with lower temperature, which is a phenomenon of heat conduction. This phenomenon of heat conduction is essentially a process of energy propagation, and the process of energy transfer in solids can be realized in three forms: free electrons, lattice vibration waves, and electromagnetic radiation. Therefore, the process of heat conduction in solids can be realized by the movement of free electrons, phonons, and photons. The carriers of the heat transfer process are different for materials with different properties; for insulators, the carrier of heat transfer is phonons, and for pure metals, the carrier is electrons.

Polymer composites mainly rely on lattice vibration to conduct heat conduction, that is, rely on phonons to conduct heat transfer. When the heat is transferred to the polymer molecular chain, the vibration of the atoms is caused, and the heat can be transferred.

At present, there are three main theories of thermal conductivity mechanism of filled thermally conductive polymer materials: thermal conduction path theory, percolation theory, and thermoelastic coefficient theory.

At present, the main point of view on the thermal conduction mechanism of polymer matrix composites is the "path theory," which believes that the thermal conductivity of polymer matrix composites is analogous to the electrical conduction process, and the thermal conductivity of polymer matrix composites can be. The increase is primarily in the ability of the filler phase to form thermally conductive pathways or networks in the matrix.

The thermal conductivity of composite materials depends on the synergy between the matrix and the filler. When the filler content is low, the polymer matrix is a continuous phase, and the filler is a dispersed phase. The matrix and the filler form an island structure, and the filler is surrounded by the polymer matrix. After the coating, with the addition of fillers, the fillers contact each other to form a local heat conduction network chain. If the filler is continued to be added, the local heat-conducting network chains are connected and penetrated, and the matrix and the fillers become continuous phases. However, this mechanism cannot explain the percolation behavior that thermal conductivity does not increase sharply with the addition of thermally conductive fillers. The percolation theory is similar to the percolation phenomenon of filled conductive polymer materials, but this theory is controversial. The percolation theory cannot explain the phenomenon that the thermal conductivity decreases with the decrease of the particle size of the filler, and it is difficult to observe the percolation phenomenon with ordinary fillers. The percolation phenomenon can only be observed for filler particles with extremely high thermal conductivity. The electrical conductivity of the polymer near the percolation transition point increases sharply, while the thermal conductivity of the polymer is not large. The electrons are due to the tunnel effect. It can pass through the polymer barrier to conduct electricity, but phonons cannot. There is a huge interfacial thermal resistance between the filler and the polymer matrixes. Even if the filler content is above the percolation transition point, the thermal conductivity is still very low. The thermoelastic coefficient theory regards thermal conductivity as the thermoelastic coefficient of the phonon heat transfer process, but the thermal conductivity of composite materials is not a path-dependent property, but a macroscopic property that depends on the overall properties of the material. The thermal conductivity enhancement of the composite system by thermally conductive fillers follows the combined enhancement mechanism of the overall thermos-elasticity. Similar to vibration and wave, the phonons transferring thermal vibration will also reflect, refract, interfere, and block at the interface of the two phases with different thermal conductivities. The phenomenon causes phonon scattering [6]. If there are defects between the matrix and the filler, the scattering of phonons at the interface will be more serious. Therefore, the interface between the matrix and the filler in the composite material will reduce the thermal conductivity of the material. Schematic illustration of interface hear transfer in interfacial layer of composite material is shown in Figure 9.2.

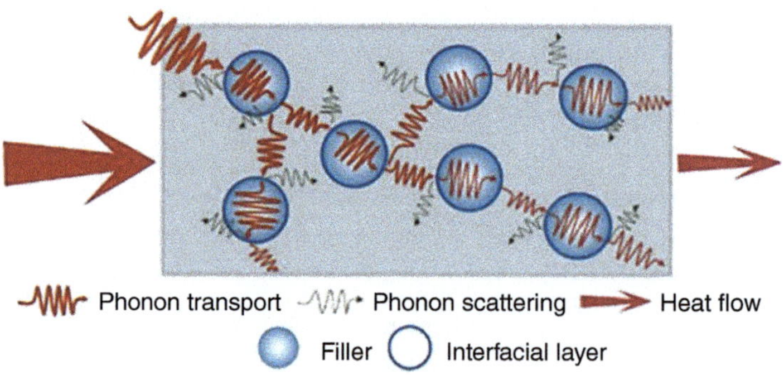

Figure 9.2 Schematic illustration of interface hear transfer in interfacial layer of composite material. Source: Reproduced with permission from Xiaze et al. [3]; © 2022 Journal of Composite Materials.

9.2.4 Thermal Conductivity Model

There are many models for estimating thermal conductivity of polymer nanocomposites based on effective medium theory. Similar thermal percolation behavior of conductive polymers at the electrical percolation threshold has also been observed from polymer nanocomposites, but the thermal properties are quite different from the electrical properties, and the electrical properties of the polymers are so poor that they can be ignored, only the conductive properties. Fillers are good for conducting electricity, while in the thermal case, there is no absolute thermal insulator, and the polymer matrix cannot be ignored. The basic reason for this behavior is that the thermal resistance of the interface or the intrinsic thermal conductivity of the composition is related to geometrical factors and the thermal overrun. There is no clear evidence of infiltration behavior. Therefore, in the heated debate, whether the reinforcement of such composites should be explained according to thermal percolation or effective medium theory remains a question. The current thermal conductivity models for polymer composites mainly include Maxwell–Eucken model, Y. Agari model, Bruggeman model, Every model, and Hamilton–Crosser model [7].

For the thermal conduction model of filled composite materials, many scholars have proposed different thermal conduction models to describe and simulate the thermal conduction process, such as Maxwell's theoretical model, Bruggeman theoretical model, Nielsen theoretical model, and Cheng–Vochen theoretical model. Among them, the earliest and most widely used is the Maxwell theoretical model. The Maxwell model can be described by the following formula:

$$\lambda_c = \lambda_m \left[\frac{2v_f\left(\frac{\lambda_f}{\lambda_m} - 1\right) + \left(\frac{\lambda_f}{\lambda_m} + 2\right)}{v_f\left(1 - \frac{\lambda_f}{\lambda_m}\right) + \left(\frac{\lambda_f}{\lambda_m} + 2\right)} \right] \tag{9.2}$$

where λ_c is the thermal conductivity of the composite material, λ_f is the thermal conductivity of the uniformly distributed filling phase, v_f is the spherical filling phase volume fraction, and λ_m is the thermal conductivity of the continuous matrix.

This model assumes that the filling phase has a spherical shape, and the volume fraction in the matrix is not large (dilution medium), but usually some scholars also apply the Maxwell model to the case where the filling volume of the filling phase is large. In addition, the thermal resistance at the interface between the matrix and the filler phase and the non-spherical filler phase particles is ignored in Maxwell's theoretical model, which is why the thermal conductivity predicted by this model is generally larger than the actual measured thermal conductivity. The thermal resistance at the interface between the matrix phase and the filling phase in the composite material is mainly derived from the poor adhesion between the two phases and the difference in the linear thermal expansion coefficient between the two materials. This thermal resistance is called Kapitsa thermal resistance [8].

Hasselman improved the Maxwell theoretical model to include the thermal resistance of the interface between the filling phase and the matrix:

$$\lambda_c = \lambda_m \frac{2\left(\dfrac{\lambda_f}{\lambda_m} - 1 - \dfrac{\lambda_f}{r \cdot G}\right) v_f + \dfrac{\lambda_f}{\lambda_m} + 2 + 2\dfrac{\lambda_f}{r \cdot G}}{\left(1 - \dfrac{\lambda_f}{\lambda_m} + \dfrac{\lambda_f}{r \cdot G}\right) v_f + \dfrac{\lambda_f}{\lambda_m} + 2 + 2\dfrac{\lambda_f}{r \cdot G}} \tag{9.3}$$

where r is the radius of the spherical filling phase particle and G is the thermal conductivity at the boundary of the two phases.

In order to be able to take into account the shape of the filler phase particles in the composite material, Hamilton and Crosser further improved the Maxwell model:

$$\lambda_c = \lambda_m \left(\frac{\lambda_f + (n-1)\lambda_m + (n-1)v_f(\lambda_r - \lambda_m)}{\lambda_f + (n-1)\lambda_m - v_f(\lambda_f - \lambda_m)}\right) \tag{9.4}$$

where n is a factor that varies with the shape of the filler phase particles, $n = 1/\Psi$, and Ψ is the filler phase particle, the sphericity of $\Psi = 1$ for spherical particles.

In addition to the Maxwell model, in 1935, Bruggeman predicted the thermal conductivity of heterogeneous composite materials and proposed the Bruggeman theoretical model:

$$1 - v_f = \frac{\lambda_f - \lambda_c}{\lambda_f - \lambda_m}\left(\frac{\lambda_m}{\lambda_c}\right)^{1/3} \tag{9.5}$$

ang.J improved the Bruggeman theoretical model and applied it to aluminum nitride/polyimide. In the prediction of thermal conductivity of amine composites, the shape of the filler particles and the thermal resistance at the boundary of the composite are comprehensively considered in the improved model. The model can be described by the following formula:

$$(1 - v_d) = \left(\frac{\lambda_m}{\lambda_c}\right)\left(\frac{\lambda_c - \lambda_f(1-\beta)}{\lambda_m - \lambda_f(1-\beta)}\right)^n \tag{9.6}$$

where $\beta = RK/R$, R is the radius of the filling phase particle, RK is the Kapitsa radius, $RK = Rb \times \lambda_m$; Rb is the thermal resistance at the boundary, and vd is the volume of the spherical particle fraction.

9.3 Composite Materials in Electronic Devices

With the rapid development of electronic and communication technology, the use of high-performance chips and large-scale and ultra-large-scale integrated circuits is becoming more and more extensive. The integration level, packaging density, and operating frequency of electronic device chips continue to increase, while the volume is gradually shrinking (for example, the feature size of microprocessors has been reduced from 0.35 μm to 0.18 μm from 1990 to 2000), which makes the chip. The heat flux density increases rapidly. Because high temperature will have a harmful effect on the performance of electronic components, for example, excessive temperature will endanger the junction of the semiconductor, damage the connection interface of the circuit, increase the resistance of the conductor, and cause mechanical stress damage. As the temperature increases, the failure rate increases exponentially, and even for some devices, the failure rate more than doubles when the ambient temperature increases by 10 °C, which is called the 10 °C rule. According to statistics, 55% of the failure rate of electronic equipment is caused by the temperature exceeding the specified value. At the same time, most electronic chips have low calorific value on standby and large calorific value during operation, and the temperature rises rapidly in an instant. This requires substrates and packaging materials with increasingly superior properties, such as high thermal conductivity, low coefficient of expansion [9], and low dielectric coefficient and thermal stability. Although metals have high thermal conductivity, it is difficult to match the thermal expansion coefficients of silicon and ceramic substrates for circuits. Therefore, the development of thermally adaptive composite materials plays an important role in the development and application of microelectronic devices.

Heat-adaptive composite material (also known as heat-matching composite material) is a kind of composite material that has the required thermal conductivity or thermal expansion coefficient through the selection and arrangement and orientation of the components and their contents.

9.3.1 Electronic Heat Dissipation and Thermal Adaptation Materials

Electronic devices and chips need to work under certain temperature conditions, and at the same time, electronic devices will generate a certain amount of heat during operation, making their temperature rise continuously. The normal operating temperature range of electronic devices is generally −5 to 65 °C, and the maximum allowable operating temperature is 100 to 120 °C. Too high or too low temperature will reduce the performance of electronic devices and even cause device damage. To control electronic products to work within a certain temperature range [10], it is necessary to preheat and dissipate heat for some overheated chips or overheated device surfaces. With the rapid development of electronic technology, the high frequency and high speed of electronic devices and the density and miniaturization of large-scale integrated circuits make the heat generation per unit volume of electronic devices increase rapidly. Taking microelectronic chips as an example, it has generally reached 60–90 W/cm^2 at present, and the highest is more than 200 W/cm^2. The heat dissipation technology of electronic devices has increasingly become a very critical technology in the development and research of electronic products. The heat dissipation performance of electronic devices is good or bad. It directly affects the reliability and performance of electronic products. Mitllal's research results show that the failure rate of electronic components can be reduced by 4% when the temperature of electronic components is lowered by 1 °C above the normal operating temperature level; if the temperature is increased by 10–20 °C, the failure rate is increased by 100%. However, the traditional convective heat transfer method and forced air cooling method relying on single-phase fluid are difficult to meet the heat dissipation requirements of many electronic devices. Therefore, new heat dissipation technologies must be researched and developed to meet the requirements of high heat flux density. In recent years, the application of heat-adaptive composite phase change materials to heat dissipation technology of electronic components has received extensive attention abroad and has been used in high-tech systems and military equipment such as aviation, aerospace, and microelectronics.

9.3.2 Preparation and Application of Thermally Adaptive Composites

Different from the traditional heat dissipation method, the heat dissipation technology based on the fast thermal response heat storage material requires not only the high heat storage density of the phase change material but also the material with high thermal conductivity and fast heat transfer rate. The main measure to improve the thermal conductivity of composites is to add fillers with good thermal conductivity, such as metals (such as Cu and Al), ceramics (such as AlN, $A1_2O_3$, Si_3N_4, and BN), graphite, or highly graphitized carbon fibers, into the composite matrix [11]. Wait. Although inorganic substances, including various low melting point alloys, have large heat storage density and high thermal conductivity relative to organic substances, inorganic substances generally have the phenomenon of supercooling and phase transition separation during the phase transition process, and inorganic substances also have high thermal conductivity. There is a problem of corrosiveness, which limits its application. Table 9.1 shows thermophysical properties of the common electronic packaging materials.

At present, in the field of heat dissipation technology of electronic components, organic substances are generally used as phase change materials, and paraffin wax, which has high heat storage density, stable performance, and low cost, is most widely used [12]. However, the low thermal conductivity of paraffin severely restricts the heat transfer performance of the heat storage device. Velraj proposed to add high thermal conductivity materials such as aluminum powder, copper powder, graphite powder, and foamed aluminum rate. However, since the density of powder materials is usually greater than that of paraffin, it is difficult to form a uniform phase change

Table 9.1 Thermophysical properties of the common electronic packaging materials.

Material	Conductivity W/(m K)	CTE (25–400 °C)/ $\times 10^{-6}$ K	Density (g/cm^3)
Si	150	4.1	2.30
GaAs	39	5.8	5.30
Al_2O_3	21	7.3	3.70
BeO	250	6.7–8.0	2.90
AlN	70–260	3.5	3.26
SiC	60–200	3.5	3.18
Kovar	17	5.2	8.10
Inovar	11	0.4	8.04
W/Cu	200	7.0	15.70
Mo/Cu	398	17.8	10.00

material during compounding; on the other hand, there is no direct force between powder (including metal foam) and paraffin, and in the region of its melting temperature, paraffin will. Its pores or powder particles are separated. Some scholars have also proposed to set a fin structure in the heat storage device of the heat dissipation system to improve the heat transfer rate by increasing the heat exchange area. However, since the thermal resistance of the phase change material itself has not changed, the improvement is limited. In addition, the above two methods will lead to the reduction of the heat storage efficiency and the increase of the mass and volume of the heat storage device of the heat dissipation system. In addition, paraffin is a solid–liquid phase change material. In order to prevent the liquid flow and leakage of paraffin during phase change, some scholars have also proposed a scheme to prepare microcapsule phase change heat storage materials, that is, to wrap the phase change material in microcapsules. In the shell, during the phase change process of the microcapsule phase change material, the inner core undergoes a solid–liquid phase transition, while the outer polymer film remains solid, so this type of phase change material appears as solid particles macroscopically. HaWlader et al. used paraffin as the phase change material, gum arabic and gelatin as the capsule materials, formaldehyde and ethanol as the solvent, and the complex coagulation technology to prepare the capsule-type phase change material. The larger the substance ratio of coating material to paraffin, the greater the fraction of paraffin encapsulated in the capsule; the hydrophilicity of the capsule paraffin phase change material mainly depends on the ratio of paraffin to the substance of the coating material. The larger the ratio, the lower the hydrophilicity. The prepared encapsulated paraffin showed a large storage (release) heat density; the thermal cycle test showed that even after 1000 thermal cycles, the encapsulated paraffin could still maintain its structural shape and heat storage density unchanged [13]. However, the shell of the microcapsule is usually a polymer material, and its thermal conductivity is very low, which reduces the heat transfer performance between the heating electronic device and the heat dissipation system (Ye Hong et al). The method of polyethylene (HDPE) solves its shaping problem, using paraffin as the organic phase change heat storage material, and blending and melting paraffin and high-density polyethylene with a higher melting point at a temperature higher than their melting point and then cooling to high density. Below the melting point of polyethylene, high-density polyethylene first solidifies and forms a spatial network structure, and the liquid paraffin is bound in it to form a fixed phase change paraffin, but HDPE also has the problem of low thermal conductivity. The above methods restrict the application of phase change heat storage materials in the field of heat dissipation technology for electronic devices [14].

Recently, the scheme of using capillary action to adsorb phase change materials into porous matrices to prepare composite phase change materials has attracted the attention of researchers. Using inorganic matter with a microporous structure with a large specific surface area as a supporting material, liquid organic or inorganic phase change heat storage materials (above the phase transition temperature) are sucked into the micropores through the capillary force of the micropores to form organic/inorganic or inorganic/organic composite phase change heat storage materials. In this composite phase change heat storage material, when the organic or inorganic phase change heat storage material undergoes a solid–liquid phase transition in the micropores, due to the capillary adsorption force, the liquid phase change heat storage material is difficult to remove from the micropores: overflow in the hole. There are many kinds of porous media with abundant pore spaces, which are ideal storage media for phase change materials. Alternative porous media include gypsum, expanded clay, expanded perlite, expanded shale, porous concrete, and the like. Using porous media as the encapsulation material of the phase change, material can make the composite material have the advantages of structure-function integration, save space in application, and have good economy. The pores inside the porous medium are very small, and the storage reliability of the phase change material in the porous medium can be improved by means of the capillary effect. The porous medium also disperses the phase change material into fine individuals, which effectively improves the heat transfer efficiency of the phase change process [15].

Electronic materials used in packaging materials, conductive frames, semiconductor support plates, etc. not only need to have high thermal conductivity but also need to have a low coefficient of thermal expansion (CTE). In recent years, a large number of such electronic materials have been developed. The general design idea is to use the high thermal conductivity of metals (alloys) such as Cu and Al and the low expansion of materials such as A1203, SiC, AlN, FeNi, w, and Mo, using various composite processes to manufacture composite materials with both high thermal conductivity and low expansion. Controlled thermal expansion composite materials have a wide range of applications in the field of electronics and can also be used in automotive, aviation, and other fields. Electronic packaging materials are required to have good thermal conductivity and coefficient of thermal expansion (CTE) that matches semiconductor materials or ceramic substrate materials such as $A1_2O_3$ and AlN, and high mechanical strength and low density.

TanF·L is filled with an alkane phase change material in the heat storage unit (HSU, heat storage unit) of electronic devices and uses the latent heat of solid–liquid phase transition to absorb heat from the chip and keep the chip at a low temperature of about 50 °C [16]. This heat storage unit (HSU) is used as a heat sink in electronic components. Solid PCM material is stored in the cavity. When the PCM absorbs heat, its state becomes liquid; in heat dissipation, it can return to solid state. This temporary thermal management can generally last about 2 hours, and its general cooling power can be about 30 W.

9.4 Thermal Functional Composites

Thermal functional composites can be roughly divided into three categories: thermally conductive composites, heat-resistant composites, and heat storage composites. With the rapid development of ultra-large-scale power transmission and transformation technology and microelectronics integration technology and assembly technology, high-power heating has become a restrictive factor in its service life, and the development of thermally conductive functional composite materials with excellent comprehensive performance has become a research hotspot: heat-proof composite materials. It has major applications in the development of aerospace science and technology and aerospace engineering such as manned spaceflight, lunar exploration and deep space exploration, and hypersonic vehicles; heat storage composite materials are one of the important research fields to improve energy utilization efficiency and protect the environment, and become the future development of green energy. It has a wide range of application prospects in the fields of solar energy utilization [17], "peak shifting and filling valley" of electric power, recycling and utilization of waste heat and waste heat, and heating of industrial and civil buildings.

9.4.1 Thermally Conductive Composites

9.4.1.1 Review of the Latest Research Progress

High thermal conductivity polymer matrix composites play an important role in heat transfer, cooling, and improving service life in many manufacturing and high-tech fields such as smart grid equipment and technology, microelectronics, and aerospace, and have become a current research hotspot. At present, the research on thermally conductive polymer composites at home and abroad mainly focuses on how to improve the thermal conductivity under low filling. In the industrial field, adding 55–70 vol% fumed silica to epoxy resin has achieved industrial production. Thermal conductivity is also related to the relative content, morphology, distribution, and interaction of each component, which is the focus of future research on improving thermal conductivity.

9.4.1.2 Comparative Analysis at Home and Abroad

In the late 1970s, foreign countries began the research of thermally conductive polymer materials, and by the mid-1980s, began to research thermally conductive insulation composites. Since then, foreign countries have high quality in research products, and the technology is generally at the international leading level. At the same time, the industrialization process of the research has been advancing, and some large group companies have formed some mature products, such as DSM, Albis, CoolPolymer, Ovation Polymers, and other companies [18]. DSM of the Netherlands has launched various high-performance thermally conductive products with PA46 as the matrix, which are mainly used in heat dissipation shells, radiators, and LED plastic packaging. China's research on polymer thermal conductive insulation composites has expanded to numerous types of thermally conductive insulation composites, and the matrix is mainly PA class, polyester class, PC, and other materials but also the choice of PPS, ABS, and other mechanical properties of better materials.

9.4.2 Heat-Resistant Composite Materials

9.4.2.1 Review of the Latest Research Progress

Thermal protection composites are key materials for the design and development of advanced thermal protection systems, which have a non-underestimated role in new power systems and loaded vehicles, reusable launchers, etc., and are one of the determinants of the advanced and reliability of the related power systems and vehicles. In thermal protection systems, thermal protection composites play functions such as thermal insulation, load bearing, and even wave-transparent and anti-destructive, involving multiple thermal responses of materials in special environments and high-temperature evolution of multi-component multi-scale structures. Therefore, the research on thermal protection composites has significant multidisciplinary crossover characteristics. High-performance, lightweight, multi-functional, low-cost, interdisciplinary basic research, and innovative research on material systems and preparation methods are the current research hotspots of thermal composite materials [19].

9.4.2.2 Comparative Analysis at Home and Abroad

Along with the development of aerospace technology, China's research and application of heat-proof composite materials has made a series of important achievements and breakthroughs, forming a comprehensive technical system covering applied basic research, product development, and quality assurance, making great contributions to China's aerospace development. In recent years, new breakthroughs have been made in the ablative heat-proof composite spalling control, carbon/carbon composite oxidation resistance mechanism, thermal wave-transparent mechanism, and basic research on thermal wave-transparent materials. Taking woven carbon1 phenolic composites as the object, we have revealed the ablative exfoliation mechanism across the sublimation zone and mastered the exfoliation control technology. Carbon/carbon composites exhibiting trace ablation above 2000 were developed, and the ablation resistance under simulated conditions was improved by more than 50% compared with pure carbon1carbon. The basic research and material development of high-temperature and high-efficiency thermal insulation composites have been carried out, and the typicality of rigid thermal insulation tiles and nano-super thermal insulation materials based on microstructure and defect control can reach the advanced level of similar materials abroad. Foreign in the development of low and medium-density carbon 1 phenolic composites to reduce the mass ratio of space probe anti-heat structure, successfully developed PICA (Phenolic Impregnated Carbon Ablator) material and applied to the heat shield of stardust returner. NASA Greene Research Center is studying enhanced SiO_2–Al_2O_3 composite aerogel systems to explore the potential application of such materials in vehicle thermal protection systems, and NASA Marshall Space Center is also exploring the new system of aerogel materials in the field of high-temperature thermal insulation [20].

9.4.3 Thermal Storage Composites

9.4.3.1 Review of the Latest Research Progress

Thermal storage materials are widely used in various fields such as electric power, electronics, and energy. It can not only recover and make secondary use of industrial waste heat and waste heat to reduce environmental pollution but also realize energy saving and emission reduction and replace non-renewable energy. At present, thermal storage technology includes sensible thermal storage and phase change thermal storage. In recent years, composite phase change thermal storage materials have emerged, which can effectively overcome the shortcomings of single inorganic or organic phase change thermal storage materials and improve the application effect of phase change materials as well as expand their application scope. Therefore, the development of composite phase change thermal storage materials is the trend and research focus of future development.

9.4.3.2 Domestic and Foreign Comparative Analysis

In recent years, the research on phase change thermal storage has become more and more extensive and profound. Foreign countries mainly study the compatibility mechanism of phase change materials and matrix materials, find suitable materials, modify phase change materials, and improve the compatibility of phase change materials and matrix materials [21]. Research on the application of nanomaterial synthesis technology and nanoscale heat transfer technology to phase change composites can overcome the serious problems of low thermal conductivity, easy phase separation, and poor stability of previous phase change materials, and make nanostructured thermal storage materials that meet the requirements of waste heat storage at all levels that will be a promising development direction and a challenging research topic.

At present, a lot of research has been carried out on refrigeration, low-temperature, and high-temperature phase change thermal storage materials at home and abroad. Due to the wide temperature range of medium-temperature phase change materials, their material forms are spread over almost all types of phase change thermal storage materials, and they have also started to receive attention. In recent years, the development of solar thermal power generation, mobile thermal storage technology, and other related fields has created a lot of space for the application of medium-temperature phase change thermal storage materials [22].

9.4.4 Application Foresight

For thermal conductive composites, they will be mainly used in smart grid equipment and technology, especially nano-composite insulation materials for high-voltage electrical equipment, which will help to promote the construction and development of China's AC/DC UHV grid. For thermal composite materials, it will focus on solving the key technologies of long-time high-temperature and high-efficiency thermal insulation material structure control and material component manufacturing, high-temperature long-time thermal insulation, and broadband thermal wave transmission structure development, which will serve the development of China's aerospace science and technology and aerospace engineering. For thermal storage composites, mainly the application field of advanced solar thermal storage materials and other applications is the key to the development of green energy.

9.4.5 Future Forecast

In the next 10–30 years, the development plan for thermal functional composites mainly focuses on the preparation of composites with high thermal conductivity and thermal functional composites with insulation and heat resistance. According to the current research status of thermal functional composites, the research focus and difficulty lie in the preparation of thermal functional composites to meet the needs of smart grid equipment and technology, microelectronics, aerospace, and many other manufacturing industries and high-tech fields, and basically meet the needs of localization in the early stage.

The period from 2030 to 2050 is roughly divided into three phases: near-term (10 years), mid-term (20 years), and far-term (30 years).

Recently, basic research on design, preparation technology, and characterization of thermo-functional composites is carried out for the requirements of different uses, and possible applications are explored. Based on the initial thermal functional composites, we will focus on the development of thermally conductive functional composites with the average value of thermal conductivity reaching above 2.0 W/m K, starting from the selection of fillers, modification, and optimization of preparation process [23].

In the medium term, it is necessary to conduct in-depth research on the design, preparation, and characterization of thermally functional composites, and to study in depth the mechanisms of thermal conductivity and thermal storage to obtain—certain applications. It is not only to improve the thermal conductivity of the composites effectively but also to study the thermal conductivity mechanism and the establishment of thermal conductivity model. During this period, we will try to reach the world level of thermal functional composites and meet the basic needs of localization.

In the long term, we will further improve the industrial economy and leadership of thermal functional composites, form a complete system for industrial production, and further develop high-end thermal functional composites with good comprehensive performance and green environment.

9.5 The Modification of Composite Materials

In order to improve the thermal conductivity of polymers, it is the most simple and effective way to introduce thermally conductive fillers with thermal conductivity into the polymer matrix. Common thermally conductive fillers include metal-based fillers, ceramic-based fillers, and carbon-based fillers. Among all fillers, graphite has the characteristics of lightweight, low cost, good chemical stability, and excellent thermal conductivity, so it is widely used in the preparation of low-density, corrosion-resistant, and easy-to-process thermally conductive composite materials. Expanded graphite (EG) is a natural flake graphite oxidized by an oxidant. The graphite intercalation compound obtained by intercalation is a new type of functional graphite material, which is characterized by rapidly expanding to form worm-like expanded graphite after absorbing heat at high temperatures. EG not only retains the inherent heat resistance and corrosion resistance of traditional graphite but also has the characteristics of large specific surface area and high surface energy. However, after expansion, EG will change from a compact scale-like structure to a loose carbon layer and a porous worm-like structure, thus forming a thermal insulation layer. The resulting graphite not only has a large specific surface area but also has a certain thermal insulation ability, so it cannot be directly used as a filler material to prepare thermally conductive composite materials [24].

There are still a large number of graphite nanosheets (GNP) on the surface of EG after expansion. GNP has excellent thermal conductivity. As we all know, air is a poor conductor of heat, and EG's lighter density and porous structure make it have certain thermal insulation. If the pores between GNPs and GNPs in EG are made smaller, the potential of EG can be fully utilized to prepare thermally conductive composites. The toughening mechanism of particle interleaving is shown in Figure 9.3.

Theoretically, the thermal conductivity of diamond is as high as 2200–2600 W/(m K), and the coefficient of thermal expansion is about $(0.86 \pm 0.1)/\times 10^6$/K. However, pure diamond is difficult and costly to obtain. The proper method is to use diamond as reinforcement to prepare metal matrix composites. Cu is a high cost-effective alternative, whose thermal conductivity could reach 398 W/(m K). Although Cu has a higher thermal expansion coefficient, $17.5/\times 10^6$/K, it can be adjusted by adding diamond. Therefore, Diamond/copper composite is a new type of thermal management material with great potential [25].

With the rapid improvement of modern science and technology, electrical and electronic equipment and devices are further developing toward the direction of miniaturization, multi-function, and high integration. The operating frequency and heat flux density have increased sharply, resulting in a large amount of heat accumulating on the equipment, thus changing the working environment. The high

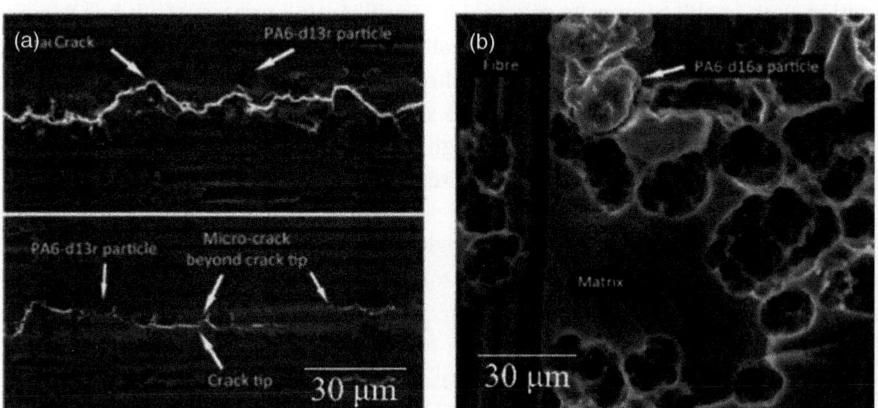

Figure 9.3 Toughening mechanism of particle interleaving: (a) Particle bridging and crack deflection; (b) microstructure of thermoplastic resin particles and thermosetting resin matrix. Source: Reproduced with permission from Wang et al. [14]; © 2020 Applied Science and Manufacturing.

temperature greatly affects the reliability, safety, and working life of microelectronic devices. Therefore, achieving efficient heat dissipation of thermal management systems has become a key issue in the development of advanced electronic devices. Epoxy resin (EP)-based composites are widely used in electronic packaging and other fields due to their excellent mechanical properties, environmental degradation resistance, wear resistance, electrical insulation, low cost, and easy processing. However, like other polymer materials, EP suffers from low thermal conductivity and poor thermal stability, which severely limits its application in thermal management systems. As we all know, it is extremely difficult to prepare a polymer matrix with high intrinsic thermal conductivity by changing the molecular structure. Therefore, researchers usually introduce high thermal conductivity fillers into EP to form a thermally conductive network structure and use the synergistic effect of the two materials to prepare: thermal interface material with high thermal conductivity and easy processing. Currently, metal-based and carbon-based materials are widely used as fillers in EP thermally conductive composites due to their excellent thermal properties, but their inherent high electrical conductivity hinders their application in the field of insulation. In contrast, the high thermal conductivity ceramic filler boron nitride (h-BN) not only has the advantages of high thermal conductivity, excellent oxidation resistance, low dielectric constant, etc. but also has a wide bandgap of 5–6 eV [26]. Electrical insulation properties, these characteristics make h-BN, have a wider application space in thermal insulation thermal management systems. Among h-BN of different dimensions, two-dimensional (2D) boron nitride nanosheets (BNNS) are a graphite-like layered material composed of two atoms of B and N alternately bonded by strong covalent bonds and arranged in six. The edge-shaped honeycomb structure and the adjacent layers are stacked and bonded by weak van der Waals forces. Because BNNS has the advantages of large specific surface area and high aspect ratio, it can effectively improve its interaction with the polymer chain and increase the thermal contact area, so that the thermal conduction path in the composite material is significantly increased, so it has become a high thermal conductivity EP composite material, one of the most promising candidates. The SEM of polyvinyl alcohol (PVA)/BNNS/PDMS composites with section perpendicular to fiber direction is shown in Figure 9.4 [27].

With the rapid development of the high-tech era, the development of thermally conductive polymer composites with excellent properties is crucial in the field

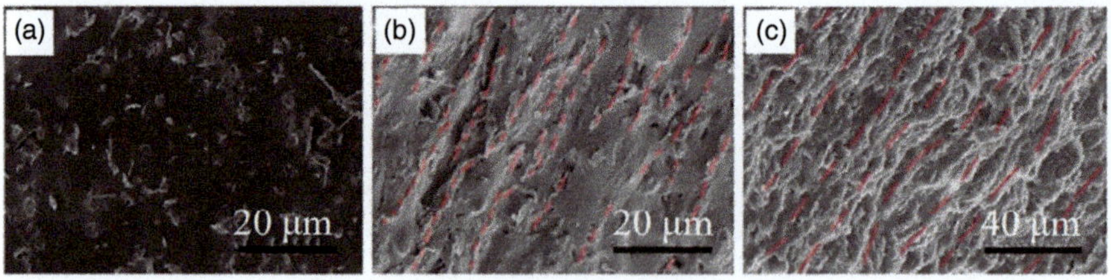

Figure 9.4 SEM of polyvinyl alcohol (PVA)/BNNS/PDMS composites with section perpendicular to fiber direction (a) and parallel to fiber direction (b, c). Source: Reproduced with permission from Chen et al. [15]; © 2017 ACS Appl Mater Interfaces.

of electronics industry. Boron nitride nanosheets (BNNS) have excellent thermal and insulating properties, and their introduction into epoxy resin (EP) matrix can make composites have a variety of excellent properties. BNNS with high specific surface area can significantly increase the thermal contact area in composites, and researchers have explored a variety of methods to prepare high-quality BNNS, including mechanical exfoliation, hydrothermal, liquid-phase exfoliation, and CVD. The modification of BNNS by covalent or non-covalent bonds is crucial to improve the interfacial compatibility between BNNS and EP. For example, silane coupling agents can link BNNS and EP through covalent bonds, thereby effectively reducing the interfacial thermal resistance in composites [28]. Combining 2D-BNNS with fillers of different dimensions or constructing 3D-BNNS thermally conductive structures can promote the formation of large thermally conductive networks. At present, a large number of studies have combined BNNS with 0D and 1D fillers, and some studies have combined three different fillers, which greatly improves the comprehensive properties of the composite material. The 3D-BNNS thermally conductive structure is usually formed by the self-assembly of the template method, in which the ice template method is widely used due to its flexibility and scalability. Some researchers have also applied the composite filler system to the ice template method to prepare a dense 3D-BNNS thermal network. The thermal conductivity of EP composites is maximized by utilizing their synergistic effects. All in all, researchers should continue to explore green processes capable of large-scale production of BNNS and conduct more in-depth studies on the modification mechanism and thermal conduction mechanism. In addition, the design and micro-control of the 3D-BNNS structure is still an important development direction for the preparation of high thermal conductivity EP composites in the future. The microstructure of SiC – GFs/Al composites is shown in Figure 9.5.

In summary, the interface plays a key role in the improvement of the thermal conductivity of the composite material, and the thermal conduction behavior of the composite material is affected by a variety of interface factors such as the number of interfaces, coupling strength, and interaction force. Although people have adopted a variety of methods to design and improve the interface, according to the existing research, no matter how high the filler content is, the thermal conductivity of the composite material has never reached or even close to the thermal conductivity of the filler itself. Therefore, for the interface structure, the research on the thermal conduction behavior of the interface remains to be further explored. In the future, it is necessary to strengthen research in the following aspects [29]:

(1) Develop interface characterization technology. Understanding the interface is a prerequisite for controlling the interface. At present, SEM, TEM, EDS, XPS, and other instrumental methods are mainly used to analyze the interface morphology and the detection of interface components. In the future, if the interaction force on the interface can be directly measured or the interface can be observed in more detail, it will undoubtedly greatly promote the interface design and interface research of composite materials;
(2) Strengthen the related theory of interface heat conduction research. Although the emergence of a new thermal conduction theoretical model considering the

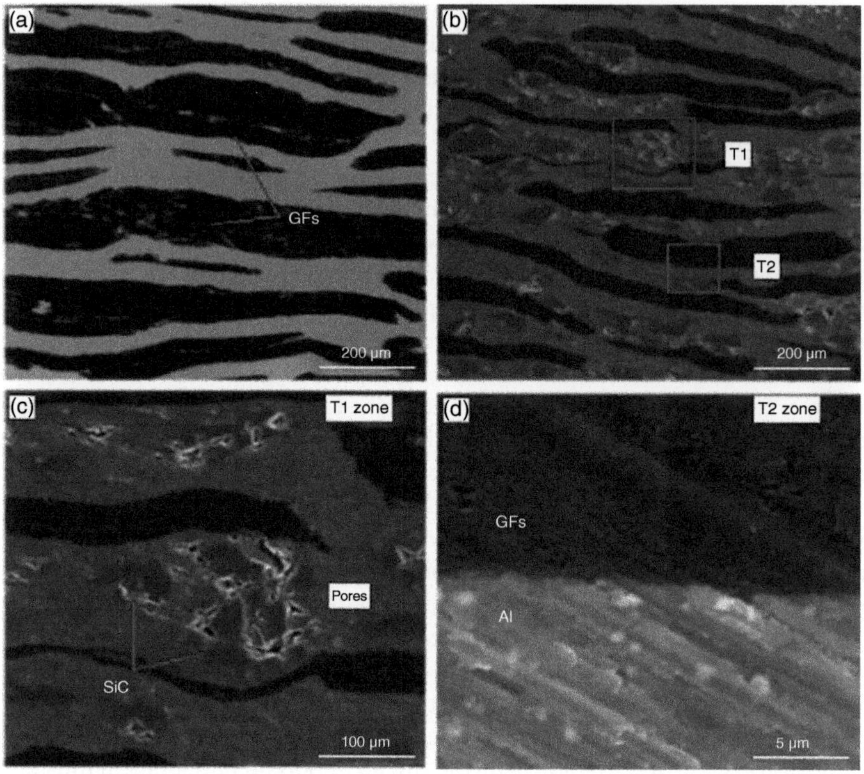

Figure 9.5 Microstructure of SiC GFs/Al composites: (a) 60Vol%Al; (b) 20Vol%SiC–40Vol%GFs/40Vol%Al; (c) T1 zone; (d) T2 zone. Source: Reproduced with permission from Fankun et al. [16]; © 2022 Journal of Composite Materials.

interface thermal resistance is closer to reality than the traditional thermal conduction model, a universal thermal conduction model has not yet appeared, and the understanding of the interface and the thermal conduction effect of the interface is still limited and needs to be further explored;

(3) The experimental measurement method of heat conduction needs to be more scientific and standardized. Existing thermal conductivity measurement methods, including laser flash method, hot wire method, and protective hot plate method, cannot avoid the influence of interface contact thermal resistance, which leads to large differences in data when measuring the same material with different measurement methods. Therefore, how to overcome the interface contact thermal resistance and reduce the experimental error is still a problem that requires great efforts to solve. The SEM images of liquid crystal epoxy resin (a–a") and SiCNWs-LCE/liquid crystal epoxy composites (b–b") is shown in Figure 9.6.

With the miniaturization, integration, and functionalization of electronic devices and the upgrading of new application technologies such as chip packaging, flexible electronic devices, and light-emitting diodes, heat dissipation has become a

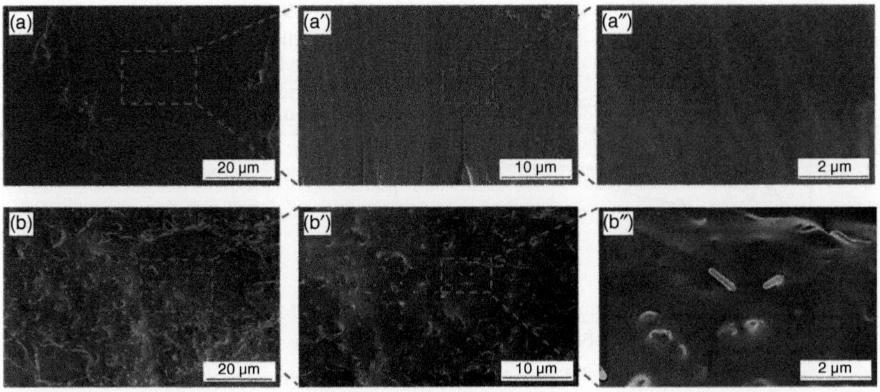

Figure 9.6 SEM images of liquid crystal epoxy resin (a–a") and SiCNWs-LCE/liquid crystal epoxy composites (b–b"). Source: Reproduced with permission from Xiaolian et al. [17]; © 2022 Journal of Composite Materials.

challenging research topic. Due to its inherent unique properties (lightweight, easy processability, excellent impact resistance, chemical resistance, and electrical insulation), polymers have set off a research boom in the field of thermal conductivity. Most polymers are poor conductors of heat with low thermal conductivity, which ranges from 0.1 to 0.5 W/(m K), which limits their wide industrial application as thermally conductive materials. Adding high thermal conductivity inorganic fillers and forming a continuous thermal network within the polymer matrix is a common strategy to improve thermal conductivity. The thermally conductive pathways increase with increasing filler content, the faster the heat diffuses through the thermal load (phonons) within the composite material along the thermally conductive network. However, the accompanying problems such as the deterioration of mechanical properties cannot be ignored, which requires the development of new high thermal conductivity fillers and processing methods. At present, researchers have made great progress in the preparation of filled thermally conductive polymer matrix composites with high thermal conductivity, especially the preparation of anisotropic thermally conductive composites. Some commercial products have a thermal conductivity of 10 W/(m K) above. At the same filler content, the orientation will form a more efficient two-dimensional thermal conduction path, and the contact thermal resistance between fillers and the interface thermal resistance between the filler and the polymer matrix will decrease, thereby achieving high thermal conductivity at low filling levels. On the other hand, fillers in composites are equivalent to newly introduced defects to the polymer, and low filling levels reduce the number of defects in the polymer matrix and avoid serious deterioration of the mechanical properties of polymer-based thermally conductive composites. It can be seen that the spatial arrangement of fillers is a key factor affecting the thermal conductivity of polymer matrix composites The quality change curves under different aging environments is shown in Figure 9.7.

By inducing orientation of fillers with non-spherical features (such as sheets, rods, or fibers) through magnetic, electric, and mechanical forces, an efficient

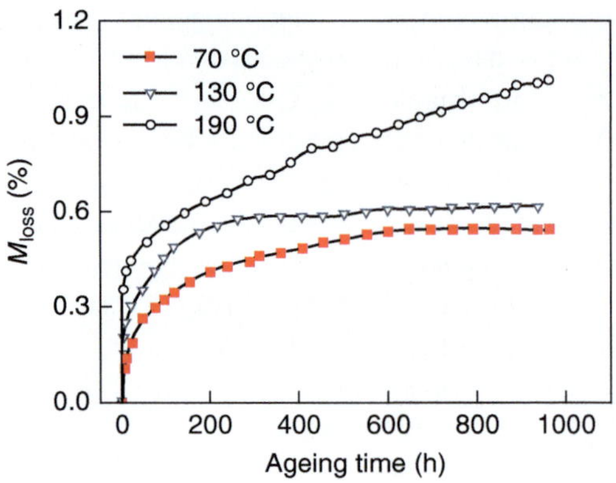

Figure 9.7 Quality change curves under different aging environments. Source: Reproduced with permission from Yaoxiong et al. [18]; © 2022 Materials Engineering.

two-dimensional thermal conduction network was constructed within the polymer matrix. These methods have their own characteristics and advantages and disadvantages:

(1) The orientation of the fillers in the magnetic field induction method and the AC (DC) electric field induction method can be controlled, but with the increase of the filler content, the viscosity of the mixture increases, and the filler is prone to agglomeration, which affects the filler.
(2) Electrostatic flocking method, ice crystal induction method, and vacuum filtration method require pre-construction of thermally conductive filler network, while other methods are the direct mixing of filler and polymer, in the polymer matrix.
(3) The composite materials prepared by the magnetic field induction method, the electric field induction method, and the mechanical force induction method all have an orientation structure and form a two-dimensional heat conduction network. The infrared spectra of T800 carbon fiber/epoxy composites under different aging environments is shown in Figure 9.8.

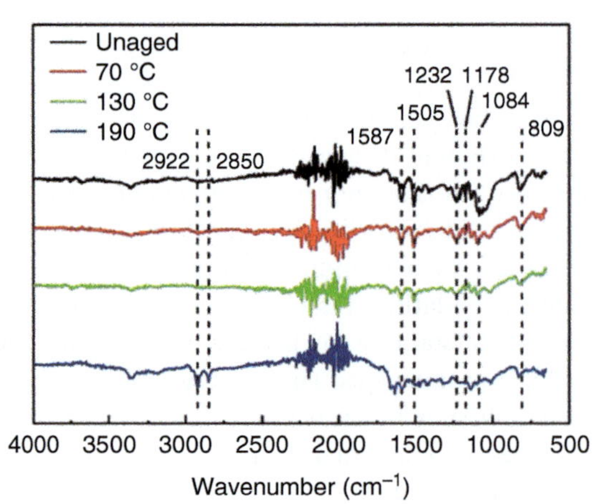

Figure 9.8 Infrared spectra of T800 carbon fiber/epoxy composites under different aging environments. Source: Reproduced with permission from Yaoxiong et al. [18]; © 2022 Materials Engineering.

Thermally conductive polymer matrix composites have been developed for nearly 30 years and are widely used in many fields. They play an irreplaceable and important role in the heat dissipation and thermal management of electronic devices. With the vigorous development of 5G technology, thermally conductive polymer matrix composites will face greater challenges. While focusing on improving thermal conductivity, electromagnetic shielding effectiveness, electrical insulation, and mechanical and mechanical properties should also be considered and developed in the direction of multi-functionality. At present, some highly thermally conductive polymer-based composite materials prepared based on external force induction have begun to be applied and play an important role in electronic equipment, 5G communications, and new energy vehicles. The future research and development also need to focus on the following aspects:

(1) Develop new thermally conductive fillers, especially to enrich the types of fillers with high thermal conductivity and non-spherical fillers, to provide more choices of high-performance fillers and the possibility of compounding for the preparation of thermally conductive composites;
(2) Control the orientation and distribution of fillers in the polymer matrix by various means (or driving forces), study the synergistic effect of different types and sizes of fillers and the surface modification technology of fillers, enhance the response of fillers to external forces, and improve fillers and polymerization. The interfacial interaction between the substrate and the substrate can be used to construct an efficient thermal conduction network to achieve the goals of low filling, high thermal conductivity, and good comprehensive performance;
(3) Carry out the theoretical prediction and thermal conduction mechanism of polymer matrix composites with anisotropic structural characteristic research, to provide theoretical reference for the design and optimization of high thermal conductivity composite materials;
(4) Imitate nature; drawing inspiration from nature, structural bionic, and method bionic is an important way to obtain high thermal conductivity composite materials;
(5) Optimize process steps, promote green preparation, reduce preparation costs, and move laboratory research to large-scale production as soon as possible. Figure 9.9 shows the schematic diagram of thermal vibration enhancement of thermally conductive filler-filled composites and Figure 9.10 shows TG curves of composite at different heating rates.

In recent years, researchers have paid great attention to the development of ceramic-filled polymer composites with high thermal conductivity and high electrical resistance. Epoxy resin (EP) is a kind of thermosetting polymer synthetic material with good corrosion resistance, insulation, and high strength. It also has chemical resistance and good dimensional stability, high bonding strength, and excellent comprehensive performance. It is cheap and widely used. However, the thermal conductivity of EP is about 0.2 W/(m K), and it is generally necessary to add inorganic fillers to improve the thermal conductivity. The main function of inorganic fillers is to reduce thermal expansion coefficient, water absorption rate,

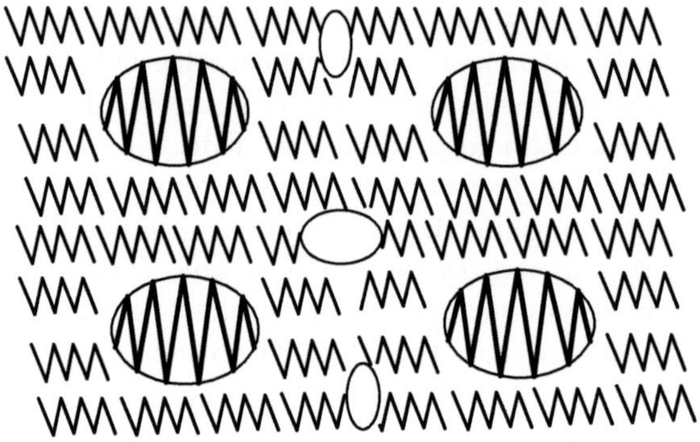

Figure 9.9 Schematic diagram of thermal vibration enhancement of thermally conductive filler-filled composites. Source: Reproduced with permission from Panmin et al. [30]; © 2018 Engineering Plastics Applications.

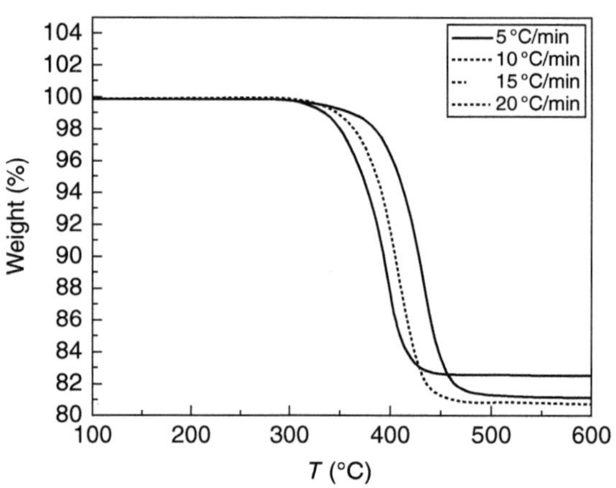

Figure 9.10 TG curves of composite at different heating rates. Source: Reproduced with permission from Chunlei et al. [31]; © 2012 FRP/Composite Materials.

molding shrinkage rate, and production cost, and can also play a reinforcing role and improve reliability. In order to improve the thermal conductivity of existing polymer packaging materials and meet the strength and insulation properties required by other electronic packaging, common inorganic fillers are silicon dioxide (SiO_2), aluminum oxide (Al_2O_3), mica, boron nitride (BN), aluminum nitride (AlN), silicon carbide, diamond, alumina fiber, aluminum nitride fiber, and glass fiber/carbon fiber. Among them, AlN, BN, and diamond fillers can significantly improve the thermal conductivity of polymer materials, but they are expensive. The large-scale application is limited, and Al_2O_3 and SiO_2 fillers have been widely used due to their low prices. Carbon materials such as graphene and carbon nanotubes, metal powders, metal wires, and other metal materials are also commonly used to improve the thermal conductivity of polymer materials.

9.6 The New Packaging Material

9.6.1 Third-Generation Packaging Material-Near-Net Shape of High-Volume-Fraction SiCp/Al Composites

Among various metal matrix composites, high-volume-fraction SiCp/Al composites were the first to attract attention and have been vigorously developed. This is because SiCp itself has excellent physical properties, and on the other hand, because the market of SiCp as an abrasive is very mature and the price is low. At the same time, the aluminum alloy used as the matrix also has excellent physical properties. The main properties of the high-volume-fraction SiCp/Al composite are shown in Table 9.1. It can be seen that its thermal conductivity and thermal expansion coefficient are close to those of the W–Cu alloy, but its density is only one-fifth of that of the W–Cu alloy. In recent years, high-volume-fraction SiCp/Al composite materials, as the third-generation electronic packaging materials, have become ideal materials with high thermal conductivity and adjustable low CTE, low density, high strength, and hardness required for electronic packaging, and will gradually replace the respective. The first- and second-generation special electronic packaging alloys represented by Kovar, W–Cu and Mo–Cu are in urgent demand, especially in the packaging of aerospace and optoelectronic devices [32].

Although SiCp/Al composites show excellent comprehensive properties, from the current market situation, they are not widely used on a large scale. This aspect is due to the high production cost of SiCp/Al, which leads to its high market price. It is limited to military use. Another more important reason is that the brittleness and hardness of the high-volume-fraction SiCp/Al composite material itself make it difficult for the material to be processed into the desired device shape by secondary machining, which seriously restricts wide application of this material. In order to meet the requirements of electronic packaging for the thermal expansion coefficient of the material, the high-volume-fraction SiCp/Al composite material usually has a volume fraction of SiCp higher than 55%, such a high volume fraction, using conventional production methods, such as stirring casting, powder metallurgy, and other methods. It is difficult to obtain composite materials with uniform structure and satisfactory properties during production. Therefore, in order to expand the scope and application of SiCp/Al, it is imperative to develop a low-cost near-net-shape fabrication process for high-volume-fraction SiCp/Al composites [33].

9.6.2 Fourth-Generation Electronic Packaging Material—Diamond/Cu(Al) Composite Material

With the continuous progress and development of information technology, the complexity and integration of electronic components are increasing, and the requirements for heat pipe materials are more stringent. Especially in recent years, the successful development of high-power microwave devices and high-power laser devices requires the support of higher-performance packaging materials. Diamond has the highest thermal conductivity of all substances (700–2000 W/(m K)); using more than 60% (volume ratio) of diamond particles and copper or aluminum composite, the thermal conductivity of the composite material is expected to exceed 500 W/(m K), and in addition, with the development of synthetic diamond technology, the price of diamond particles has been greatly reduced (<2000 yuan/kg), and the price per unit volume is close to or even lower than that of refractory metals such as W and Mo. From Table 9.1, it is also possible to improve performance of diamond/Cu(Al) composite material. It can be seen that the thermal conductivity of the diamond/Cu(Al) composite material is more than three times that of the third-generation packaging material. Therefore, the diamond/Cu(AI) composite material has a very broad application prospect in the field of high-performance

Table 9.2 Main properties of metal matrix composites for electronic packaging.

Material	Conductivity W/(m K)	CTE (25–400 °C)/ $\times 10^{-6}$ K	Density (g/cm^3)
W–Cu	190–200	6–7.5	7.3
SiCp/Al	185–200	6.5–8	2.85–3.02
Diamond/Al	550–600	7–7.5	3–3.1
Diamond/Cu	600–1200	7–7.5	5.7–6

electronic packaging [34]. In recent years, diamond/Cu(AI) composite materials are attracting extensive attention from domestic and foreign researchers as the fourth-generation electronic packaging materials.

The main research hotspots are as follows:

① Improve the interface bonding between Diamond/Cu(Al) and reduce the interface thermal resistance and ② Development of the near-net-shape process of this material.

Developed countries such as the United States, Japan, and the United Kingdom have begun to use them in high-performance servers, military portable computers, plasma displays, aerospace electronic equipment, and optoelectronic systems. Diamondp/Cu.Diamondp/Al and other fourth-generation packaging materials, while most domestic researches focus on depositing thick diamond films on copper substrates to obtain diamond–copper layered composite materials, and research on bulk composite materials with high thermal conductivity for electronic packaging is relatively weak; in this context, Beijing University of Science and Technology took the lead in developing diamond, particle-reinforced copper (Diamondp/Research on high-performance composite materials of Cu). In view of the difficulty of machining and forming diamond/Cu materials, the porous diamond preform skeleton was prepared by powder injection molding net-shaping technology, and the metal copper or aluminum melt was infiltrated into the pores of the skeleton by the infusion process, so as to obtain diamond–copper (aluminum) composites or net final shape parts. At present, the project has made progress in stages, and the thermal conductivity of the prepared composites exceeds 500 W/(m K). Table 9.2 shows main properties of metal matrix composites for electronic packaging [30].

9.7 Thermal Management of Electronic Devices

Many electronic devices work in instantaneous high-power mode, which requires a large amount of heat dissipation generated by high-power components during operation to be dissipated only in a very short time and a limited heat dissipation area.

The ways in which temperature causes electronic devices to fail can be divided into direct and indirect effects. The direct effect refers to the occurrence of chemical reaction and accelerated diffusion under certain temperature conditions, which directly leads to the failure of the chip; the indirect effect refers to the fatigue damage of the device under the conditions of thermal stress and vibration (such as automobiles and military electronic products). In addition, the presence of temperature changes and temperature gradients can also lead to device failure. Combining phase change heat storage materials with traditional heat dissipation technology of electronic devices can effectively control the temperature rise and temperature gradient of electronic devices. And it has been used in high-tech systems such as aviation, aerospace, and microelectronics, as well as military equipment.

9.7.1 Electronic Device Heat Dissipation Technology

The problem of efficient heat dissipation of electronic devices is closely related to the application of heat transfer, fluid mechanics, and other principles. The purpose of heat dissipation of electronic devices is to control the operating temperature (or thermal control) of electronic devices to ensure the stability and reliability of their work, which involves many aspects such as heat dissipation or cooling methods and materials related to heat transfer content. At present, the cooling technology of electronic devices mainly includes two categories: air cooling technology and liquid cooling technology.

Air cooling technology is currently the most widely used electronic cooling technology, including natural convection air cooling technology and forced convection air cooling technology. The natural convection air cooling technology is mainly used in electronic devices with small-volume heating power. Natural convection depends on the density change of the fluid, and the required driving force is not large, so it is easily affected by obstacles and resistance in the flow path and reduces the flow rate and cooling rate of the fluid. The forced heat dissipation or cooling method mainly uses a fan and other equipment to force the air around the electronic device to flow, so as to take away the heat emitted by the device. This is a simple and effective heat dissipation method [35].

With the development of heat transfer technology today, the design optimization of forced air cooling radiators has been very mature. Combined with heat pipe technology, thermoelectric refrigeration technology, air-jet technology, etc., the cooling capacity of air cooling technology has been greatly strengthened. Air-jet impingement technology is a research hotspot in the field of high-efficiency air cooling technology in recent years, and it is an effective means to greatly improve air convection heat transfer coefficient. Jet impingement is different from traditional forced air cooling technology and produces an extremely high convection heat transfer effect locally. The experimental results show that the convective heat transfer coefficient of jet impingement reaches 900 W/(m^2 K), which is almost an order of magnitude higher than the traditional forced air cooling technology and is comparable to the forced liquid cooling heat transfer coefficient.

Liquid cooling technology was developed in the 1980s to solve the cooling problem of large electronic computers at that time. Because of its high efficiency and compactness, liquid cooling has been widely used in the cooling of ultra-heat flux chips, thus becoming a research hotspot in various countries. Liquid cooling can be single-phase or two-phase. The cooling effect of gas–liquid phase transition is better due to the use of the latent heat of phase transition of the coolant. Liquid cooling mainly includes direct cooling or indirect cooling, gas–liquid phase change cooling, liquid jet impingement cooling, dripping, and spray cooling [31].

9.7.1.1 Direct Liquid Cooling

In direct liquid cooling, the cooling liquid is in direct contact with the heat-generating electronic components for heat exchange. The heat source directly transfers the heat to the cooling liquid, and then the cooling liquid takes the heat away, so the convection and phase change of the fluid working medium are the main heat exchange methods between the fluid and the heat source. For example, in a fluorocarbon immersion evaporative cooling method of liquid cooling, the heat of each chip in the module is transferred to the fluorocarbon evaporation liquid immersed in them and then transferred to the inner fins provided on the inner wall of the module through the evaporative liquid, and finally the heat is carried away through a water cooling plate or radiator on the outside of the wall. The experimental effect of direct liquid cooling can reach 800 W/cm^2. Direct liquid cooling must take into account the electrical insulation and compatibility between the working fluid and electronic components, the thermal expansion and contraction of the working fluid, the sealing of the system, and the maintainability of the system. High or those components must operate in high-temperature environments with small temperature gradients between the device and the surface to be cooled, and 2D or 3D packaging of highly encapsulated or high-power electronics. Yokouchi et al proposed a low-cooling direct immersion cooling method, which can not only prevent air bubbles from gathering at the top of the module to generate a bubble layer and affect the heat generation effect but also can significantly improve the cooling effect of the module.

9.7.1.2 Indirect Liquid Cooling

Direct liquid cooling is gradually replaced by indirect liquid cooling due to the thermal shock wave phenomenon caused by thermal hysteresis and the inconvenience of system maintenance in direct liquid cooling. The indirect liquid cooling method is that the liquid coolant does not directly contact the electronic components but first transfers the heat to the heat exchanger through heat conduction, and then the heat is taken away by the cooling medium in the heat exchanger. This type of heat exchanger is characterized by a working fluid, which is usually called a cold plate. As an advanced and industrialized heat dissipation equipment, the cold plate has the advantages of uniform heat load, small temperature gradient, compact structure, high heat transfer efficiency, and large heat transfer area. With the continuous improvement of the heat generation of electronic devices, the application of cold plates in the heat dissipation of electronic devices will be further popularized. The cold plate adopts a hollow structure, usually in the form of a honeycomb or convoluted structure, and the liquid used is usually water, fluorocarbon, silicone grease, or hexylene glycol.

9.7.1.3 Liquid Jet Cooling and Spraying, Drop Cooling

In recent years, liquid jet cooling technology has received extensive attention; its principle is basically the same as that of air-jet cooling, but the cooling effect is better (especially liquid jet cooling jet evaporative cooling technology). Jet cooling has two forms: free surface jet and immersion jet. Research at home and abroad mainly focuses on jet speed, nozzle shape, refrigerant selection (such as water, FC-72, FC-77, and FC-87), microstructure rough surface, and other influencing factors.

Spray cooling is to atomize the cooling liquid into a large number of tiny droplets through a special nozzle and spray it onto the high-temperature wall surface to cool the wall surface. Since the atomized tiny droplets have great surface energy, it is very easy to vaporize (phase change) after spraying on the high-temperature wall, so spray cooling can often produce extremely high heat transfer coefficients. Spray cooling and jet cooling are two very similar cooling methods. The difference is that in jet cooling, the refrigerant liquid is only sprayed to the stagnant area, while spray cooling droplets directly affect a larger area, so spray cooling. The temperature distribution of the chip is more consistent, and the cooling effect is better.

Droplet impingement cooling is another development of liquid evaporative cooling. Carnegie Mellon University in the United States has developed an embedded integrated chip droplet cooling, which uses micro-electromechanical systems (MEMS) technology to integrate a cooling device into a chip-scale package. The small droplets ejected by the multi-nozzle evaporate evenly on the surface of the chip, avoiding the uneven temperature distribution of the chip. Heat is carried to the edge of the system by the steam and dissipated to the environment through a condenser or heat sink. EDIFICE can achieve a cooling effect of 70–100 W/cm^2 on desktop or laptop computers.

9.7.1.4 Microchannel Heat Transfer Microchannel

Microchannel heat transfer microchannel is defined as a channel or pipe with a hydraulic diameter between 1 and 1000 μm, with high surface area/volume ratio, low thermal resistance, low flow rate, etc., and is an effective heat dissipation solution. Microscale channels are fabricated on oriented silicon wafers or on substrates using techniques such as anisotropic etching, and the liquid flows through the micro-channels by vaporizing or directly carrying heat away. It uses the particularity of microscale heat transfer to achieve the purpose of efficient cooling and is currently a research hotspot in various countries. The substrates of microchannel heat sinks include metal heat sinks and semiconductor heat sink materials. The selection of materials mainly considers the thermal conductivity and linear expansion coefficient of the materials. At present, the most used materials are silicon and copper, and the liquid is absorbed in the microchannel. Heating rapidly develops into nucleate boiling, where the liquid is in a highly unbalanced state with great heat transfer capacity and channel wall superheat much less than would be the case with conventional dimensions.

There are still some difficulties to be solved in microchannel heat transfer. Due to the small cross-sectional area of the microchannel, the liquid single-phase flow through the microchannel will be accompanied by a large temperature rise, which can cause serious problems such as high thermal stress or chip thermoelectric mismatch. The temperature rise can be reduced by raising the pressure head and increasing the flow rate, but the flow rate cannot be large enough due to noise and other factors, so the temperature rise problem cannot be solved fundamentally. The use of gas–liquid phase change can solve the problem of high-temperature gradient, but this will bring some other problems, such as complex structure, and the flow requires a larger pressure drop.

9.7.2 Phase Change Temperature Control

Phase change temperature control is to use the phase change process of phase change materials to store or release heat, so as to achieve the temperature control of the object. Phase change temperature control has the advantages of compact structure, reliable performance, economic energy saving, etc., as early as the 1960s was applied to the aerospace electronic equipment temperature control. In recent years, with the development of various portable electronic devices in the direction of miniaturization and high integration, phase change temperature control has been applied to the temperature control of these electronic devices, and the application research in this area has gradually become a research hotspot in the field of temperature control.

Phase change material (PCM) is a material with a specific function, which can develop a biological phase change at a specific temperature (phase change temperature) and is accompanied by the phase change process to absorb or emit a large amount of heat. Using this property of phase change energy storage materials to store or emit heat, it is possible to regulate or control the temperature of the working source or the surrounding environment of the material to achieve its specific application function. The ideal phase change energy storage material should have the following characteristics: the phase change temperature is the specific temperature to be controlled; a sufficiently large phase change heat; good reversibility of the phase change; low expansion and contraction during the phase change; high specific gravity and specific heat of the two phases; non-toxic and non-corrosive; low cost; and easy manufacturing. Phase change energy storage materials are mainly divided into solid–solid phase change and solid–liquid phase change energy storage materials according to the phase change form and phase change process; high-, medium-, and low-temperature energy storage materials according to the phase change temperature range; and inorganic and organic materials (including polymers), energy storage materials according to their composition. Usually, phase change materials are composed of multiple components, including main heat storage agent, phase change point adjusting agent, anti-subcooling agent, anti-phase separation agent, phase change accelerator, and other components. Figure 9.11 shows microscopic morphology of graphene, pure PP and graphene/PP composites with different mass fractions.

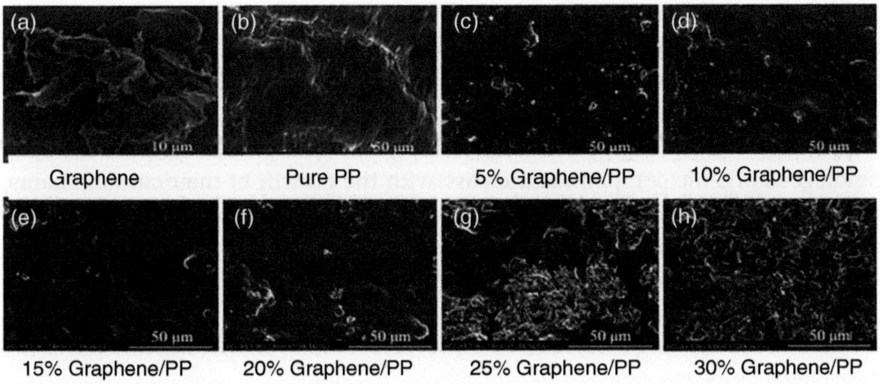

Figure 9.11 Microscopic morphology of graphene, pure PP, and graphene/PP composites with different mass fractions. Source: Reproduced with permission from Yi et al. [36]; © 2022 China Testing.

9.7.2.1 Inorganic Energy Storage Materials

Inorganic hydrated salts have a large heat of melting and fixed melting point, which is an important class of low- and medium-temperature phase change materials, mainly crystalline hydrated salts, molten salts, etc. The most typical is crystalline hydrated salts, and these materials have the advantages of large heat of melting, high thermal conductivity, small volume change during the phase change, etc. The more used are mainly alkali and alkaline earth metal halides, nitrates, sulfates, phosphates, carbonates, and acetates.

However, such materials are prone to subcooling and phase separation phenomena. The phenomenon of subcooling means that the material does not crystallize when it condenses to the condensation point but needs to reach a certain temperature below the condensation point before it starts to crystallize, resulting in fluctuations in the phase change temperature. The phenomenon of supercooling is related to the nature of the material, the cooling rate, and the type and content of impurities. To prevent the phenomenon of supercooling, we often choose compounds with a small tendency to supercool, a slightly higher melting point than the phase change material, and a composition and nature close to the phase change material as the anti-supercooling agent. Phase separation phenomenon refers to the repeated phase change process, often leading to brine separation; some salts do not dissolve in crystalline water and sink to the bottom, no longer combined with crystalline water, the formation of layered phenomenon, resulting in a substantial reduction in energy storage capacity, shortening the use of cycle. For this reason, anti-phase separation agents need to be added, often choose thickening agents, crystal structure change agents, etc.

9.7.2.2 Organic Energy Storage Materials

Organic phase change materials commonly used are advanced aliphatic hydrocarbons, fatty acids or their esters or salts, alcohols, aromatic hydrocarbons, aromatic ketones, amides, freons, polyhydroxy carbonates, etc. In addition, polymers include polyolefins, poly polyols, polyalkenols, polyalkenoic acids, polyamides, and some other polymers, of which typical ones are urea, C_nH_{2n+2}, C_nH_{2n+2} C_nH_{2n}, $C_{10}H_8$, CFC, PE, PEG, PMA, PA, etc.

Generally speaking, the phase change temperature and phase change enthalpy of homologous organic compounds increase with the growth of their carbon chains, so that energy storage materials with a range of phase change temperatures can be obtained, but the value of the increase in phase change temperature decreases as the carbon chains grow. Since the phase change material of polymer compound class is a mixture with a certain molecular weight distribution and the crystallization is not complete due to the long molecular chain, it has a melting temperature range for the phase change process.

Organic phase change materials have the following advantages: better formability in the solid state, generally less prone to subcooling and phase separation, less corrosive materials, more stable performance, less toxicity, and low cost; at the same time, these materials also have the following disadvantages: small thermal conductivity and low density, and thus the unit volume of energy storage capacity is small, and organic materials generally have a low melting point, not suitable for high-temperature applications. In addition, they are volatile, easily burned, or even exploded or aged by the slow oxidation of oxygen in the air. In order to obtain phase change materials with appropriate phase change temperature and superior performance, it is often necessary to compound several organic phase change materials to form binary or multiphase change materials, and sometimes also compound organic phase change materials with inorganic phase change materials to make up for the shortcomings of both, to obtain better performance of phase change materials and thus better applications. Relationship between thermal conductivity and content of composites is shown in Figure 9.12.

At present, in order to overcome the shortcomings of the mobility of solid–liquid phase change thermal storage materials, a class of shape-stable solid–liquid phase change materials has emerged, that is, the shape can maintain the solid shape during the phase change without making it flow. This type of material is mainly in the organic class (work material) energy storage materials by adding polymer resin types

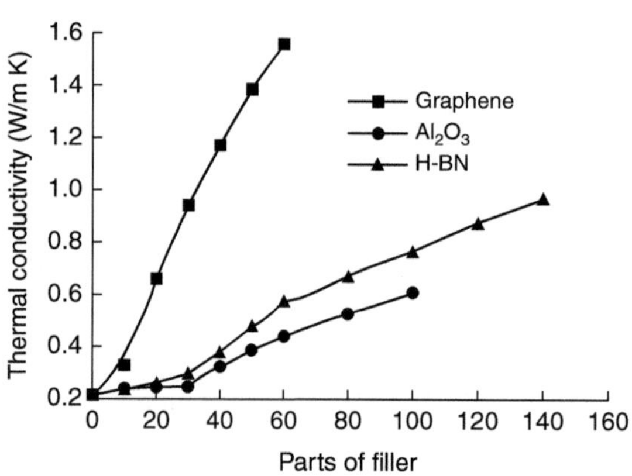

Figure 9.12 Relationship between thermal conductivity and content of composites. Source: Reproduced with permission from Liu et al. [34]; © 2022 Chemical Engineering.

(carrier matrix), such as polyethylene, polymethacrylic acid, and polystyrene, so that they are molten together or prepared by using physical blending method and chemical reaction method to infuse the work material into the carrier.

Thermally adapted composites (also known as thermally matched composites) are the other composites that have a suitable thermal conductivity or thermal expansion coefficient through the selection of components and their contents and the design of alignment orientation. The fast thermal response phase change thermal storage material with high thermal conductivity is one of the research hotspots for the application of thermally adapted composites in the field of electronic heat dissipation technology. Bending strength of the E51/Al_2O_3 composites is shown in Figure 9.13.

Recently, the technique of preparing composite phase change materials by adsorption of phase change materials into porous matrices using capillary action has attracted the attention of researchers. Inorganic materials with large specific surface area microporous structures are used as support materials, and the capillary force of microporous pores is used to draw liquid organic or inorganic phase changes thermal storage materials (above phase change temperature conditions) into the microporous pores to form organic/inorganic or inorganic/organic composite phase change thermal storage materials. In this composite phase change thermal storage material, when the organic or inorganic phase change thermal storage material in the microporous solid a liquid phase change, due to the capillary adsorption force, the liquid phase change thermal storage material is difficult to overflow from the microporous. The porous medium has a wide variety and a richly varied pore space, which is an ideal storage medium for phase change materials. The available porous media include gypsum, expanded clay, expanded perlite, expanded shale, and porous concrete. Adopting porous media as the encapsulation material of phase change substances can make the composite material have the advantages of structural–functional integration, which can save space and have good economy in application. The pores inside the porous medium are very small, which can improve the storage reliability of the phase change substances in the porous medium

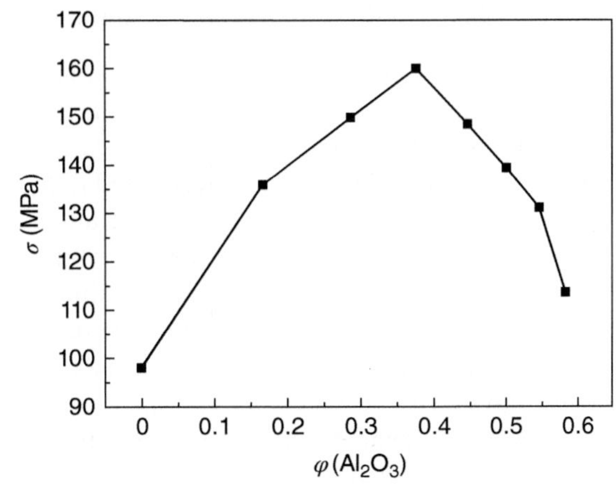

Figure 9.13 Bending strength of the E51/Al_2O_3 composites. Source: Reproduced with permission from Panmin et al. [35]; © 2011 Electronic Components and Materials.

with the help of capillary effect. The porous medium also disperses the phase change material into fine individuals, which effectively improves the heat transfer efficiency of its phase change process. Xavierpy et al. pressed expanded graphite into blocks with different bulk densities and immersed it in melted paraffin wax to form a paraffin wax/graphite composite phase change material, using the high thermal conductivity of graphite to improve the thermal conductivity of paraffin wax. The thermal conductivity of the prepared composite phase change material was comparable to that of a single graphite (4–70 W/(m K)), while the thermal conductivity of pure paraffin wax was 0–24 W/(m K), which shows a significant increase in thermal conductivity [37].

9.8 Methods for Improving Thermal Conductivity of Composite Materials

For composite materials widely used in electronic industrial equipment and devices, it is necessary to improve their thermal conductivity. By improving the thermal conductivity of materials, the thermal resistivity can be reduced and the heat dissipation performance of the device can be improved, which is beneficial to reduce thermal damage to the circuit board and prolong the service life of the device. Figure 9.14 shows SEM images of composite powders with different contents of graphene oxide.

The improvement of thermal conductivity of polymer matrix composites mainly depends on the filling phase, and the condition of the filling phase forming thermal conduction paths in the matrix will determine the thermal conductivity of the final composite material. There are many factors that affect the thermal conductivity of composite materials. The thermal conductivity of composite materials is related to the volume fraction of each component, the thermal conductivity of the filling phase, and the arrangement, shape, and orientation of each component in the composite material. The methods to improve the thermal conductivity of composite materials mainly include the following aspects.

9.8.1 Choose a Reasonable Filling Amount

The thermal conductivity of the composite material is improved compared with that of a single matrix material, mainly due to the thermal conduction path formed by the filling phase in the matrix. When the filling amount of the second phase is small, the filling phase is mostly uniformly dispersed in the entire matrix, it is difficult for the particles of the filling phase to contact or interact, and it is difficult to form a thermal conduction path, so it does not contribute much to the thermal conductivity of the entire material system. When the filling amount of the second phase reaches a certain level greater than a certain penetration value, the particles of the filling phase begin to contact each other to form a thermal conduction path, and the thermal conductivity of the composite material will be significantly improved at this time. After the penetration value is exceeded, with the continuous increase of the filling phase, the density of the thermal conduction path formed increases, and the

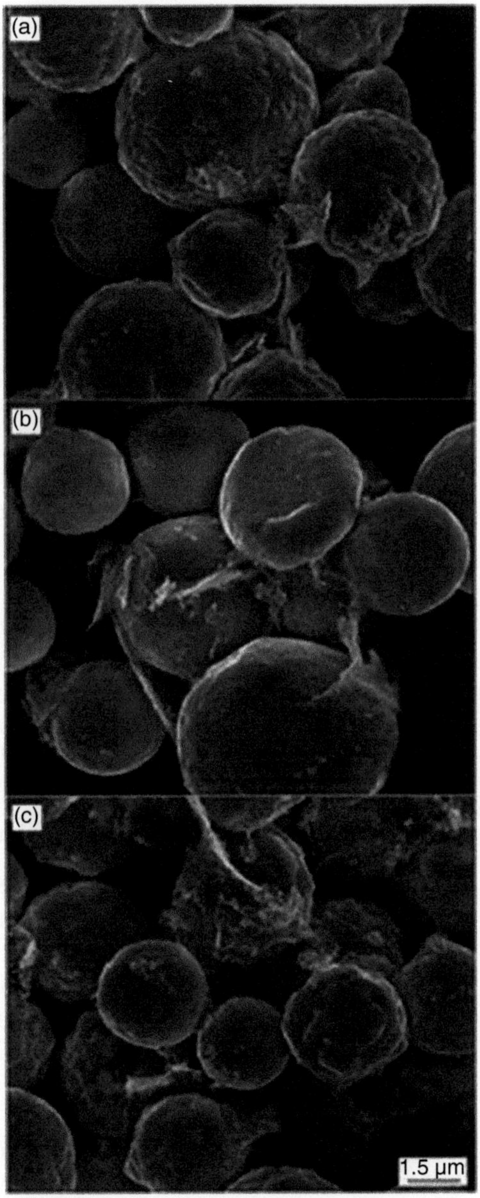

Figure 9.14 SEM images of composite powders with different contents of graphene oxide: (a) 0.1%; (b) 0.25%; (c) 0.5%. Source: Reproduced with permission from Caizi et al. [38]; © 2021 Chinese Journal of Nonferrous Metals.

thermal conductivity of the composite material will also be enhanced, but when the filling amount of the filling phase in the matrix reaches saturation or the thermal conductivity network, when the density of the composite material cannot be further increased, the thermal conductivity of the composite will no longer improve with the increase of the filling amount. However, increasing the thermal conductivity by increasing the filling volume will pose new problems for the filling process. Glass transition temperature is shown in Figure 9.15.

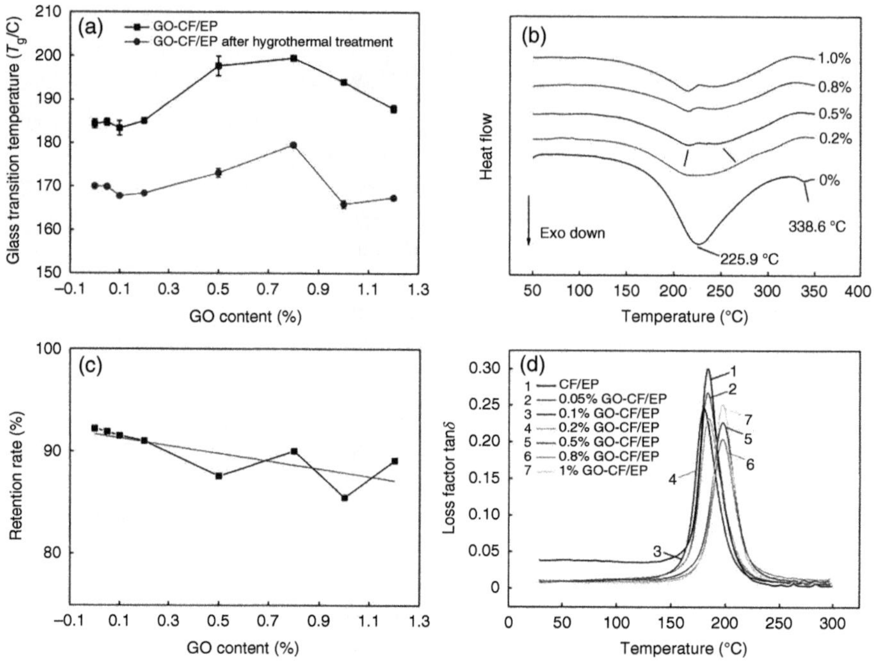

Figure 9.15 Glass transition temperature (a), DSC curves (b), Tg retention rate curves (c), and dynamic loss factor (d) of GO-carbon fiber/epoxy composites with different contents of GO. Source: Reproduced with permission from Wenjun et al. [39]; © 2021 Journal of Composite Materials.

9.8.2 Change the Structure and Morphology of the Filling Phase

The dispersion effect of fillers with different morphological structures in the matrix is different, which has an important influence on the formation of the thermal conductive network. Chu Luxuan found that when the same content of whisker ZnO and powdered ZnO with a particle size of 1 μm were added to EVA, when the mass fraction of the filling amount was 0.2, the thermal conductivity of ZnO/EVA filled with whisker ZnO was 0.82 W/(m K), the thermal conductivity of powdered ZnO/EVA is 0.56 W/(m K), and the reason for this difference may be that ZnO whiskers can more effectively form a "thermal conduction network" of interconnecting pathways in the EVA matrix, thereby further improving the thermal conductivity of ZnO/EVA. Figure 9.16 shows SEM images of TGO under different temperature.

9.8.3 Change the Surface Morphology of the Filling Phase

The thermal conductivity of composites depends not only on the thermal conductivity of the filling phase itself but also on the degree of wetting between the surface of the filling phase and the matrix, which is due to the wettability between the filling phase and the matrix. Moisture affects some factors that directly affect the thermal conductivity of composite materials, such as the bonding degree between the filling

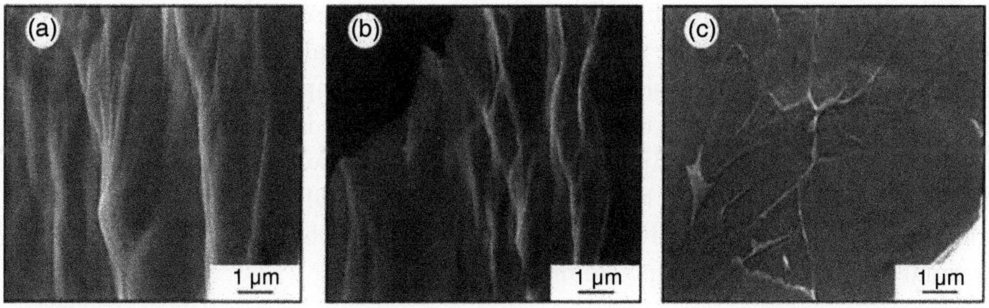

Figure 9.16 SEM images of TGO under different temperatures: (a) 600 °C, (b) 800 °C, (c) 1000 °C. Source: Reproduced with permission from Fanglan et al. [37]; © 2019 New Carbon Materials.

phase and the matrix, the thermal resistance of the interface between the filling phase and the matrix, the uniform dispersion of the filling phase, and the maximum filling amount of the filling phase in the matrix. Therefore, it is of great significance to modify the surface of the packed phase particles. The surface treatment methods of the filling phase can be divided into physical surface modification and chemical surface modification. Physical surface modification includes liquid coating method and surface adsorption coating method, and chemical surface modification includes plasma surface modification method, surface grafting modification method, and mechanochemical modification method. Kondo Toshiki et al. finally obtained a composite material with a thermal conductivity of 1.16 W/(m K) by blending, granulating, and injecting a certain amount of MgO with polyamide resin.

9.8.4 Improving the Dispersion Form of Filling Phase

Nanomaterials have attracted more and more attention in recent years. For composite materials using nanomaterials as filling phase, due to nanoscale quantum effects and surface and interface effects, microscale and macroscale, there is a clear difference in the heat transfer under the interfacial properties, which play an important role in thermal conduction at the nanoscale, for example, the thermal conductivity of CNTs (carbon nanotubes)/epoxy composites is smaller than that of CNTs, because the interfacial thermal resistance between CNTs and the surrounding matrix hinders the acoustic heat conduction of the sub. In the laboratory process, it was found that the nano-sized filled phase also differs in the thermal conductivity of the composite due to the difference in the structure and shape of the filled phase, dispersion, etc. Generally, the granular filling phase mostly forms an "island structure" in the matrix and is covered by the matrix, and it is difficult to form a thermal conduction network between them. The interface between the matrix and the filling phase increases, and the thermal resistance of the interface increases, so the thermal conductivity increases. For fibrous and tubular filled phases, such as CNTs, due to the huge difference in thermal conductivity between transverse and longitudinal directions, it is difficult to make the filling phase oriented during the filling process, and the thermal conductivity of the composite material is improved and limited. For sheet-like materials, such as graphite, the spreadability is better than that of granular and tubular materials, and it is easier to form a thermally conductive network to improve thermal conductivity. In addition, the dispersion of the filler phase in the matrix is an important influencing factor. Due to the nano-size effect, the nanoscale filling phase is prone to agglomeration. The agglomeration of the filling phase in the matrix will cause a large thermal resistance, and the phonon heat conduction process will be hindered, which will affect the thermal conductivity. It will be an effective method to improve the thermal conductivity of composite materials [36].

9.9 The Application of Composite Materials

Potting is a process of rationally arranging, assembling, bonding, connecting, isolating, and protecting the various components that constitute an electrical device according to the prescribed requirements, so that the electronic device can prevent the intrusion of moisture, dust, and harmful gases, while slowing down the vibration, preventing damage from external force and stabilizing the effect of internal parameters of the element. For example, circuit boards used outdoors need to be potted to prevent corrosion of circuits by moisture, salt spray, and condensation; in order to improve performance, all transformers in electronic equipment working at sea require potting. In the field of airborne and aerospace, some circuit boards need to be partially reinforced or encapsulated as a whole; in order to avoid corrosion or breakage of the solder joints of some cable sockets, they also need to be encapsulated, and so on. With the rapid development of high-performance and high-density packaging technology of potting devices in the field of electronics today, higher requirements are also placed on potting materials.

9.9.1 Classification of Potting Materials

The composition of potting materials is complex, with a wide variety, different characters, and wide uses. Therefore, there are various classification methods. The general classification methods of potting materials are shown in Table 9.2.

9.9.2 Research Status of Potting Materials

Potting materials are the basis of potting technology. Understanding the characteristics of various potting materials and selecting appropriate potting materials according to potting requirements are the conditions of potting technology. There are many kinds of potting materials. At present, the most studied are mainly epoxy resin, polyurethane elastomer, and silicone polymer-based potting materials.

① The characteristics of epoxy resin general epoxy potting glue:
   a. Good performance, long pot life, and strong productivity;
   b. Low viscosity and strong impregnation;
   c. Component sedimentation is small, with no stratification;
   d. The curing exothermic peak is low, and the curing shrinkage is small;

Epoxy potting includes one-component epoxy resin potting glue and two-component epoxy resin potting glue. One-component epoxy potting compound belongs to the type of heat curing and is a new variety developed abroad in recent years. Its development makes up for the shortage of two-component heat-curing potting glue. The required potting equipment is simple, easy to use, and has small requirements for equipment and technology; however, the cost is high and the material storage conditions are strict [38].

Epoxy potting glue can be divided into two types: heat curing and room temperature curing in terms of curing conditions. The heat-curing two-component epoxy potting compound is the most widely used and the most used variety. It is characterized by low viscosity, stable process, long pot life, good impregnation effect, and excellent comprehensive performance. Room temperature curing epoxy potting glue is generally two-component, which can be cured at room temperature after potting. It has low equipment requirements and is easy to use. However, the potting material has high viscosity, poor impregnation effect, and short pot life, and automatic production is difficult. The cured product has poor electrical insulation and heat resistance and is generally used in applications that are not suitable for heating and curing.

② Features of polyurethane elastomergeneral polyurethane potting:
   a. Molecular structure with hydrophobicity, excellent weather resistance, and low temperature;
   b. Polyurethane has low viscosity and good fluidity;
   c. Easy to cast molding and can be cured at room temperature;
   d. Low curing stress, small shrinkage, and less heat release;
   e. The processing equipment and process are simple, and the degree of automation and continuity is high.

The bonding performance of polyurethane is between epoxy and silicone, and the temperature resistance generally cannot exceed 100 °C, and it has melted and softened at 170–190 °C. However, polyurethane materials are toxic and harmful to human health. Low-toxicity polyurethane materials should be used when selecting them, which are generally used in the packaging of electrical equipment and electronic components with complex internal structures. At the same time, the thermal conductivity of polyurethane material is extremely poor, even lower than the thermal conductivity of air, less than 0.023 W/m K. It roughly includes two-component polyurethane potting glue and one-component polyurethane potting glue.

③ Silicone polymer
   a. It can withstand high temperatures for a long time and has good insulation and good stability;
   b. Liquid before vulcanization, easy to perfuse, and easy to use;
   c. No low molecular weight is released during potting, no stress shrinkage, deep vulcanization, and no corrosion;
   d. Affordable and color adjustable.

Silicone resin encapsulants include room temperature vulcanized silicone rubber, two-component addition silicone rubber encapsulants, and two-component condensation silicone rubber encapsulants. Room temperature vulcanized silicone rubber or silicone gel is used for potting electronic and electrical components, which can play the role of sealing, shock absorption, dust removal, moisture resistance, corrosion resistance, and improved performance and stability parameters. And because it is liquid before vulcanization, it is easy to pour and use. When using silicone gel for potting, it does not release low molecules, has no stress shrinkage, can be deeply vulcanized, and has no corrosion. The transparent silicone becomes a transparent elastomer after vulcanization, and the components encapsulated in the adhesive layer are clearly visible. Use a needle to pierce the inside to measure the parameters of the components one by one, which is convenient for inspection and repair. There are also opaque gray or black, and different colors are used in different ranges. The room temperature vulcanized foamed silicone rubber is used for the magnetic core board of the internal memory of the electronic computer. The flame-resistant potting compound prepared on the basis of addition-type RTV silicone rubber is used for the molding of TV high-voltage caps, high-voltage cable sheaths, and other products. When airtight packaging is not required or impregnation and potting protection are inconvenient, one-component room temperature vulcanized silicone rubber can be used as a surface coating protection material. Generally, the surface protection coating of electronic components is made of room temperature vulcanized silicone rubber, and the inner coating is carried out with addition-type silicone gel.

9.9.3 Research Status of Thermally Conductive Potting Composite Materials

At present, the research of high thermal conductivity composite materials is mainly based on the blending type, that is, a certain high thermal conductivity inorganic filler is blended into the matrix material with specific requirements. If inorganic thermal conductive fillers are used to replace part of the polymer matrix, the result will improve the thermal conductivity of the entire system, that is, to replace relatively disordered lattice structure materials with ordered lattice structure materials, resulting in the thermal conductivity of the entire system and performance improvements. Figure 9.17 shows schematics of hot-pressing preparations and heat conduction along the plane direction of composites.

Since the field of potting materials is mainly based on epoxy resin, polyurethane elastomer, and silicone polymer, the research on its thermal conductivity application is also carried out in these three categories of materials, while the research on oil-based potting materials and inorganic potting materials rarely reported. Wang Zheng used silicone-modified epoxy as the matrix and mixed thermally conductive and insulating fillers to prepare a thermally conductive potting compound with good thermal conductivity, high temperature resistance, and small internal stress, which is suitable for thermal conduction of military high-power circuits (DC/DC power supply). Chen Yuanzhang made thermally conductive and insulating silicone potting by compounding alumina and silicone rubber, and studied its physical

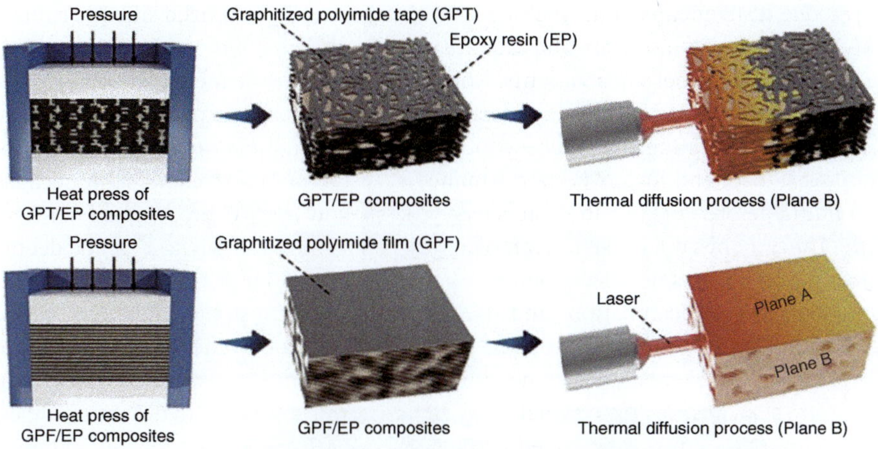

Figure 9.17 Schematics of hot-pressing preparations and heat conduction along the plane direction of composites. Source: Reproduced with permission from Li et al. [33]; © 2021 New Carbon Materials.

and mechanical properties, environmental aging resistance, and actual heat dissipation effect. The results show that the potting compound meets the needs of military electronic potting. Zhang Wenjie studied the silicone potting compound filled with $A1_2O_3$ and A1N, and obtained a potting compound with a thermal conductivity of 0.89 W/(m K). Xu developed an ultra-fine SiO_2/silane mixture paste with high thermal conductivity and high specific heat.

9.9.4 Research on Fillers

Thermally conductive fillers are the main factors affecting the thermal conductivity of thermally conductive composites. The type, particle size, structural shape, surface wettability, doping fraction, and their own thermal conductivity of fillers are important to thermally conductive composites.

9.9.4.1 The Effect of Filler Thermal Conductivity on Thermal Conductivity

Since the thermal conductivity of different types of fillers varies greatly, the thermal conductivity of the composite material obtained is higher when the filler with high thermal conductivity is used under the same filling amount. However, when the fillers with low thermal conductivity and high thermal conductivity are used for filling at the same time, the thermal conductivity of the composites does not change much. This is mainly because when the amount of fillers is small, the fillers are in the matrix package, and the difference between the fillers and the fillers is small. The thermal resistance is very large, and the thermal conductivity is mainly determined by the thermal conductivity of the matrix. Therefore, only when the filling amount of the filler reaches a certain critical value, the thermal conductivity of the material can be affected. Xu filled epoxy with BN and AlN particles, respectively. When the dosage of BN was 57 vol%, the thermal conductivity of the system reached

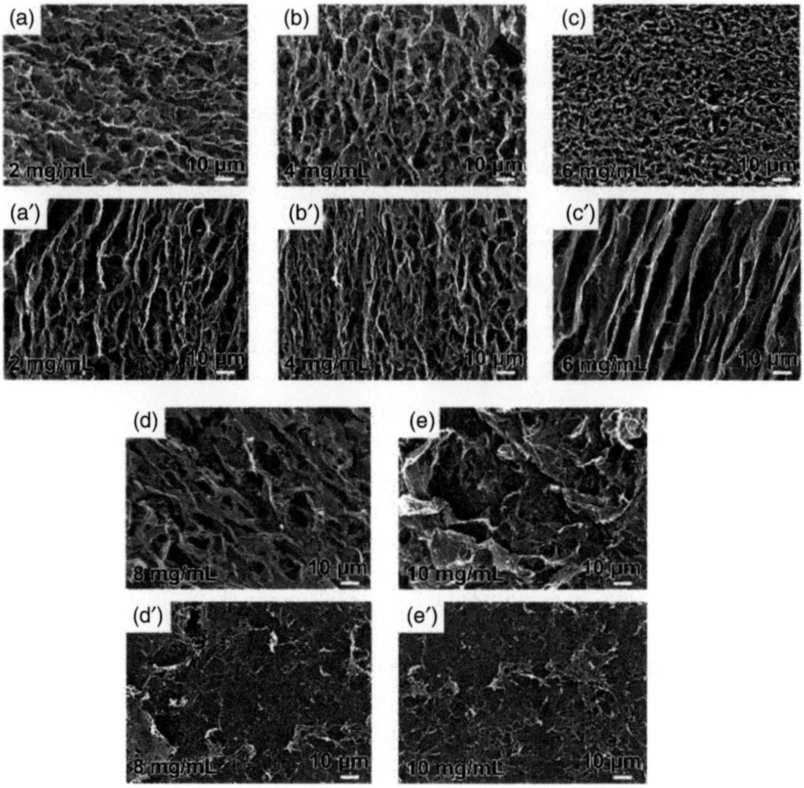

Figure 9.18 SEM images of TAGA with different GO concentrations. Source: Reproduced with permission from Yifan et al. [19]; © 2022 Functional Materials.

10.3 W/(m K), and when the dosage of 60 vol% A1N was used, the thermal conductivity of the system reached 11.0 W/(m K) [39]. The phenolic resin was filled with AlN to prepare a thermally conductive electronic packaging material. When the maximum filling amount of AlN was 78.5 vol%, the thermal conductivity was 32.5 W/(m K). SEM images of TAGA with different GO concentrations is shown in Figure 9.18.

9.9.4.2 The Effect of Filler Particle Size on Thermal Conductivity

Use fillers with less thermal conductivity to form the largest stacking degree between the fillers and obtain higher thermal conductivity. Ideally, the thermal conductivity of the composite material can reach 20 times that of the matrix. BemdW systematically studied the thermal conductivity of polypropylene composites filled with different thermally conductive fillers and different densities and found that the thermal conductivity of composites has a lot to do with the distribution of fillers, rather than the thermal conductivity of the filler itself. However, at a high filling amount, the effect of particle size on the thermal conductivity will be weakened, because at a high filling amount, a thermal network chain has been formed inside the matrix, and the effect of particle size can be ignored.

9.9.4.3 Effect of Filler Surface Modification Treatment on Thermal Conductivity

Due to the poor compatibility between the inorganic filler particles and the organic resin matrix interface, the particles are easily aggregated into agglomerates in the matrix and cannot be effectively dispersed. In addition, due to the difference in surface tension between inorganic particles and organic resins, it is difficult for the resin to wet the surface of the particles, resulting in voids at the interface between the two, increasing the interfacial thermal resistance of the composite. In order to improve the combination of the two interfaces, the thermal conductive particles must be surface treated. The dispersion state of the filler, the degree of bonding between the filler and the matrix, and the thermal barrier of the matrix and the filler interface are all affected by the surface wetting degree of the filler, especially the nano-filler. There will be no nanometer-size dispersion in the matrix. When the thermally conductive fillers are formed in a "separated distribution state" in the matrix through a special process, even a small amount will impart high thermal conductivity to the material. Therefore, it is important to modify the filler particles to improve their distribution in composites. Liu Daolong et al. found that after irradiating HDPE with γ-rays, oxygen-containing polar groups such as carbonyl groups were introduced into its molecular chain, and the polarity was improved. Using it as the matrix and Al_2O_3 with high thermal conductivity as the filler, the thermal conductivity of the prepared thermally conductive plastic is obviously improved, the interfacial interaction of the system is enhanced, and the mechanical properties are also significantly improved.

9.9.4.4 Effects of Mixed Particle-Size Fillers on Thermal Conductivity

When a variety of particle-size thermal conductive fillers are mixed and filled, the thermal conductivity is improved by improving the mutual cooperation between the fillers. Four kinds of AlN powders with different particle sizes are mixed with epoxy according to a certain proportion. When the filler content is 80%, the thermal conductivity reaches 4.1 Wl (m K). Three different particle sizes of Al_2O_3 were mixed with epoxy according to a certain proportion. When the filler content was 73%, the thermal conductivity reached 4.05 W/(m K). Mixing a series of BN powders with different particle sizes and polymers, the thermal conductivity of the material is 18.3 w/(m k).

9.10 Conclusion

Composite material has attracted widespread attention due to its unique performance advantages, and it has continued to develop with the rapid development of electronic packaging technology. And among them A1, sic. composite materials have become ideal packaging materials for modern microelectronic components due to their advantages in performance and cost, especially in high-density packaging, military, and aerospace industries. This kind of material has achieved mass production abroad, but it is still among the prohibited products. Compared with foreign countries, there is still a certain gap in domestic research and production levels, but the development direction is clear. After a period of research, further development will be achieved in product molding and performance. At the same time, domestic material developers and packaging designers should strengthen mutual contacts and exchanges, promote each other, and jointly promote the development of electronic packaging technology in China.

References

1 Jiawei, Y., Ruixuan, F., Yifan, N. et al. (2022). Research progress of nano-carbon material/thermoplastic resin interlayer toughened thermosetting resin-based composites. *Journal of Composite Materials* 39 (2): 528–543.

2 Junyu, Z., Chunbo, W., Chengyang, W. et al. (2022). Preparation and properties of thermally conductive expanded graphite/polyetherimide composite materials. *Chemical Journal of Chinese Universities* 43 (4): 187–196.

3 Xiaze, L. and Bianying, W. (2022). Influence of interface effect on thermal conduction behavior of functional composite materials. *Journal of Composite Materials* 39 (4): 1498–1510.

4 Yiteng, Z. (2021). Application of polyimide composite materials in the field of microelectronics. *China Broadband* 3: 91.

5 Fei, W. (2020). Application status and development trend of composite materials in the field of military electronic equipment. *Fiber Composite Materials* 37 (3): 105–108. 104.

6 Chao, Z., Ruiqin, B., Yong, M. et al. (2019). Research progress of epoxy resin-based composites for electronic packaging. *Engineering Plastics Application* 47 (10): 144–149.

7 Zhu Xiaomin, Y. and Jiakang, W.X. (2012). Microstructure and properties of Al/Si/SiC composite materials for electronic packaging. *Chinese Journal of Nonferrous Metals (English Edition)* 22 (7): 1686–1692.

8 Jing, Z., Chaoqun, P., Richu, W. et al. (2015). Research progress of metal matrix composites for electronic packaging. *Chinese Journal of Nonferrous Metals* 25 (12): 3255–3270.

9 Moore, A.L. and Li, S. (2014). Emerging challenges and materials for thermal management of electronics. *Materials Today* 17 (4): 163–174.

10 Chen, H.Y., Ginzburg, V.V., Yang, J. et al. (2016). Thermal conductivity of polymer-based composites: fundamentals and applications. *Progress in Polymer Science* 59: 41–85.

11 He, J., Wang, H., Qu, Q.Q. et al. (2020). Three-dimensional network constructed by vertically oriented multilayer graphene and SiC nanowires for improving thermal conductivity and operating safety of epoxy composites with ultralow loading. *Composites Part A: Applied Science and Manufacturing* 139: 106062.

12 Han, G.J., Zhao, X.Y., Feng, Y.Z. et al. (2021). Highly flame-retardant epoxy-based thermal conductive composites with functionalized boron nitride nanosheets exfoliated by one-step ball milling[J]. *Chemical Engineering Journal* 407: 127099.

13 Xie, B.S., Li, C.C., Chen, J. et al. (2020). Exfoliated 2D hexagonal boron nitride nanosheet stabilized stearic acid as composite phase change materials for thermal energy storage. *Solar Energy* 204: 624–634.

14 Wang, W.-T., HaNa, Y., Potter, K., and Kim, B.C. (2020). Effect of the characteristics of nylon microparticles on Mode-I interlaminar fracture toughness of carbon-fibre/epoxy composites. *Composites Part A: Applied Science and Manufacturing* 138: 106073.

15 Chen, J., Huang, X., Sun, B. et al. (2017). Vertically aligned and interconnected boron nitride nanosheets for advanced flexible nanocomposite thermal interface materials. *ACS Appl Mater Interfaces* 9 (36): 30909–30917.

16 Fankun, Z., Zhenghua, M., and Wei, G. (2022). Thermophysical properties of graphite/aluminum composites reinforced with silicon carbide particles. *Journal of Composites* 39 (10): 4918–4926.

17 Xiaolian, X., Zhonglei, M., Lin, S. et al. (2022). Preparation and thermal conductivity of functionalized silicon carbide nanowire/liquid crystal epoxy composites. *Journal of Composites* 39 (8): 3747–3756.

18 Yaoxiong, J., Liang, X., Qingyang, A. et al. (2022). Effect of different thermal oxygen environments on the mechanical properties of T800 carbon fiber/epoxy resin composites. *Materials Engineering* 50 (4): 156–161.

19 Yifan, Z., Chengzhi, L., Xueqin, Y. et al. (2022). High-temperature thermally reduced graphene aerogel/epoxy resin electromagnetic shielding composites. *Functional Materials* 53 (3): 3186–3192.

20 Alok, A. and Alok, S. (2013). Development of a heat conduction model and investigation on thermal conductivity enhancement of AlN/Epoxy composites. *Procedia Engineering* 51: 573–578.

21 Tekce, H.S., Kumlutas, D., and Tavman, I.H. (2007). Effect of particle shape on thermal conductivity of Copper reinforced polymer composites. *Journal of Reinforced Plastics and Composites* 26: 113–123. Kondo Toshiki. Polyamide Composition for Housing of Electronic.

22 Jia-Jun, W. and Xiao-Su, Y. (2004). Effects of interfacial thermal barrier resistance and particle shape and size on the thermal conductivity of AlN/PI composites. *Composites Science and Technology* 64: 1623–1628.

23 Guo, H., Wang, G.Z., Jia, C., and Zhang, X.M. (2014). Low-temperature heat conduction characteristics of diamond/Cu composite by high pressure infiltration. *Acta Materiae Compositae Sinica* 31 (3): 550–555.

24 Wang, Y.L., Duan, K.Y., Wang, K. et al. (2018). Structure and thermal properties of layered Ti-clad diamond/Cu composites prepared by SPS and HP. *Rare Metal Materials and Engineering* 47 (7): 2011–2016.
25 Chang, G., Sun, F.Y., Duan, J.L. et al. (2018). Effect of Ti interlayer on interfacial thermal conductance between Cu and diamond. *Acta Materialia* 160: 235–246.
26 Bai, G.Z., Zhang, Y.J., Liu, X.Y. et al. (2020). High-temperature thermal conductivity and thermal cycling behavior of Cu–B/diamond composites. *IEEE Transactions on Components, Packaging, and Manufacturing Technology* 10 (4): 626–636.
27 Chen, Z.R., Ma, D.J., Wang, S.M. et al. (2020). Wear resistance and thermal stability enhancement of PDC sintered with Ti-coated diamond and cBN. *International Journal of Refractory Metals and Hard Materials* 92: 105278.
28 Geiger, A.L., Hasselman, D.P.H., and Donaldson, K.Y. (1993). Effect of reinforcement particle size on the thermal conductivity of a particulate silicon carbide-reinforced aluminum-matrix composite. *Journal of the American Materials Science Letters* 12: 420–423.
29 Ye, W.T., Wei, Q.P., Zhang, L. et al. (2018). Macroporous diamond foam: a novel design of 3D interconnected heat conduction network for thermal management. *Materials & Design* 156: 32–41.
30 Yuning, X., Hua, L., and Qian, S. (2018). Research progress of thermally conductive epoxy resin-based composites for electronic packaging. *Engineering Plastics Applications* 46 (12): 143–147.
31 Chunlei, H., Yunhua, Y., Xiaochao, L. et al. (2012). Research on accelerated thermal oxidative aging of carbon fiber/epoxy resin composite materials. *FRP/Composite Materials* 2: 25–29.
32 Liu, X.Y., Sun, F.Y., Wang, L.H. et al. (2020). The role of Cr interlayer in determining interfacial thermal conductance between Cu and diamond. *Applied Surface Science* 515: 146046.
33 Li, W.-L., Xuan-Ke, L., Shen, K. et al. (2021). Preparation and property characterization of high thermal conductivity polyimide graphite film/epoxy resin composites. *New Carbon Materials* 36 (5): 971–979.
34 Liu, S.H. (2022). Preparation and properties of epoxy resin-based thermally conductive composites. *Chemical Engineering* 50 (7): 22–26.
35 Panmin, L., Chaowei, Z., Qiming, T. et al. (2011). Preparation and thermal conduction model of epoxy resin/alumina composite materials. *Electronic Components and Materials* 30 (11): 26–29.
36 Yi, Z., Wei, L., Xiaojuan, F. et al. (2022). Measurement and analysis of thermal conductivity of graphene/polypropylene composite materials. *China Testing* 48 (2): 66–74.
37 Fanglan, G., He, X., and Zhoufan, Z. (2019). Thermal conductivity of graphene/cellular alumina/epoxy resin composites. *New Carbon Materials* 34 (6): 587–592.

38 Caizi, X., Duosheng, L., Yin, Y. et al. (2021). Effect of thermally reduced graphene nanosheets on properties of copper matrix composites. *Chinese Journal of Nonferrous Metals* 31 (6): 1516–1525.

39 Wenjun, L., Jianlong, Y., Chuan, Z. et al. (2021). Hygrothermal properties and microscopic morphology of graphene oxide modified carbon fiber/epoxy resin composites. *Journal of Composite Materials* 38 (5): 1416–1425.

Index

Thermal Management Materials for Electronic Packaging: Preparation, Characterization, and Devices, First Edition. Edited by Xingyou Tian.

r

s

t